27 $\frac{50}{}$
NEW

EVOLUTION'S RAINBOW

Evolution's Rainbow

Diversity, Gender, and Sexuality
in Nature and People

Joan Roughgarden

UNIVERSITY OF CALIFORNIA PRESS   Berkeley   Los Angeles   London

University of California Press
Berkeley and Los Angeles, California

University of California Press, Ltd.
London, England

© 2004 by Joan Roughgarden

Library of Congress Cataloging-in-Publication Data

Roughgarden, Joan.
    Evolution's rainbow : diversity, gender, and sexuality
in nature and people / Joan Roughgarden.
        p. cm.
Includes bibliographical references and index.
    ISBN 0-520-24073-1 (alk. paper)
    1. Biological diversity. 2. Sexual behavior in animals.
3. Gender identity. 4. Sexual orientation. I. Title.
QH541.15.B56.R68 2004
305.3—dc22

                                    2003024512

Manufactured in the United States of America
13 12 11 10 09 08 07 06 05 04
10 9 8 7 6 5 4 3 2 1

The paper used in this publication meets the minimum
requirements of ANSI/NISO Z39.48-1992(R 1997) (Per-
manence of Paper).

To my sisters on the street
To my sisters everywhere
To people everywhere

Contents

Diversity Denied

On a hot, sunny day in June of 1997, I attended my first gay pride parade, in San Francisco. The size of the crowd amazed me. As I marched from Civic Center up Market Street to San Francisco Bay, a throng of onlookers six persons deep on both sides shouted encouragement and support. For the first time, I felt the sheer magnitude of the gay community.

I stored this impression in the back of my mind. How, I wondered, does biology account for such a huge population that doesn't match the template science teaches as normal? When scientific theory says something's wrong with so many people, perhaps the theory is wrong, not the people.

It wasn't just the number of gay people that astonished me, but the diversity of personal expression in the parade. A drag queen or two were featured in the newspapers, but many other, less flamboyant presentations with different mixtures of gendered symbols were evident as well. I was intrigued, and resolved to investigate further if I ever got the chance. During the next few months I intended to transition into a transgendered woman.[1] I didn't know what the future held—whether I'd be fired as a biology professor, whether I'd become a nightclub waitress, whether I'd even stay alive. I couldn't make long-term plans.

Still, I found my mind leaping from one question to another: What's the real story about diversity in gender and sexuality? How much diversity exists in other vertebrate species? How does diversity evolve in the animal kingdom? And how does diversity develop as individuals grow up: what role do genes, hormones, and brain cells play? And what about diversity in other cultures and historical periods, from biblical times to our own? Even more, I wondered where we might locate diversity in gender expression and sexual orientation within the overall framework of human diversity. Are these types of diversity as innocent as differences in height, weight, body proportion, and aptitude? Or does diversity in gender expression and sexuality merit special alarm and require careful treatment?

A few years after the 1997 parade, I was still alive and still employed. I had been forced to resign from my administrative responsibilities, but found myself with more time for research and writing. I was able to revisit the questions that had flooded my mind as I walked in the parade on that lovely day. This book is the result.

I found more diversity than I had ever dreamed existed. I'm an ecologist—diversity is my job—and yet I was still astonished. Much of this book presents the gee-whiz of vertebrate diversity: how animal families live, how animal societies are organized, how animals change sex, how animals have more than two genders, how species incorporate same-sex courtship, including sexual contact, as regular parts of their social systems. This diversity reveals the evolutionary stability and biological importance of expressions of gender and sexuality that go far beyond the traditional male/female or Mars/Venus binary. I also found that as we develop from tiny embryos to adults, our genes make decisions. Our glorious diversity is the result of our "gene committees" passing various biochemical resolutions. No gene is king, no body type reigns supreme, nor is any template universal in a cacophonous cellular democracy.

I studied how some cultures value transgender people, found where in the Bible transgender people occur, and learned that people from various cultures organize categories of identity differently. Although all cultures span the same range of human diversity, they have different ways of distinguishing gay, lesbian, and transgender identities.

All these facts were new to me, and even now seem utterly engaging, leading to page after page of I-didn't-know-that, wow, and really. This book, then, is a memoir of my travels though the academic spaces of ecology and evolution, molecular biology, and anthropology, sociology, and theology. My general conclusion is that each academic discipline has its own means of discriminating against diversity. At first I felt that the book's main message would be a catalogue of diversity that biologically validates divergent expressions of gender and sexuality. This validating catalogue is indeed important. But as I reflected on my academic sojourn, I increasingly wondered why we didn't already know about nature's wonderful diversity in gender and sexuality. I came to see the book's main message as an indictment of academia for suppressing and denying diversity. I now conclude that all our academic disciplines should go back to school, take refresher courses in their own primary data, and emerge with a reformed, enlarged, and more accurate concept of diversity.

In ecology and evolution, diversity in gender and sexuality is denigrated by sexual selection theory, a perspective that can be traced to Darwin. This theory preaches that males and females obey certain universal templates—the passionate male and the coy female—and that deviations from these templates are anomalies. Yet the facts of nature falsify Darwin's sexual selection theory. In molecular biology and medicine, diversity is pathologized: difference is considered a disease. Yet the absence of a scientific definition of disease implies that the diagnosis of disease is often a value-loaded exercise in prejudice. And in the social sciences, variation in gender and sexuality is considered irrational, and personal agency is denied. Gender- and sexuality-variant people are thought to be motivated by mindless devotion to primitive gods, or compelled by far-fetched psychological urges, or brainwashed by social conventions, and so on: there is always some reason to avoid taking gender- and sexuality-variant people seriously.

The fundamental problem is that our academic disciplines are all rooted in Western culture, which discriminates against diversity. Each discipline finds its own justification for this discrimination. This book blows the whistle on a common pattern of disparaging gender and sexuality variation in academia and predicts foundational difficulties for each discipline.

Although criticism is valuable in its own right, and a critic has no responsibility to suggest solutions, I do suggest improvements when I can. I offer alternatives for interpreting the behavior of animals, interpretations that can be tested and will lead ultimately to more accurate science. I suggest new perspectives on genetics and development that may yield a more successful biotechnology industry. I show that mathematical criteria for the rarity of a genetic disease point to possibly overlooked advantages for genes presently considered defective. I suggest new readings of narratives recorded from gender-variant people across cultures. I call attention to overlooked aspects of the Bible that endorse gender variation.

I do *not* argue that because gender and sexuality variation occur in animals, this variation is also good for humans. People might anticipate that as a scientist I would say, "Natural equals good." I do *not* advocate any version of this fallacy that confuses fact with value. I believe the goodness of a natural trait is the province of ethical reasoning, not science. Infanticide is natural in many animals but wrong in humans. Gender variation and homosexuality are also natural in animals, and perfectly fine in humans. What seems immoral to me is transphobia and homophobia. In the extreme, these phobias may be illnesses requiring therapy, similar to excessive fear of heights or snakes.[2]

I also do not suggest that people are directly comparable to animals. Indeed, even people in different cultures have life experiences that may not be comparable, and comparing people to animals is even riskier. Still, parallels can sometimes be found between cultures. Rugby is a counterpart to American football but located in a different sports culture. Some aspects of American football, like the way play begins by hiking the ball, are comparable to rugby. Similarly, parallels can sometimes be drawn between how people behave and how animals behave, as though animals offered biological cultures resembling ours. I'm quite willing to anthropomorphize about animals. Not that animals are really like people, but animals are not just machines either. We make an error if we attribute too much human quality to animals, but we underestimate them if we think they're mechanical robots. I've tried to strike a balance here.

I've borrowed the word "rainbow" for the title of the book and use it throughout. The word "rainbow" signifies diversity, especially of racial and cultural minorities. The Reverend Jesse Jackson ran for president with the Rainbow Coalition. The rainbow also symbolizes gay liberation.

You probably work with or supervise biologically diverse people. You may be the parent or relative of an unusual child. You may be a teacher, Scout master, coach, minister, legislator, policy analyst, judge, law enforcement officer, journalist, or therapist wondering why your colleagues, clients, or constituencies are so different from the norms we were indoctrinated with as children. You may be a student in college or high school trying to understand diverse classmates. You may be taking a deep breath before coming out yourself, or you may have come out years ago and wish to connect with your roots. You may be studying gender theory and wondering where science fits in, or you may be a woman scientist wondering how to contribute to feminist theory. You may be a conservation biologist wondering how to make biodiversity more relevant to human affairs. You may be a medical student with a professional need for more information about diversity than medical school teaches. You may belong to a discussion group in your place of worship trying to understand how to be inclusive. You may be a young doctoral student shopping for a thesis topic. This book is for all of you.

In Part 1, Animal Rainbows, I begin with my own discipline of ecology and evolution. I've written previously on the evolution of sex: why organisms have evolved to reproduce sexually rather than simply by budding, fragmentation, parthenogenesis, or some other nonsexual means.[3] Reproduction that uses sex rather than bypassing it is better because species need a balanced portfolio of genes to survive over the long term, and sex continually rebalances a species' genetic portfolio. Yet, even though this benefit of gene pool mixing is universal, the means of implementing sexual reproduction are incredibly diverse, spanning many styles of bodies, family organizations, and patterns of bonding between and within sexes, each with its own value and its own internal logic.

Part 1 reviews the body plans, genders, family organizations, female and male mate choices, and sexualities of animals, leading to the conclusion that Darwin's theory of sexual selection is false. I find that competitive tooth-and-claw narratives about nature have been greatly exaggerated, that all sorts of friendships occur among animals, many mediated by sexuality, and that many social roles are signaled by gendered bodily symbols. The great difference in size between an egg and a sperm (a ratio in mass of usually one million to one) is not present to the same degree at the levels of body, behavior, and life history. When a gender binary does exist, the difference is usually slight and sometimes reverses gender stereotypes. Furthermore, there are often more than two genders, with multiple types of males and females. This real-life diversity in gender expression and sexuality challenges basic evolutionary theory.

Darwin is known for three claims: that species are related to one another by sharing descent from common ancestors, that species change through natural selection, and that males and females obey universal templates—the males ardent and the females coy. This third claim results from Darwin's theory about sexual selection, and this claim, not the first two, is what is specifically under challenge. The picture conveyed by Darwin's sexual selection theory is both inaccurate in detail and inadequate in scope to address real-world animal diversity. Darwin's theory of sexual selection is perhaps valid for species like the peacock, whose males have showy ornaments directly used in courtship, but it isn't a general biological theory of gender roles. Twisting Darwin's original theory to conform with today's knowledge renders the theory a tautology. Instead, I submit that the time has come to acknowledge the historical value of Darwin's theory of sexual selection and move on.

I've suggested a new theory that I call "social selection." This new theory accommodates variation in gender and sexuality. It envisages animals as exchanging help in return for access to reproductive opportunity, producing a biological "labor market" for mutual assistance by employing reproductive opportunity as currency. This theory proposes that animals evolve traits that qualify them for inclusion in groups that control resources for reproduction and safe places to live and raise offspring. These traits, called social-inclusionary traits, are either possessed only by females and unexplained by any theory, such as the penis of female spotted hyenas, or possessed only by males and interpreted as a secondary sex characteristic even though they are not actually preferred by females during courtship.

Part 2, Human Rainbows, deals with the areas of biology focused on human development. I tell the story of human embryogenesis as a first-person narrative ("when my sperm part met my egg part") to emphasize that agency and experience function throughout life, before birth and after. I also wish to destabilize the primacy of individualism, to emphasize how much cooperation takes place during development, from the mother who chemically endorses some sperm and not others as competent to fuse with one of her eggs, to genes that interconnect to produce gonads, tissues that touch each other and direct each other's development, and hormones from adjacent babies in utero that permanently influence each other's temperament. Therefore, what we become arises more from our relationships than from our atomic genes, just as a piece of coal's atomic bonds differ from a diamond's, even though both consist solely of carbon atoms.

I've coined the term "genial gene" to distinguish my conception from the popular notion of the selfish gene, which is imagined to single-handedly control development for its own ends. Instead, I emphasize that genes must cooperate lest the common body they inhabit sink like a lifeboat filled with squabbling sailors. I dwell at length on genetic, physiological, and anatomical differences among people. We are as different from each other under the skin as we are on the surface. Although biological differences can be found between the sexes and between people of differing gender expression and sexuality, biological differences can also be found between any two people. For instance, musicians who are string players have been discovered to have brains that differ from those of people who don't play strings. Part 2 shows how medicine seizes on the often tiny anatomical differences between people, and on differences in life experience, to differentiate them from an artificial template of normalcy and deny a wide range of people their human rights by defining them as diseased. Meanwhile, in our society we face not only persecution of people with diverse expressions of gender and sexuality, but also the prospect of doing permanent harm to the integrity of the gene pool of our species, thereby damaging our species for posterity. Part 2 concludes with a summary of the dangers inherent in attempts by genetic engineers to "cleanse" diversity from our gene pool.

In Part 3, Cultural Rainbows, the book progresses from biology to social science, offering a survey and new reading of gender and sexuality variation across cultures and through history. Many tribes of Native Americans accommodated gender and sexuality variation by identifying people as "two-spirits" and including them within social life to an extent that is inspirational to those persecuted in modern society. In Polynesia, the *mahu,* comparable to the Native American two-spirits, are experiencing cultural tension as a result of the introduced Western concept of transgender. Across the globe in India, we find a large castelike group of transgender people called *hijra;* there are over one million hijra in a total population of one billion Indians. The hijra enjoy an ancient pedigree and provide an Asian counterpart to the European history of gender variation that extends from Cybelean priestesses in the Roman empire to the transvestite saints of the Middle Ages, including Joan of Arc (called here Jehanne d'Arc), a transgender man. Early transgender people in Europe were classed as eunuchs, a large group similar to the hijra, with whom they may share a common origin. The Bible, in both Hebrew and Christian testaments (including a passage from Jesus), explicitly endorses eunuchs for baptism and full membership in the religious community. Gender variation was recognized in early Islamic writings as well.

Early Greece enforced a gender binary for techniques of sexual practice: certain practices were considered appropriate for between-sex sexuality and others for same-sex sexuality. Approved practices were called "clean" and those disapproved called "unclean." The Bible is relatively silent on same-sex sexuality, in spite of the centuries-old belief that the Bible condemns homosexuality. I suggest the Bible's clear affirmation of gender variation and its relative silence on same-sex sexuality reflect different ages of gender- and sexuality-variant categories of identity. The category of eunuch extended to the time of Christ and beyond into prehistory, whereas homosexuality as a category of personal identity originated relatively recently in Europe, during the late 1800s. Thus, when the Bible was written, there existed a language for categories of gender variance but not for categories of sexuality variance.

My focus then shifts to anthropologists working in Indonesia, who describe coming reluctantly to acknowledge a legitimate element of masculine gender identity in lesbian expression, although they at first believed that lesbian sexual orientation should not include a masculine presentation. In contrast, an investigator studying Mexican *vestidas* (transgender sex workers) never moves beyond pejorative descriptions. Also, an interesting situation has occurred in the Dominican Republic, where enough intersexed people lived in several villages to have produced a special social category, the *guevedoche*. I wind up the cultural survey by returning to the contemporary United States to discuss the politics of transgender people and their growing alliance with gay and lesbian organizations, and conclude by stating a political agenda for transgendered people. Part 3 demonstrates that our species manifests the same range of variation across cultures and through time, but shows great variation in how we package people into social categories.

In Part 3, I've discussed affirming diversity from a religious standpoint. I believe that ignoring religion, and the Bible specifically, is to work with tunnel vision. Regardless of what science tells us, if people believe that the Bible disparages lesbian, gay, and transgender people, then the cause of inclusion is jeopardized because many would choose religion over science. In fact, I find that the Bible is mostly silent about sexual orientation and that the passages about eunuchs that directly affirm transgender people have been largely ignored. Overall, the Bible gives no support to the religious persecution of gender and sexuality variation. Moreover, the well-known story of Noah's ark imparts a moral imperative to conserve all biodiversity, both across species and within species.

As an appendix, I offer concrete policy recommendations. I suggest strengthening the undergraduate curriculum in psychology and improved education for premedical and medical students to prepare them better to understand natural diversity. I propose new institutional processes to prevent continuing medical abuse of human diversity under the guise of treating diseases. I demand that genetic engineers take an oath of professional responsibility and that they be licensed to practice genetic engineering only after having passed a certifying examination. Finally, I float the idea that our country should construct a large statue and plaza, called the Statue of Diversity, which would be to the West Coast what the Statue of Liberty is to the East Coast.

This book is my first "trade book," a term publishers use for books intended for a wide audience rather than specifically for classroom use—my previous books have been specialized textbooks, monographs, or symposium proceedings.[4] In this type of book I'm free to express opinion and to adopt an informal style. In this book, I freely declare where I'm coming from. Being up front about my position automatically raises the question of objectivity; I've told the truth, and the whole truth, as best I can. Yet I offer my own interpretation of the facts, as if I were a lawyer for the defense opposing lawyers for the "persecution." You, my readers, are a jury of friends and neighbors, and you will make up your own mind. Please consider that everyone writing on these topics is writing from a particular perspective and with a vested interest. Some benefit from the biological excuse for male philandering that Darwin's sexual selection theory provides. Others find validation in Darwin's reinforcement of their aggressive worldview. Still others enjoy the genetic elitism of sexual selection theory, confident that their own genes are superior. I find that refuting sexual selection theory imbues female choice with responsibility for decisions about power and family far more sophisticated than what Darwin envisioned, and empowers varied expressions of gender and sexuality.

At times I've loved writing this book; at other times I've felt afraid of what I have to say. The view of our bodies, of gender and sexuality, that emerges is strikingly new. But I've carried on because I've found the message to be positive and liberating. I hope you enjoy this book. I hope it betters your life.

I thank the staff of the Falconer Biology Library at Stanford University for extensive help with research. I am deeply grateful for reviews from Blake Edgar, Patricia Gowaty, Scott Norton, Robert Sapolsky, and Bonnie Spanier, together with editorial improvements from the staff of the University of California Press, especially Elizabeth Berg and Sue Heinemann. I've been blessed by love from my closest friend, Trudy, and my sisters at Trinity Episcopal Church in Santa Barbara, especially Terry.

PART ONE

ANIMAL RAINBOWS

1

Sex and Diversity

ll species have genetic diversity—their biological rainbow. No exceptions. Biological rainbows are universal and eternal. Yet biological rainbows have posed difficulties for biologists since the beginnings of evolutionary theory. The founder of evolutionary biology, Charles Darwin, details his own struggle to come to terms with natural variation in his diaries from *The Voyage of the Beagle*.[1]

In the mid 1800s, living species were thought to be the biological equivalent of chemical species, such as water or salt. Water is the same everywhere. Countries don't each have water with a unique color and boiling temperature. For biological species, though, often each country does have a unique variant. Darwin saw that finches change in body size from island to island in the Galápagos. We can see that robins in California are squat compared to robins in New England, and lizards of western Puerto Rico are gray compared to the brownish ones near San Juan. Darwin recognized that the defining properties of biological species, unlike physical species, aren't the same everywhere. This realization, new and perplexing in the mid 1800s, remains at times perplexing today.

In Darwin's time, the Linnaean classification system, which is based on phyla, genera, species, and so forth, was just becoming established. Naturalists mounted expeditions to foreign places, collecting specimens for museums and then pigeonholing them into Linnaeus's classification system. At the same time, physicists were developing a periodic table for elements—their classification scheme for physical species—and chemists were classifying recipes for various compounds on the basis of chemical bonds. But the biological counterpart of physical classification didn't work very well. If Boston's robin is different San Francisco's, and if intermediates live at each gas station along Route 80, what do we classify? Who is the "true" robin? What does "robin" mean? Biological names remain problematic in zoology and botany today. Biological rainbows interfere with any attempt to stuff living beings into neat categories. Biology doesn't have a periodic table for its species. Organisms flow across the bounds of any category we construct. In biology, nature abhors a category.

Still, a robin is obviously different from a blue jay. Without names, how can we say whether it is a robin or a blue jay at the bird feeder? The work-around is to collect enough specimens to span the full range of colors in the species' rainbow. Then specialists in biological classification, taxonomists, can say something like, "A robin is any bird between six and seven inches in length with a red to orange breast."[2] No single robin models the "true robin"; *all* robins are true robins. Every robin has first-class status as a robin. No robin is privileged over others as the exemplar of the species.

DIVERSITY—GOOD OR BAD?

Rainbows subvert the human goal of classifying nature. Even worse, variability in a species might signify something wrong, a screwup. In chemistry a variation means impurity, a flaw in the diamond. Doesn't variability within a species also indicate impurity and imperfection? The most basic question faced by evolutionary biology is whether variation within a species is good in its own right or whether it is simply a collection of impurities every species is stuck with. Evolutionary biologists are divided on this issue.

Many evolutionary biologists are positive about the rainbow. They view it as a reservoir of genes that can come to the forefront at different times and places to guarantee a species' survival under changing conditions. The rainbow represents the species' genetic assets.[3] According to this view, the rainbow is decidedly good. This view is optimistic about the capability of species to respond to ever-changing environmental conditions. This view affirms diversity.

Other evolutionary biologists are negative about the rainbow, believing that all gene pools—even our own—are loaded with deleterious mutations, or bad genes. During the 1950s, studies claimed that every person has three to five lethal recessive genes that would surface if they chose the wrong marriage partner, causing their children to die.[4] This view is pessimistic about the future, suggesting that evolution has already reached its pinnacle and all variation is useless or harmful.[5] This school of evolutionists believed in a genetic elite, advocating artificial insemination from sperm banks stocked with genes from great men. This view represses diversity.

Darwin himself was ambivalent on the value of rainbows. Darwin argued that natural selection is the mechanism that causes species to evolve. On the one hand, because natural selection depends on variation, Darwin viewed the rainbow as a spectrum of possibilities constituting the species' future. A species without variability has no evolutionary potential, like a firm with no new products in the pipeline. On the other hand, Darwin viewed females as shopping around for mates with desirable genes while rejecting those with inferior genes. This view demeans the variation among males and implies a hierarchy of quality, suggesting that female choice is about finding the best male rather than the best match. Darwin both affirmed and repressed diversity at different times within his career.

The philosophical conflict over whether to affirm or to repress diversity is still with us today, permeating everything from the way biologists interpret motives for an animal's choice of a particular mate to how medical doctors handle newborn babies in the hospital.

THE COSTS VERSUS THE BENEFITS OF SEX

How, then, are we to decide whether rainbows are good or bad? Who is correct, the diversity affirmers or the diversity repressers? To answer this most fundamental question of evolutionary biology, let's compare species with full rainbows to species with very limited rainbows. Species who manage to reproduce without sex have limited rainbows. By sex, I mean two parents mixing genes to produce offspring. Lots of species propagate without sex. In such species, everyone is female and offspring are produced without fertilization. In addition, in many species offspring may be produced either with or without fertilization, depending on the season.

If you go to Hawaii, look at the cute geckoes on the walls. You're seeing an asexual species—all these geckoes are female.[6] Females in all-female species produce eggs that have all the needed genetic material to begin with. In sexual species, like humans, an egg has only half the genetic material needed to produce a baby; a sperm has the other half, so combining these yields the required material. In addition, eggs from an all-female species don't need fertilization by a sperm to trigger the cell divisions that generate an embryo. Females in all-female species clone themselves when they reproduce.

The Hawaiian all-female geckoes are locally abundant and widespread throughout the South Pacific, from the lovely Society Islands of French Polynesia to the Marianas Islands near New Guinea. More all-female species live in Mexico, New Mexico, and Texas—all varieties of whiptail lizards These small, sleek tan and brown-striped animals dart quickly along the ground looking for food. The all-female species of whiptail lizards live along streambeds, while sexually reproducing relatives typically live up-slope from the streams in adjacent woods or other vegetation. Every major river drainage basin in southwestern North America is a site where an all-female whiptail lizard species has evolved. More than eight all-female species are found in this area. Still more all-female species of lizards are found in the Caucasus Mountains of Armenia and along the Amazon River in Brazil. All-female fish occur too. Indeed, all-female animal species are found among most major groups of vertebrates.[7]

Also, some species have two kinds of females: those who don't mate when reproducing and those who do mate. Examples include grasshoppers, locusts, moths, mosquitoes, roaches, fruit flies, and bees among insects, as well as turkeys and chickens.[8] Fruit flies grow easily in the laboratory and are especially well studied. Over 80 percent of fruit fly species have at least some females that reproduce entirely asexually. Although the majority of females in these species reproduce through mating, selection in the laboratory increased sixtyfold the proportion of females not needing to mate, yielding a vigorous all-female strain.[9]

Thus all-female species are well known among animals. So why don't even more all-female species exist? Indeed, why aren't *all* species all-female? To answer this question, let's look at the costs and benefits of reproducing with and without sex.

Sexual reproduction cuts the population's growth rate in half—this is the cost of sex. Only females produce offspring, not males. If half the population is male, then the speed of population growth is half that of an all-female population. An all-female species can quickly outproduce a male/female species, allowing an all-female species to survive in high-mortality habitats where a male/female species can't succeed. (This result is also true in hermaphrodite species, in which the fifty-fifty allocation of reproductive effort to male and female function reduces the female allocation used to make eggs by half.)

The potential for doubling production in an all-female species hasn't escaped the attention of agricultural scientists. In the 1960s, turkeys and chickens were bred to make all-female strains.[10] Indeed, the cloning of a sheep in Scotland reflected a fifty-year-old aspiration to increase agricultural production by taking the sex out of reproduction. However, despite the big advantage in population growth rate that all-female species enjoy and the many examples of all-female species that do occur, clonally reproducing species remain a tiny minority. Far and away most species are sexual. Nature has experimented many times and keeps experimenting with clonal species, but with little success. Sex does work. Why?

The benefit of sex is survival over evolutionary time. Lacking sex, clonal species are evolutionary dead ends. On an evolutionary time scale, almost all clonal species are recently derived from sexual ancestors. On the family tree of species, asexual species are only short twigs, not the long branches.[11] The advantages of sex are also demonstrated by species who can use sex or not, depending on the time of year. Aphids (tiny insects that live on garden plants) reproduce clonally at the beginning of the growing season, switching to sexual reproduction at the end of the season. Aphids benefit from fast reproduction when colonizing an empty rose bush, but the anticipated change of conditions at the end of the season makes sexual reproduction more attractive.[12]

Clonally reproducing species are "weeds"—species specialized for quick growth and fast dispersal, like plants that locate and colonize new patches of ground. The common dandelion of North America is a clonal reproducer whose sexual ancestors live in Europe.[13] Weeds eventually give up their territory to species who are poorer colonizers but more effective over the long term.[14] The geckoes who colonized the South Pacific and the whiptail lizards of New Mexico streambeds make sense in these contexts, where dispersal is at a premium or the habitat is continually disturbed.

Clonal reproduction is a specialized mode of life, not recommended for any species that fancies itself a permanent resident of this planet. But we haven't answered *why* sexual reproduction is good over the long term. Two theories have been offered for why sex benefits a species, one diversity-affirming, the other diversity-repressing. Both theories agree that asexual species are short-lived in evolutionary time relative to sexual species and that sex guarantees the longer species survival. Both theories therefore agree that sex is beneficial to a species. Both theories also agree that the purpose of sex isn't reproduction as such, because asexual species are perfectly capable of reproducing. But the theories have different perceptions of why sex is good. The diversity-affirming theory views diversity itself as good and sex as maintaining that diversity. The diversity-repressing theory views diversity as bad and sex as keeping the diversity pruned back.[15] Let's start with the diversity-affirming theory.

THE DIVERSITY-AFFIRMING THEORY

According to the diversity-affirming theory for the benefit of sex, sex continually rebalances the genetic portfolio of a species. Think of a savings account and jewelry—a rainbow with two colors. How much can both colors earn together? When demand for jewelry is low, one can't sell jewelry, even to a pawnshop, and earning 2 percent from a bank account looks great. When jewelry is hot, interest on a bank account looks cheap and selling jewelry turns a good profit. The overall earnings are the total from both investments.

A species earns offspring instead of money from its investments. The long-term survival of a species depends on being sufficiently diversified to always have some offspring-earning colors. Although biologists may talk about the rainbow as a source of genes for new environments, it is in fact more important for surviving the regular fluctuations between hot and cold, wet and dry, and the arrival and departure of new predators, competitors, and pathogens like the bubonic plague or AIDS.[16]

The social environment within a species is always changing too. Concepts of the "ideal" mate change through time. Among humans, men have sometimes preferred the amply proportioned Mama Casses among us, at other times the skinny Twiggys, as recorded in the portraits of women from art museums. Other aspects of our social environment have also changed over the centuries, like the fraction of time spent with others of the same sex or the opposite sex, or the number of sex partners a person has. Changes in the social setting within a species, as well as changes in the ecological and physical environment, all affect which colors of the rainbow shine the brightest at any one time.

A clonal species can accumulate diversity through mutation, or it may have multiple origins, thereby starting out with a limited rainbow. In fact, several genetically distinct clones have been detected among the South Pacific geckoes and dandelions. Still, these mutation-based and origin-based rainbows are nearly monochromatic.[17]

Furthermore, even the limited rainbow of a clonal species is continually endangered. The colors that shine brightly are always crowding out the colors that don't, causing diversity to contract over time. Recall the jewelry and the savings account. If diamonds are valuable for a long time, their value grows and comes to overshadow the savings account. If profits are automatically reinvested in the most immediately successful venture, the portfolio gradually loses its diversity. Then when the demand for jewelry drops—say because a new find of diamonds floods the market—the portfolio takes a big hit. This progression is similar to that of the clonal reproducer, which courts danger by concentrating on only a few investments. Instead, one should redistribute some earnings each year across the investments. If jewelry has a good year, sell some and put the proceeds in the savings account. If interest is high one year, then withdraw some funds and buy jewelry. Shuffling money across investments in this way maintains the portfolio's diversity, and a bad year for one investment doesn't cause disastrous losses in the portfolio. Wall Street investors call this shuffling "rebalancing a portfolio." This is the strategy of the sexual reproducer. Every generation when sexually reproducing animals mate, they mix genes with one another and resynthesize the colors in short supply. Thus, according to the diversity-affirming theory, sex serves to maintain the biological rainbow, which conserves the species.[18]

THE DIVERSITY-REPRESSING THEORY

According to the diversity-repressing theory for the benefit of sex, sex protects the genetic quality of the species. The diversity-repressing theory envisions that asexual species accumulate harmful mutations over time and gradually become less functional, as though asexual lizards gradually lost the ability to run fast or digest some food. Sex supposedly counteracts this danger by allowing family lines that have picked up harmful mutations to recombine, producing offspring free of bad mutations. According to this theory, some offspring will possess both families' mutations and will die even more quickly, but other offspring will have none of the mutations, and will prosper on behalf of the species. According to this theory, without sex each and every family line inexorably accumulates mutations, leading eventually to species extinction.

ENDING THE DEBATE

Although both the diversity-affirming and diversity-repressing views have a long history, the time has come for closure. The time has come to reject the diversity-repressing view as both theoretically impossible and empirically vacuous. The scenario envisioned by the diversity-repressing theory can't exist. In an asexual species, when a bad gene arises, the line where the mutation originated is lost to natural selection, whereas the lines without the mutation prosper. The entire stock never deteriorates, because natural selection doesn't look the other way while a bad gene spreads. Instead, natural selection eliminates a bad gene when it first appears, preserving the overall functionality of the species. No evidence whatsoever shows asexual species becoming extinct because of a progressive accumulation of disabilities and loss of functionality. A bad gene never gets going in an asexual species, and sex's supposed pruning of the gene pool is unnecessary and mythical.

On the other hand, the environment does change from year to year, and individuals who don't do well one year may shine when conditions change, and vice versa. Butterflies whose enzymes work at cold temperatures thrive in dark, damp years, while butterflies whose enzymes function best at hot temperatures do better in sunny drought years. All butterflies are perfectly good butterflies, even if the abilities of some don't match the opportunities currently supplied by the environment.

I don't see any grounds for dignifying the diversity-repressing view for the benefit of sex as a viable alternative to the diversity-affirming view. To be agreeable, one might say both theories are valid. But this compromise isn't true. Conceding, even slightly, that one function of sex is to prune diversity puts forth a view that hasn't earned its place scientifically. Accepting a diversity-repressing view of sex simply to be polite admits through the back door a philosophical stance that may later be used to justify discrimination.

Therefore, I accept as a working premise that a species' biological rainbow is good—good because diversity allows a species to survive and prosper in continually changing conditions. I further accept that the purpose of sex is to maintain the rainbow's diversity, resynthesizing that diversity each generation in order to continually rebalance the genetic portfolio of the species. I reject the alternative theory that sex exists to prune the gene pool of bad diversity.

Darwinists have to take a consistent stand on the value of diversity. They can't maintain on the one hand that most variation is good because it's needed for natural selection and on the other hand also maintain that females must continually shop for males with the best genes as though most genes could be ranked from good to bad. Instead, I argue that almost all diversity is good and that female choice is more for the best match than for the best male.

How then should we assess the rainbows in our own species? We should be grateful that we do reproduce sexually, although we probably take this gift for granted. I feel too that we should conserve and embrace our rainbows. Affirming diversity is hard, very hard. We must come to accept ourselves and love our neighbors, regardless of color in the rainbow.

Overall, sex is essentially cooperative—a natural covenant to share genetic wealth. Sexual reproduction is not a battle.

2

Sex versus Gender

To most people, "sex" automatically implies "male" or "female." Not to a biologist. As we saw in the last chapter, sex means mixing genes when reproducing. Sexual reproduction is producing offspring by mixing genes from two parents, whereas asexual reproduction is producing offspring by one parent only, as in cloning. The definition of sexual reproduction makes no mention of "male" and "female." So what do "male" and "female" have to do with sex? The answer, one might suppose, is that when sexual reproduction does occur, one parent is male and the other female. But how do we know which one is the male? What makes a male, male, and a female, female? Indeed, are there only two sexes? Could there be a third sex? How do we define male and female anyway?

"Gender" also automatically implies "male" and "female" to most people. Therefore, if we define male and female biologically, do we wind up defining gender as well? Similarly, for adjectives like "masculine" and "feminine," can we define these biologically? Moreover, among humans, is a "man" automatically male and a "woman" necessarily female? One might think, yes, of course, but on reflection these key words admit lots of wiggle room. This chapter develops some definitions for all these words, definitions that will come in handy later on.

When speaking about humans, I find it's helpful to distinguish between social categories and biological categories. "Men" and "women" are so-cial categories. We have the freedom to decide who counts as a man and who counts as a woman. The criteria change from time to time. In some circles, a "real man" can't eat quiche. In other circles, people seize on physical traits to define manhood: height, voice, Y chromosome, or penis. Yet these traits don't always go together: some men are short, oth-ers are tenors, some don't have a Y chromosome, and others don't have a penis. Still, we may choose to consider all such people as men anyway for purposes like deciding which jobs they can apply for, which clubs they can join, which sports they may play, and whom they may marry.

For biological categories we don't have the same freedom. "Male" and "female" are biological categories, and the criteria for classifying an organism as male or female have to work with worms to whales, with red seaweed to redwood trees. When it comes to humans, the biological criteria for male and female don't coincide 100 percent with present-day social criteria for man and woman. Indeed, using biological categories as though they were social categories is a mistake called "essentialism." Es-sentialism amounts to passing the buck. Instead of taking responsibility for who counts socially as a man or woman, people turn to science, try-ing to use the biological criteria for male to define a man and the bio-logical criteria for female to define a woman. However, the definition of social categories rests with society, not science, and social categories can't be made to coincide with biological categories except by fiat.

MALE AND FEMALE DEFINED

To a biologist, *"male" means making small gametes, and "female" means making large gametes.* Period! By definition, the smaller of the two gametes is called a sperm, and the larger an egg.* Beyond gamete size, biologists don't recognize any other universal difference between male and female. Of course, indirect markers of gamete size may exist in some species. In mammals, males usually have a Y chromosome. But

---

*A gamete is a cell containing half of its parent's genes. Fusing two gametes, each with half the needed number of genes, produces a new individual. A gamete is made through a spe-cial kind of cell division called meiosis, whereas other cells are made through the regular kind of cell division, called mitosis. When two gametes fuse, the resulting cell is called a zygote. A fertilized egg is a zygote.

whether an individual is male or not comes down to making sperm, and the males in some mammalian species don't have a Y chromosome. Moreover, in birds, reptiles, and amphibians, the Y chromosome doesn't occur. However, the gamete-size definition is general and works throughout the plant and animal kingdoms.

Talk of gamete size may seem anticlimactic. Among humans, for example, centuries of poetry and art speak of strength and valor among men, matched by beauty and motherhood among women. Saying that the only essential difference between male and female is gamete size seems so trivial. The key point here is that "male" and "female" are biological categories, whereas "man" and "woman" are social categories. Poetry and art are about men and women, not males and females. Men and women differ in many social dimensions in addition to the biological dimension of gamete size.

Yet, biologically, the gamete-size definition of "male" and "female" is far from anticlimactic. In fact, this definition is downright exciting. One could imagine species whose members all make gametes of the same size, or several gamete sizes—small, medium, and large—or a continuum of gamete sizes ranging from small to large. Are there any such species? Almost none. Some species of algae, fungi, and protozoans have gametes all the same size. Mating typically occurs only between individuals in genetic categories called "mating types." Often there are more than two mating types.[1] In these cases, sex takes place between the mating types, but the distinctions of male and female don't apply because there is only one gamete size.[2] By contrast, when gametes do come in more than one size, then there are generally only two sizes, one very small and the other very large. Multicellular organisms with three or more distinct gamete sizes are exceedingly rare, and none is known to have a continuum of gamete sizes.

More than two gamete sizes occur in some colonial single-celled organisms, the protozoans. In the green ciliate *Clamydomonas euchlora,* the cells producing gametes may divide from four to sixty-four times. Four divisions result in relatively big gametes, whereas sixty-four divisions produce lots of small gametes. The cells that divide more than four times but less than sixty-four make various intermediate-sized gametes. Another ciliate, *Pandorina,* lives in colonies of sixteen cells. At reproduction, some cells divide into eight big gametes and others into sixteen small gametes. However, any two of these can fuse: two big ones, one big and one small, or two small ones.[3] These species are at the borderline between single-celled and multicellular organisms.

In the fruit fly *Drosophila bifurca* of the southwestern United States, the sperm is twenty (yes, twenty) times longer than the size of the male who made it! These sperm don't come cheap. The testes that make these sperm comprise 11 percent of the adult male's weight. The sperm take a long time to produce, and males take twice as long to mature as females. The sperm are so expensive that males conserve them, "offering" them to females in small amounts, leading to a one-to-one gamete ratio.[4] So much for the vision of one huge egg surrounded by zillions of tiny sperm. Although giant sperm are a marvelous curiosity, the important finding is that some species of *Drosophila* have three sperm sizes—one giant type and two smaller varieties that overlap somewhat, totaling four gamete sizes (three sperm sizes plus one egg size). In *Drosophila pseudoobscura* from Tempe, Arizona, the tail of the big sperm is 1/3 millimeter long, and the tail lengths of the two small types are 1/10 and 1/20 millimeter.

Female *Drosophila* in some species can store sperm for several days or even up to a month after mating. About one-third of the sperm are the giant type; the remaining two-thirds are small. Females preferentially store the large sperm, although some small sperm are stored too. Females also control which sperm are used for fertilization and preferentially select the large sperm. Whether the small sperm are ever used for fertilization has been hard to demonstrate. The amount of material in a large sperm is about one hundred times that of a small sperm. Therefore, to break even, the fertilization rate for small sperm needs to be only 1/100 of the fertilization rate of large sperm, and this low rate would be hard to detect.[5]

If different individuals made the different-size gametes, we could have as many as four sexes in *Drosophila,* one for each gamete size. In this species, however, every male apparently makes all three of the sperm sizes in the same ratio, so all the males are apparently the same. If further research reveals that the sperm makers differ in the ratio of sperm sizes they produce, we will have discovered a species with more than two sexes. Such a discovery would not violate any law of nature, but it would be very rare and would certainly make headlines. So, for practical purposes, male and female are universal biological categories defined by a binary distinction between small and large gametes, sperm and egg.

Why are two gamete sizes practically universal in sexually reproducing species? The current theory imagines a hypothetical species starting with two mating types that produce gametes of the same size. These gametes fuse with each other to produce a zygote, and each gamete contributes half the genes and half the cytoplasm needed by the zygote. Then the gamete in one of the mating types is hypothesized to evolve a smaller size to increase quantity while sacrificing quality. The gamete in the other mating type responds by evolving a larger gamete size to compensate for the lowered quality of the small gametes now being made by its counterpart. Overall, this back-and-forth evolutionary negotiation between the mating types with respect to gamete size culminates in one mating type making the tiniest gametes possible—gametes that provide genes and nothing else, whereas the other mating type makes gametes large enough to provide genes as well as all the cytoplasm the zygote needs to start life.[6]

This little story of how the gamete binary originates is completely conjectural and untested, and points to the need for much further thought on such an important issue. This story also leaves unexplained why some groups, such as fungi, persist with only one gamete size, and why rare groups such as *Drosophila* occur with multiple sperm sizes.

GENDER DEFINED

Up to now, we've come up with two generalizations: (1) Most species reproduce sexually. (2) Among the species that do reproduce sexually, gamete size obeys a near-universal binary between very small (sperm) and large (egg), so that male and female can be defined biologically as the production of small and large gametes, respectively. Beyond these two generalizations, the generalizing stops and diversity begins!

The binary in gamete size doesn't extend outward. The biggest error of biology today is uncritically assuming that the gamete size binary implies a corresponding binary in body type, behavior, and life history. No binary governs the whole individuals who make gametes, who bring them to one another for fertilization, and who interact with one another to survive in a native social context. In fact, the very sexual process that maintains the rainbow of a species and facilitates long-term survival automatically brings a cornucopia of colorful sexual behaviors. Gender, unlike gamete size, is not limited to two.[7]

"Gender" usually refers to the way a person expresses sexual identity in a cultural context. Gender reflects both the individual reaching out to cultural norms and society imposing its expectations on the individual. Gender is usually thought to be uniquely human—any species has sexes, but only people have genders. With your permission, though, I'd like to widen the meaning of gender to refer to nonhuman species as well. As a definition, I suggest: *Gender is the appearance, behavior, and life history of a sexed body.* A body becomes "sexed" when classified with respect to the size of the gametes produced. Thus, gender is appearance plus action, how an organism uses morphology, including color and shape, plus behavior to carry out a sexual role.

Now we're free to explore the zoological (and botanical!) counterpart of human gender studies. So, we may ask: How much variety occurs in gender expression among other species? Let's take some favorite stereotypes and see. We'll look mostly at vertebrates; even more variety occurs with invertebrates and plants.

An organism is solely male or female for life. No, the most common body form among plants and in perhaps half of the animal kingdom is for an individual to be both male and female at the same, or at different times during its life. These individuals make both small and large gametes during their lives.

Males are bigger than females, on the average. No, in lots of species, especially fish, the female is bigger than the male.

Females, not males, give birth. No, in many species the female deposits the eggs in the pouch of the male, who incubates them until birth. In many species, males, not females, tend the nest.

Males have XY chromosomes and females XX chromosomes. No, in birds, including domesticated poultry like chickens, the reverse is true. In many other species, males and females show no difference in chromosomes. In all alligators and crocodiles, some turtles and lizards, and the occasional fish, sex is determined by the temperature at which the eggs are raised. A female can control the sex ratio among her offspring by laying eggs in a shady or a sunny spot.

Only two genders occur, corresponding to the two sexes. No, many species have three or more genders, with individuals of each sex occurring in two or more forms.

Males and females look different from one another. No, in some species, males and females are almost indistinguishable. In other species, males occur in two or more forms, one of which resembles a female, while the others are different from the female.

The male has the penis and the female lactates. No, in the spotted hyena, females have a penislike structure externally identical to that of males, and in the fruit bat of Malaysia and Borneo, the males have milk-producing mammary glands.[8]

Males control females. No, in some species females control males, and in many, mating is a dynamic interaction between female and male choice. Females may or may not prefer a dominant male.

Females prefer monogamy and males want to play around. No, depending on the species, either or both sexes may play around. Lifelong monogamy is rare, and even within monogamous species, females may initiate divorce to acquire a higher-ranking male.

One could tick off even more examples of gender stereotypes that are often thought to be "nature's way" but that have no generality within biology. Instead, let's look closer at the lives of these organisms to see whether what they do makes sense to us. Be prepared, though, to shrug your shoulders and wonder about the mystery of life.

Note that by defining gender as how an organism presents and carries out a sexual role, we can also define masculine and feminine in ways unique to each species. "Masculine" and "feminine" refer to the distinguishing traits possessed by most males and females respectively. Cross-gender appearance and behavior are also possible. For example, if most females have vertical stripes on their bodies and males do not, then a male with vertical stripes is a "feminine male." If most males have antlers and females do not, then a doe with antlers is a "masculine female."

Politically, locating the definition of male and female with gamete size keeps society's gender categories at arm's length from biology's sex binary. We don't have to deny the universality of the biological male/female distinction in order to challenge whether the gender of whole organisms also sorts into a male/female binary. In humans specifically, a gender binary for whole people is not clear-cut even though the difference between human sperm and egg is obvious—a size ratio of about one million to one.

3

Sex within Bodies

Although the binary in gamete size is practically universal, the way male and female functions are packaged into individual bodies does not fit into any consistent polarity. We tend to think that males and females must be in separate bodies because most of us are, as are most of the animals we live with, such as our pets, domesticated stock, and the birds and bees around our parks. However, many species have other ways of organizing sexual functions.

An individual body who makes *both* small and large gametes at some point in life is called a hermaphrodite. An individual who makes both sizes at the same time is a simultaneous hermaphrodite, and one who makes them at different times is a sequential hermaphrodite. Most flowering plants are simultaneous hermaphrodites because they make pollen and seeds at the same time. Pollen is the male part of a plant and the ovule is the female part. A pollinated ovule turns into a tiny embryo that detaches, to be blown away by the wind or carried away by an animal.

Among animals, hermaphrodism is common in the ocean.[1] Most marine invertebrates, such as barnacles, snails, starfish, fan worms, and sea anemones, are hermaphroditic. Many fish are too. If you go snorkeling at a coral reef in Hawaii, the Caribbean, Australia, or the Red Sea, chances are that about a quarter of the fish you see will be hermaphroditic. Or take a look at some of the colorful fish popular in tropical aquaria—they are often sequential hermaphrodites. Most species of wrasses, parrot fish, and larger groupers are hermaphroditic, as are some damselfish, angelfish, gobies, porgies, emperors, soapfishes, dottybacks, and moray eels (all from shallow waters), and many deep-sea fish as well.[2]

Hermaphrodism is a successful way of life for many species; my guess is that hermaphrodism is more common in the world than species who maintain separate sexes in separate bodies (called gonochorism). The separate-sex/separate-body state is often viewed as "normal," suggesting that something unusual favors hermaphrodism in plants, on coral reefs, and in the deep sea. Alternatively, hermaphrodism may be viewed as the original norm, prompting us to ask what there is in mobile organisms in the terrestrial environment that favors separate sexes in separate bodies.

WHAT FISH CAN TELL US

FEMALES CHANGING TO MALE

Sex change is only one of several interesting aspects of coral reef fish society. The bluehead wrasse is named for the blue head of the largest males. When small and just entering sexual maturity, fish of both sexes look similar. Later three genders develop. One gender consists of individuals who begin life as a male and remain so for life. Another gender consists of individuals who begin as females and later change into males. These sex-changed males are larger than those who have been male from the beginning. The third gender consists of females who remain female. We'll call the two male genders the "small unchanged males" and the "large sex-changed males," respectively. The large sex-changed males are the biggest individuals of the three genders, and they attempt to control the females. In some species, the large sex-changed males maintain and defend the females, and in others they defend locations that females appear to prefer.

Fertilization is external—a female releases eggs into the water and a male then releases a cloud of sperm around the eggs to fertilize them. The unfertilized eggs are out in the open and can potentially be fertilized by any male in the vicinity.

The small unchanged and large sex-changed males are hostile to each other. The large sex-changed males chase the small unchanged males away from the territory or from females they control. The small un-changed males are more numerous than the large sex-changed males and may form coalitions to mate with females that a large sex-changed male is trying to control. The small unchanged males mate by darting in and fertilizing the eggs that a large sex-changed male was intending to fertil-ize. Some small unchanged males keep the large sex-changed male busy while others are mating.

Different ecological circumstances favor unchanged and sex-changed males. The wrasses live both on coral reefs and in the seagrass beds nearby. In seagrass, females nestled among grass blades can't be guarded very well, and the balance of hostilities tips in favor of the small un-changed males. This situation leads to only two genders, unchanged males and females. On the coral reef, clear water and an open habitat structure permit the large sex-changed males to control the females, and the balance tips in their favor.[3] This situation encourages the presence of all three genders. Simple population density also shifts the gender ratios. At high densities females are difficult to guard and small unchanged males predominate, whereas at low densities a large sex-changed male can control a "harem."[4] Whether females prefer either type of male isn't known.

The sex changes are triggered by changes in social organization. An-other type of wrasse is the cleaner wrasse, named for its occupation of gleaning ectoparasites from other fish. When a large sex-changed male is removed from his harem, the largest female changes sex and takes over. Within a few hours, she adopts male behavior, including courtship and spawning with the remaining females. Within ten days, this new male is producing active sperm. Meanwhile the other females in the harem remain unchanged.[5] I haven't been able to find out whether any female can turn into a male if she is the biggest female when the existing male dies, or whether females divide into two groups—those who remain female no matter what and those who change sex when circumstances are right.

Does this animal society seem oh-so-bizarre? It isn't. Aspects of this system appear again and again among vertebrates, especially the themes of male control of females or their eggs, multiple male genders, hostility among some of the male genders, flexible sexual identity, and social or-ganization that changes with ecological context. Still, if you think the coral reef fish scene is bizarre, you're not alone—so did the biologists who first witnessed it. We're only just realizing that the concepts of gender and sexuality we grew up with are seriously flawed.

MALES CHANGING TO FEMALE

Sex changes from male to female also occur. A group of damselfish are called clown fish because their bold white strips remind one of the white makeup used by clowns. These fish live among the tentacles of sea anemones, which have cells in their tentacles that sting any animal who touches them. To survive in this lethal home, a clown fish secretes a mucus that inhibits the anemone from discharging its stinging cells. Although living within the anemone's tentacles provides safety for the clown fish, the size of its house is limited by how big its sea anemone grows. An anemone has space for only one pair of adult clown fish and a few juveniles.

The female is larger than the male. If she is removed, the remaining male turns into a female, and one of the juveniles matures into a male.[6] The pair is monogamous. Female egg production increases with body size. A monogamous male finds no advantage in being large because he's not controlling a harem of females. The advantage for males of remaining small and for females of becoming large may account for the developmental progression from male to female.[7]

MALE AND FEMALE SIMULTANEOUSLY

Hamlets, which are small coral reef basses, don't have to worry about choosing their sex: they are both sexes at the same time. However, they cross-fertilize and must mate with a partner to reproduce. These simultaneous hermaphrodites change between male and female roles several times as they mate. One individual releases a few eggs and the other fertilizes them with sperm. Then the other releases some eggs, which the first fertilizes with sperm, and so on, back and forth.[8]

No one has offered any suggestions about why hamlets are simultaneous hermaphrodites. Deep-sea fish also tend toward simultaneous hermaphrodism, which for these species is viewed as an adaptation to extremely low population density.[9] Hamlets don't have a strange appearance, nor do any other hermaphroditic fish. Hermaphroditic fish look like, well, just fish. Hamlets are not particularly rare, nor are they derived from ancestors who were rare or lived in the deep sea. So just why hamlets are simultaneously hermaphroditic remains mysterious.

MALE AND FEMALE CRISSCROSSING

Changing sex once may seem a big deal, but some fish do it several times during their life span. An individual may change from an unsexed juvenile to a female, then to a male, and then back to a female. Or it may change from a juvenile to a male, then to a female, and then back to a male. In certain species, sexual identity can be changed as easily as a new coat.

Sex crisscrossing was first discovered in a species of goby, which is the largest family of fish. Gobies are tiny and often live on coral reefs—in this case, on the Pacific island of Okinawa.[10] These gobies live as monogamous pairs on branching coral, and the males care for the eggs. About 80 percent of the juveniles mature female, and the rest mature male. Some of the females later switch to male, and some of the males later switch to female. Of those that have switched once, a small fraction later switch back again—the crisscrossers.

Why go to the expense of changing one's sexual wardrobe? One theory envisages pair formation in gobies as resulting when two larvae drop out of the plankton together onto a piece of coral.[11] They awake after metamorphosis to discover that they are both the same sex. What to do? Well, one of them changes sex. Changing sex has been suggested as a better way of obtaining a heterosexual pairing than moving somewhere else to find a partner of the opposite sex when traveling around is risky. Thus this theory comes down to a choice: switch or move. This theory is rather heterosexist, though. As the hamlets show, a heterosexual pair is not necessary for reproduction, because both could be simultaneously hermaphroditic and not have to bother with crisscrossing.

A species of goby from Lizard Island on Australia's Great Barrier Reef has recently been discovered to crisscross, but in a way that is interestingly different from the Okinawan goby.[12] In the Australian goby, all the juveniles mature into females, with some later becoming males. The males, however, can change back into females. In fact, the meaning of male is ambiguous here. The investigators defined a male to be any fish with at least some sperm production. All males, however, contain early-stage oocytes—cells that develop into eggs—in their gonads. So all the males remain part female. The species therefore consists of two genders at any one time: all-female fish and part-male-part-female fish.

Among flowering plants, populations with hermaphrodites and females are common,[13] more so than populations with males and females. These mixed hermaphrodite/single-sex species contrast with most plant species, which are entirely hermaphroditic. (Perhaps as more gobies are investigated, a species will be found consisting of females and hermaphrodites, just as in plants.)

Plants also offer the most amusing examples of crisscrossing sex changes. In a tropical ginger from China, some individuals are male in the morning, making pollen, while others are female in the morning, receiving pollen. Then they switch sexes in the afternoon. This phenomenon, called flexistyly, is known in eleven families of flowering plants.[14] The ginger's diurnal sex change is not too different from how hamlets mate, where members of a mating pair switch back and forth between male and female once a minute.

These examples of sequential, simultaneous, and crisscrossing hermaphrodism show that male and female functions don't need to be packaged into lifelong distinct bodies. Hermaphroditic vertebrate species are successful and common.

INTERSEXES IN MAMMALS

Can mammals be hermaphroditic too, or have we been left out? Not entirely. Mammals described as hermaphrodites are often reported, although the word "hermaphrodite" is misleading.

Let's work out some definitions. The reproductive system in mammals consists of gonads—the place where eggs and sperm develop—and plumbing, which transports gametes from the gonads to their destination. The plumbing consists of internal pipes and external valves. The internal pipes are fallopian tubes, muellerian ducts, and so forth. External valves include the penis, clitoris, scrotum, labia, and so on. An "intersexed" individual has gonads to make both eggs and sperm and/or combinations of sperm-related and egg-related plumbing parts. With so many parts in the overall system, many combinations are possible.

To be more specific, we can distinguish intersexed gonads, with some combination of ovarian and testicular tissue, from intersexed genitals, with some combination of egg- and sperm-related plumbing.[15] We could even distinguish internal genitally intersexed and external genitally intersexed to pinpoint where the combined plumbing is located. Although the gamete-size binary implies that only two sexed functions exist, many body types occur, ranging from all-sperm parts, through various combinations of both sperm- and egg-related parts, to all-egg parts.

To manufacture a hermaphrodite using mammalian components on a vertebrate chassis, two entire sets of gonadal and plumbing parts are needed, one for eggs and one for sperm. Mammals show many partial combinations of sperm- and egg-related parts. All the partial combinations could be stirred together into a putty from which evolution might someday mold a full mammalian hermaphrodite if selection pressure for that arose, a pressure such as those to which coral reef fish have already responded. In some mammalian species, intesexed bodies are a minority; in others, the majority.

Antlers offer easy-to-see clues for possible intersexed individuals. White-tailed deer *(Odocoileus virginianus)* possess a male body type, called a velvet-horn because these deer retain the special velvet skin over the antlers that is usually shed after the antlers have aged. Velvet-horn males have small antlers, doelike body proportions and facial features, and small testes; they are said to be infertile. Females typically don't have antlers, but there is a type of female deer with hard, bony antlers and extensively combined plumbing parts, which is believed to be infertile. A distinct fertile antlerless male morph and a distinct fertile antlered female morph occur as well.

The mention of infertility plays to the prejudice that something is "wrong" with intersexes. But the story is more complicated. The frequency of velvet-horns in white-tailed deer is around 10 percent in some areas and can reach as high as 40 to 80 percent.[16] Numbers this big contradict the idea that velvet-horns represent a deleterious mutation.

Similarly, a male morph in black-tailed deer *(Odocoileus hemionus)* called a cactus buck may be a form of intersex as well. Elk *(Cervus ela-*

phus, also called red-tailed deer), swamp deer *(Cervus duvauceli)*, Sika deer *(Cervus nippon)*, roe deer *(Capreolus capreolus)*, and fallow deer *(Dama dama)* all have a male morph with velvet-covered antlers, called a peruke, that is described as nonreproductive. Moose *(Alces alces)* have males with velvet-covered antlers, called velericorn antlers, as well as perukes and a small number of velvet-antlered females.[17]

Because female kangaroos incubate their embryos in a pouch rather than a uterus, an intersexed individual might have both a penis and a pouch, mammary glands and testes. Intersexed kangaroos are known among eastern gray kangaroos *(Macropus giganteus)*, red kangaroos *(Macropus rufus)*, euros *(Macropus robustus)*, tammar wallabies *(Macropus eugenii)*, and quokkas *(Setonix brachyurus)*.[18]

Kangaroo rats are small mammals that are not marsupials at all, but rather rodents native to the American Southwest. Kangaroo rats hop around on their hind feet, reminding one of real kangaroos. Not to be outdone by the better-known kangaroos, kangaroo rats *(Dipodomys ordii)* have lots of intersexes. About 16 percent of the animals have both sperm- and egg-related plumbing, including a vagina, a penis, a uterus, and testes in the same individual.[19]

Pigs in the South Pacific islands of Vanuatu (formerly the New Hebrides) have been bred for their intersex expressions. Typically, these pigs have male gonads and sperm-related internal plumbing, intermediate or mixed external genitalia, and tusks like boars. In Vanuatu cultures, the pigs are prized as status symbols, and among the people of Sakao, seven distinct genders are named, ranging from those with the most egg-related external genitalia to those with the most sperm-related external genitalia. The indigenous classification of gradations in intersexuality is said to be more complete than any system of names yet developed by Western scientists and was adopted by the scientist who wrote the first descriptions of the culture. In the past, 10 to 20 percent of the domesticated pigs consisted of intersexed individuals.[20]

Bears, including the grizzly bear (*Ursus arctos,* also called the brown bear), the American black bear *(Ursus americanus)*, and the polar bear *(Ursus maritimus),* have long been symbols of gender mixing for Native American tribes. The Bimin-Kuskusmin and Inuit peoples have stories of bears who are "male mothers," giving birth through a penis-clitoris.[21] Indeed, 10 to 20 percent of the female bears in some populations have a birth canal that runs through the clitoris, rather than forming a separate vagina. An intersex female bear actually mates and gives birth through the tip of her penis.[22]

This form of intersexed plumbing is found in *all* females of the spotted hyena *(Crocuta crocuta)* of Tanzania—in which the females have penises nearly indistinguishable from those of the males.[23] Aristotle believed these animals to be hermaphrodites, but he was only half right. The first scientific investigation in 1939 showed that a spotted hyena makes only one-size gamete throughout its life, either an egg or sperm.[24] Thus these hyenas are not hermaphrodites. Rather, female spotted hyenas are intersexed, like some female bears. The females have a phallus 90 percent as long and the same diameter as a male penis (yes, somebody measured, 171 millimeters long and 22 millimeters in diameter). The labia are fused to form a scrotum containing fat and connective tissue resembling testicles. The urogenital canal runs the length of the clitoris, rather than venting from below. The animal can pee with the organ, making it a penis. Completing the picture, the female penis contains erectile tissue *(corpus spongiosum)* that allows erections like those of a male penis.

A female spotted hyena mates and gives birth through her penile canal. When mating, a female retracts the penis on itself, "much like pushing up a shirtsleeve," and creates an opening into which the male inserts his own penis. The female's penis is located in the same spot as the male's penis, higher on the belly than the vagina in most mammals. Therefore, the male must slide his rear under the female when mating so that his penis lines up with hers. During birth, the embryo traverses a long and narrow birth canal with a sharp bend in it. About 15 percent of the females die during their first birth, and they lose over 60 percent of their firstborn young.[25] These obvious disadvantages lead us to the question of why female spotted hyenas have this penis instead of a clitoris.

Female spotted hyenas have a dominance hierarchy, and the erect penis is a signal of submission. When two females interact with each other in a struggle for dominance, the one who wants to back down signals by erecting her penis.[26] No one knows why female hyenas evolved this method of signaling, but then signals always seem arbitrary in themselves. Why are traffic lights red, yellow, and green? The female penile erection of hyenas is an "honest signal." Erections occur in the "meeting ceremony" when animals greet after having been apart. The animals approach each other and stand alongside one another, head-to-tail, one or both lifting her hind leg to allow inspection of her erect penis. When only one member of a greeting pair displays an erection, she is normally the subordinate. Each hyena puts her reproductive organs next to powerful jaws. Greetings between captive females that have been separated for a week are tense and frequently wind up in a fight that starts when one bites the genitals of the other, doing occasional damage to the reproductive capability of the injured party.

The masculinized genitals of female hyenas are an example of what I call a social-inclusionary trait, which allows a female hyena access to resources needed for reproduction and survival. If a female were not to participate in social interactions using her penis for signaling, she would not be able to function in hyena society and presumably would either die or fail to breed.

It has been suggested that the enlarged clitoris is a side effect of high testosterone levels in female spotted hyenas.[27] Social life among female spotted hyenas involves lots of aggression, possibly caused by elevated blood testosterone. This testosterone might produce incidental "excess" masculinization during development. I don't buy this theory. Aggressiveness doesn't require testosterone. We're not talking about a slightly larger clitoris, but a full-fledged replica of male genital anatomy, complete with scrotal sacs and fat bodies resembling testicles. This structure can't develop from a few extra splashes of testosterone in the blood. I believe this case demonstrates that mammalian genitals have a symbolic function. In fact, displaying genitals is a mammal thing. Fish, frogs, lizards, snakes, and birds rarely have external genitals pigmented with bright colors to wave around at one another. Mammals do.

Penises can be seen in various female primates, such as bush babies, nocturnal squirrel-like primates from central Africa. Among the dozen or so known species of bush babies, all the females have a penis—that is, a long pendulous clitoris with a urethra extending through the tip so that they can pee through it.[28] The males have a bone in their penis called a baculum. Copulation is unusually slow in these primates, lasting one to two hours.[29]

Field guides to spider monkeys of South America refer to a pendulous and erectile clitoris long enough to be mistaken for a penis.[30] Over half a dozen species of these monkeys exist, named for their spectacular ability to hang from prehensile tails and move around the treetops using their hands, feet, and tails as though they were five-legged creatures. Because the clitoris looks like a penis, the presence of a scrotum is used as a field mark to indicate whether the subject is male. Scent-marking glands may also be present on the clitoris of spider monkeys.

In woolly monkeys, close relatives of the spider monkeys, the clitoris is actually longer than the penis.[31] In still another close relative, the muriqui, nipples are located along the sides, under the arms. Thus, even in primates, a gendered body can be assembled on a vertebrate chassis in many ways.

One reason the public presentation of genitals is such an emotionally charged issue for us humans is that primates use their genitals in displays even more than other mammals do. Picture books about animals often feature baboons called drills and mandrills, showing the male's colorful snout. A full-body photo, rather than just a head shot, would reveal that the color extends to the genitals. Both males and females have bright red genitals. The male displays a crimson-red penis riding astride a snow-white scrotum, and an estrous female displays large red bulbous swellings surrounding her vagina. The drills provocatively present these areas to one another's view.[32] Our own practice of covering the genitals with clothes except in particular evocative situations bespeaks the symbolic power of genital design and decoration for us too. Medicine's peculiar history of assigning gender based on genital anatomy can undoubtedly be traced to our primate dependence on genitals as symbols.

How about feminized male genitals? Spotted hyenas, bush babies, and spider monkeys offer cases of masculinized female genitals. What about the reverse? The genitals of male dolphins and whales apparently represent a different type of intersex. For the purposes of hydrodynamic streamlining, male dolphins and whales don't have external genitals. Instead, paired testes are located within the body cavity. The penis is cradled inside a "genital slit" and covered by flaps unless it is erect. Male cetaceans have no scrotum.

What would be the easiest way to develop this genital architecture for males, using mammalian body parts and a vertebrate chassis? Some of the steps ordinarily taken by terrestrial mammalian males when their genitals are developing could simply be omitted. On land, a male mammal's testes descend from the body cavity into the scrotum, whereupon they become testicles. The scrotum is derived by fusing the tissues that in females become the labia covering the vagina and clitoris. By not bothering to fuse the labial tissue into a scrotum and leaving the testes in the abdominal cavity, a developing male dolphin or whale keeps his testes protected, using the labial tissues as protective flaps. The clitoris continues to develop into a penis, as the urethra becomes included along its axis. If these steps took place on land, a mammalian male would be classified as intersexed. Thus, we might speculate that male dolphins and whales have achieved their genital architecture by making a norm out of what would otherwise be considered an exceptional intersex morphology.

Both genital and gonadal intersexes are documented in wild cetaceans. The striped dolphin *(Stenella coeruleoalba)* has some individuals who display external female genitals along with testes and internal male plumbing. The bowhead whale *(Balaena mysticetus)* has individuals with female external genitalia and mammary glands combined with male chromosomes, testes, and male internal plumbing. A fin whale *(Balaenoptera physalus)* has been described with both male and female reproductive organs, including uterus, vagina, elongated clitoris, and testes. A beluga whale *(Delphinapterus leucas)* in the St. Lawrence seaway had male external genitals combined with a complete set of two ovaries and two testes.[33]

Although a recent report on intersexes among cetaceans raises the specter of pollution causing genital deformity, the early reports on intersexes predate dangerous levels of pollution. Perhaps cetaceans are on their evolutionary way to the state that hermaphroditic fish have already attained.

The examples so far have focused on intersexed genital plumbing. What about intersexed gonads? In four species of burrowing mammals from Europe called old world moles, males have testes typical of other mammals, whereas *all* the females have ovotestes, containing both ovarian and testicular tissue. The females make eggs in the ovarian part of their ovotestes, whereas the testicular portion has no sperm, although the testicular portion does actively secrete hormones. These species come close to being hermaphroditic.[34]

Thus a number of mammalian species have recombined genital plumbing and gonads in surprising and successful ways. More generally, we see that among vertebrates, from fish through mammals, the binary distinction in gamete size does not generally extend to the entire body. Many body plans include production of both sizes of gamete at different times or the same time, as well as various genital sculptures and mixtures of genital plumbing—all as a way of serving social functions important in the society of the species.

4

Sex Roles

Even species thought of as typical, with one gender per sex and individuals who maintain a single sex throughout life, often have gender roles quite different from the traditional template. Indeed, in some species, males (apart from making sperm) look and behave much as females do in other species, and females (apart from making eggs) look and behave much as males do in other species. If these species could express their thoughts about us, they would describe our gender distinctions as reversed.

BODY SIZES REVERSED

Anglerfish are deep-sea fish who have what looks like a tiny fishing pole attached to their head. A spine projects out in front of the fish, and somewhat upward, with a frilly or luminescent bulb at its tip to lure prey. When prey comes near, the anglerfish lunges forward, "angling" and then gobbling it up.

Predators catch prey in countless tricky ways. The anglerfish's fishing pole is a neat curiosity, but what is more interesting is that the anglerfish just described are all female—fisherwomen, not fishermen. Is the anglerfish another example of an all-female species? Nope. Anglerfish males exist, but they are tiny and are called "dwarf males." These anglerfish males are incapable of independent existence. They have large nostrils for homing in on perfumes released by the females and pinchers, instead of teeth, to grasp little projections on the female. After a male attaches to the back or side of a female, their epidermal tissues fuse and their circulatory systems unite, and the male becomes an organ of the female. Multiple males may attach to one female, a case of polyandry. They thereby turn into two or more genetically distinct individuals in one body, a colony.[1]

These fish were discovered in 1922 by an Icelandic biologist who observed two small fish attached by their snouts to the belly of a large female. He thought the small fish were juveniles being suckled by their mother—which mammals do all the time, but which would be big news for fish.[2] Three years later, the small attached fishes were discovered to be reproductively mature males.

An attached male was called a "parasite," by analogy to the small ectoparasites on the outside of large individuals, such as the barnacles attached to whales or leeches that cling to people who bathe in tropical streams. The terminology is unfortunate, because here the relationship is presumably reciprocal. The anglerfish male is "merely an appendage of the female, and entirely dependent on her for nutrition. . . . [S]o perfect and complete is the union of husband and wife that one may almost be sure that their genital glands ripen simultaneously, and it is perhaps not too fanciful to think that the female may possibly be able to control the seminal discharge of the male and to ensure that it takes place at the right time for fertilization of her eggs."[3]

Over one hundred species of anglerfish are distributed throughout the world at depths below one mile. For all anglerfish, the females are much larger than the males. In other respects, though, anglerfish are diverse, exhibiting a rainbow of their own. Some species have attaching dwarf males that fuse with the body of a female, as just described; others have both free-living males and attaching males; and still other species have males who are exclusively free-living. Indeed, whenever one looks deeply into any biological category, a rainbow is revealed. The living world is made of rainbows within rainbows within rainbows, in an endless progression.

SEX ROLES REVERSED

The pipefish is a small pencil-like fish with a circular mouth that resembles a small musical pipe, like a flute. In some species of pipefish, the embryos are "glued" to the male's underside. The young fish develop there and swim away when they are mature. In other pipefish species, the males have protective skin flaps that partially cover the fertilized eggs.

In their close relative the seahorse, the skin flap is elaborated into a pouch that fully encloses the developing embryos. A female seahorse places eggs in a male's pouch. The eggs are fertilized there, forming embryos, and the male becomes "pregnant." The male provides oxygen, maintains the right salt balance, and nourishes and protects the embryos in his sac.[4]

We might think that seahorses and pipefish reverse male and female roles relative to mammals. To determine whether this impression is correct, we must consider the "parental investment" made by males and females in the raising of young. A male contributes a sperm to the embryo, which provides little energy or nutrients. A female contributes a full-sized egg to the embryo. A female therefore starts out by putting a larger investment in the embryo than the male does.

Biologists define "sex-role reversal" as occurring when the total parental investment by males in raising the young exceeds that of females.[5] Male seahorses and pipefish provide a great deal of parental investment in terms of time spent rearing the young in their pouches or glued to their undersides. Does what the males do for the young by the time the embryos mature add up to more than what the females do, given that females invested more at the beginning?

Simply having males provide some care for the young doesn't qualify as sex-role reversal. Species showing some male parental care are too numerous to mention. Many male fish watch over and nourish eggs in nests on the sea floor or lake bottom, and others even store the eggs in their cheeks (called mouth brooders). The specific way males provide parental care depends on the species, and the seahorse's pouch is one of many curious delivery styles. The style of care doesn't matter, the amount does. So, are seahorses sex-role reversed? That is, does the cumulative parental investment by male seahorses exceed that of females?

How could one tell which sex was contributing the most overall to the raising of offspring? An indication comes from the supply and demand of each sex at the time of mating. We're all familiar with supply and demand during courtship. A belle at an Alaskan mining camp has men entreating her with bags of gold dust and promises of trips to Paris. A bachelor on a love-boat cruise is entertained by women offering duty-free Cuban cigars and football lore memorized from the 49ers playbook. Let's extend this idea.

If one sex, say A, is providing most of the parental care, then few are receptive to mating at any particular time because most are occupied with raising offspring. Conversely, the other sex, B, is not very involved with raising offspring and has many individuals ready and willing to mate. This asymmetry in the supply and demand of mates leads to a dynamic tension between the sexes. The Bs compete for access to and control of the As. Provided their freedom of choice is not thwarted by the Bs' control, the As choose which B they wish to mate with.

Biologists call the ratio of receptive females to willing males the "operational sex ratio." The operational sex ratio isn't fifty-fifty because the sex with the higher parental investment is occupied with raising the offspring and is relatively unavailable for mating compared with the other sex.[6]

Returning to the seahorses and pipefish, we can ask which sex is relatively unavailable for mating because of their efforts in raising offspring. Swedish investigators found two nearby North Sea pipefish species that are indeed sex-role reversed. The females from both these species produce enough eggs for about two males during the time it takes for one male to raise his young. In the wild, the number of females with ripe eggs consistently exceeds the number of receptive males. Females in these species are polyandrous, with a harem of males. In addition, these females are larger than the males and develop bright colors at courtship time, presumably for the males to choose among, reversing Darwin's classic peacock story. Furthermore, the females, not the males, compete with one another, forming dominance hierarchies for access to the males who will tend their eggs. Nine other pipefish species in which the females alone have sexual coloration and/or grow larger than the males are thought to be sex-role reversed as well.

On the other hand, seahorses and certain other pipefish species are not sex-role reversed; they follow the model of Darwin's peacocks. Male seahorses can raise their young and get ready for the next embryos faster than female seahorses can produce egg batches. The result is a net surplus of males wanting eggs compared to females offering eggs. Males aggressively tail-wrestle and snout-snap one another for access to females, whereas females don't have any specific aggressive behaviors among themselves. Male seahorses tend to be larger, more colorful, and more distinctly patterned than females.[7]

Thus sex-role reversal definitely occurs in nature. Many feel that the concept of an operational sex ratio effectively extends Darwin's theory of sexual selection to cover sex-role reversed species—after all, the logic is the same for the mating strategies in both sex-role-typical species and sex-role-reversed species, with the identities of the excess sex and rate-limiting sex simply flipped. But no theory has been proposed to explain why sex-role reversal occurs in the first place.

Sex-role reversal is found in birds, especially aquatic and sea birds. When sex-role reversal occurs, the double standard can reverse too. Wattled jacanas from the Chagres River in Panama are large, squat black birds with white wing tips, a red face, and a long, yellow probing bill used to feed among shallow freshwater plants like hyacinths. The raucous, beefy females spend their days jousting with one another at the borders of their territories. Within these territories, harems of smaller males tend the eggs and chicks.

DNA fingerprinting has shown that males raise eggs laid by the female who controls their harem, even when the eggs were fathered by males outside the harem. The females clearly went outside their harem to obtain matings and yet burdened the males within their harem with the job of raising the young. The investigators, themselves male, were outraged, asserting that male jacanas were being "cuckolded" in spite of contributing so much parental care. One investigator stated, "It's about as bad as it can be for these guys."[8]

The converse probably wouldn't have provoked such outrage. A female in a harem controlled by a male might raise a chick fathered by that male and placed there by a female from a neighboring harem. We could imagine many reasons for such an adoption. The female might find it advantageous to raise the chick in return for the controlling male's provisioning and protection of the young she has mothered herself. Similarly, a male jacana might find it advantageous to raise a chick mothered by the controlling female in return for the controlling female's provisioning and protection of the young he has fathered himself. Thus, sex-role reversal implies that the double standard also reverses. This idea takes some getting used to.

Other birds showing sex-role reversal include two shore birds, Wilson's phalarope and the spotted sandpiper.[9] Apparently, no mammals exhibit sex-role reversal, presumably because of the very high parental investment by mammalian females. In addition to the egg, a mammalian female supplies milk to the embryo and carries the young to term, either in a placenta or a pouch. For a mammalian male, this act is hard to follow. To exceed this already high parental investment by a female, a male would require a social system allowing him to care for his offspring well beyond the age of weaning, as may be approached in humans.

The evolution of the mammalian placenta and pouch is usually presented as a physiological advance, an adaptation for nurturing embryonic development in a climate that has cooled globally since the time of dinosaurs. Alternatively, the evolutionary force behind the placenta and pouch may have been for females to assume control of their offspring. A side effect is that males then acquire an incentive to control females.

5

Two-Gender Families

Let's move on now to species with two genders that don't change sex, do not have intersexual body parts, and aren't sex-role reversed. Are such animals "normal"? Have we come at last to the familiar gender roles performed by ordinary bodies, as depicted on nature shows? Or are nature shows perhaps not telling the whole story? What goes on in two-gender animal families, and how are such families organized?

Many of us were raised to admire the nuclear family as a norm and were taught that single-parent families, families of same-sex couples, or communes were second-best alternatives or, even worse, wrong. Yet the meaning of a human family is in flux. In the United States, public attention has focused on the problem of how to define "family" as a result of a recent Supreme Court case about the rights of grandparents to visit grandchildren despite parental objection. The thirty-million-member American Association of Retired People (AARP) states that grandparents are the primary caretakers of 1.4 million children because many nuclear families have dissolved.[1]

The American Center for Law and Justice, which represents the Christian right, claims that "the traditional family, consisting of married parents and their children" is the building block of society. A leader of the Nation of Islam similarly declares, "Whenever . . . 50% of those who marry get divorced within the first three years, these are signs of the decline of a civilization."[2]

Meanwhile, the Lambda Legal Defense and Education Fund, a gay rights organization, argues that neither side, grandparents or parents, sufficiently protects children being reared in nontraditional families, affirming the primary importance of "the quality and security of the relationship between individual children and adults rather than blood ties or labels."[3] Indeed, in June 2002, the California Supreme Court ruled unanimously that a man who cared for a six-year-old boy could be considered the boy's parent, even though the man was not the boy's biological father and was never married to the boy's mother. The man was given custody of the child over the mother's objections, showing that "parenthood could be achieved through love and responsible conduct."[4]

With so much controversy about the meaning of family and parenthood, asking how animals raise their young may be helpful. What is an animal family? Does any family organization emerge as a particularly efficient way to raise young? And does biology support the belief that the nuclear family should have a privileged status in our society?

SEX AND POWER

Oh, I wish the simplest of animal families were a blissfully pair-bonded male and female. Alas, males and females negotiate over power in even the most elementary of animal families. Feminist writings call attention to a power differential between the sexes: "The image of the cage helps convey . . . the nature of oppression. 'Why can't I go to the park; you let Jimmy go!' 'Because it's not safe for girls.' "[5] Well, how safe is it outside the cage? Why does a cage exist at all? If we look at squirrels, we can see what biologists call "mate guarding," a male caging a female.

The Idaho ground squirrel *(Spermophilus brunneus)* lives in populations of one hundred to three hundred individuals in short-grass meadows.[6] The squirrels hibernate in burrows most of the year, only becoming active from late March to early August. The males wake up from hibernation about two weeks before the females. Females become sexually active for about three hours in the afternoon on the first day after hibernation. Three hours on this one day is all the family life a squirrel has.

A successful date for a male squirrel means walking behind a female while sniffing and licking her genitals, going with her into a burrow, mating there for five minutes, and then reemerging. As a sign of successful mating, the female acquires a "sperm plug" that can be seen by a human observer and probably by other squirrels as well. The male then stays within one meter of her and keeps her in a small area by "herding." About every forty minutes, he follows her back into the burrow, where they mate again. She acquires a fresh sperm plug, and he turns around to block the entrance with his body. The male rebuffs an average of four other males who try to mate with the female. If a male is displaced, he is likely to take about an hour and twenty minutes to locate another female, and she is usually guarded by some other male.

A litter typically consists of five pups. If the female has been guarded by only one male for these three hours, then the whole litter is sired by him. If more than one male has guarded the female, paternity analysis shows that her litter is sired mostly by the male who was the last to guard her or by the male who guarded her the longest.

Thus the family life of the Idaho ground squirrel consists of three hours per year. The Idaho ground squirrel family is solely a unit for reproduction. The male doesn't hang around and help raise the young. The male is said to guard the female to protect his "investment" of sperm by making sure that his female doesn't mess around. He doesn't mess around himself because if he stopped guarding he would lose his investment, and in any case finding another female is nearly impossible.

A close relative, Belding's ground squirrel (*Spermophilus beldingi*) of the midwestern United States, does not mate-guard. In this species, twenty-five minutes are all a male needs to locate and initiate courtship with another female. Female Belding's squirrels typically mate with three to five males. The female's first mate sires most of the litter, followed by the second mate, then the third mate, and so on, in contrast to the Idaho squirrel, where the last mate gets most of the sires. A Belding's male doesn't waste time guarding a female because his investment of sperm is secure—as the first mate, he's already guaranteed siring most of the young. Instead, these males hurry to find more females to mate with, and with so many nearby, why not?

A cage, then, is not biologically universal. These two closely related species of squirrels have completely different power relationships between males and females. Male Idaho ground squirrels guard the females, while male Belding's ground squirrels don't. Why? The power relationship is probably not as simple as it seems. Are the female Idaho squirrels being caged against their will? Maybe, maybe not. If asked, an Idaho male squirrel might say lovingly that he was "protecting" his female during their brief marriage from the unwanted advances of rival males. And she might agree, happy to have him there. The courtship that precedes mating suggests her acquiescence. And she is capable of physically repelling a male if she wishes: the day after her three-hour burst of sexuality, an Idaho female squirrel constructs a nest and excludes all other squirrels from a small territory around it. So a female Idaho squirrel may want to be guarded.

Why does the last male sire most of the young in the Idaho squirrel, and the first male sire most of them in Belding's squirrel? Could female squirrels control whose sperm fertilizes their eggs?[7] Could a female manipulate a male to guard, or not guard, by controlling his sperm? Can a female select the first male's sperm to do the fertilizing if she wants him to have no incentive for staying, and select the last male's sperm if she wants him to hang around and guard? Presenting mate guarding as a tactic by which males protect their investment ignores the female perspective. Females are viewed as land in which males plant seeds and which they guard if necessary. Yet females are probably active players in whether they're guarded or not.

The Idaho and Belding's squirrels may have evolved to experience pleasure differently. A female Idaho squirrel may like being squeezed into a burrow, and a male may enjoy having a female behind him as he stuffs the burrow's entrance, like a guy taking a girl for a spin in his red convertible—fun for both. Yet for a female Belding's ground squirrel, squeezing into burrows could be a total turnoff and the reason she doesn't permit the guarding. Species differences in how power is eroticized make it difficult to discern whether animals have freedom of choice during mate selection and in their family lives.

Among primates, the amount of sexual coercion varies greatly from species to species, as does the overall level of both between-sex and within-sex aggression.[8] Outside the breeding period, male mountain gorillas aggress against females one to four times a day, olive baboons about one time per day, and red howler monkeys only 0.04 times per day. The context for the aggression varies from competing for food, coming to the aid of one female against another, and breaking up fights between females. These aggressions are not directly sexual coercion, but rather reflect a social atmosphere of violence.

Within the breeding period, male rhesus monkeys attack females who consort with male rivals. Similarly, among chimpanzees, one of our closest primate relatives, males attack females who consort with lower-ranking male rivals, rather than attacking the low-ranking males themselves. Using "a fair amount of brutality,"[9] chimpanzee males persuade females to accompany them away from the group where they were living. One-third of all conceptions result from matings between pairs separated from the group for several days to over a month. Chimpanzee males also intimidate females into submitting to their advances later on. Lack of resistance does not necessarily mean willing participation, but may reflect experience with the male's previous aggression.

The record-holders for male sexual coercion are the orangutans, in which most copulations by subadult males and nearly half of all copulations by adult males occur after a female's fierce resistance has been violently overcome.[10] Other primate species showing lots of male aggression against females include white-fronted and wedge-capped capuchins, black spider monkeys, and brown lemurs. Females' counterstrategies include avoiding areas where males are found, joining a male's territory or harem to gain his protection, and forming coalitions to fight off the males.

Yet other primate species enjoy a peaceful life. Male aggression against females is rare in bonobos, a primate species as closely related to humans as the relatively violent chimpanzee is. Male sexual coercion is also rare in patas monkeys, red-backed squirrel monkeys, brown capuchins, woolly spider monkeys, and black-and-white ruffed lemurs. As with the bonobo/chimpanzee contrast, closely related species differ in the prevalence of male sexual aggression.

No explanation exists for why some societies develop coercive power relations between the sexes, whereas others form equitable power relations.[11] Although some species resemble the tough-talking television show "NYPD Blue," others resemble the peaceful Mr. Rogers. How power relates to sex is not a biological universal. We may choose to live like some species and not others.

MONOGAMY AND DIVORCE

As we turn to long-lasting two-gender families—beyond the three-hour marriages of ground squirrels—the plot thickens. Ninety percent of bird species are monogamous in the sense that a family consists of one male and one female who cohabit a nest and raise young together in that nest during at least one breeding season. In contrast, 90 percent of mammal species are polygynous, with one male for many females.[12] Among mammals, females typically occur in groupings of two or more who are serviced by one male.

But what do these statistics mean? Biologists have been slow to distinguish between economic monogamy and reproductive monogamy. Birds are identified as "monogamous" simply if they are living in the same nest and feeding nestlings together—an economic criterion. Biologists have unquestioningly assumed that the nestlings are the offspring of the couple in the nest. When distributed parenthood is discovered, biologists feel something is amiss—that one or the other member of the pair "cheated," straying outside the marriage bond. However, birds have decoupled reproductive and economic monogamy; in some species these go together, and in others not.

Black-capped chickadees *(Parus atricapillus)* of eastern Ontario are monogamous.[13] During the summer, pairs settle in territories to raise their families—a suburbanite's dream. During the winter, the chickadees cease living as couples and live in flocks of about ten birds. Because a chickadee's average life spans several years, the birds are aware of each other through both the over-wintering period and the breeding period. During the winter, males and females sort themselves into separate dominance hierarchies.

New couples form during the winter social period and settle as pairs for the summer to raise a family. As a couple, they forage together, excavate their nest cavity together, mate with one another, and defend their territory together. The male feeds the female although they both forage, and the male prevents other males from approaching the female. When forming couples, the highest-ranking female pairs with the highest-ranking male. Most couples remain together for more than one breeding season. Females may "divorce" their mate and/or mate with males other than their nest mate. A mating outside the pair-bond is called an extra-pair copulation, or EPC in biology jargon. Female chickadees divorce to obtain a male who is higher in the male dominance hierarchy than their present mate.

In one study,[14] seven females paired to seven high-ranking males were removed, temporarily creating seven very desirable widowers. Over the next two days, some of the remaining females deserted their partners and took up with the higher-ranking widowers. When the females who had been removed were restored to the site, they quickly chased the social climbers back to the males they had deserted. These hen-pecked husbands took their wayward spouses back. To complete the experiment, six females were also temporarily removed from six low-ranking males, but none of the remaining females left their mates to join these sad sacks. These males were saved from eternal loneliness only when their mates were returned to the site to rejoin them. Not only are female chickadees willing to desert their mate to acquire a higher-ranking male when one is available, but they are willing to mate with higher-ranking males in EPCs, resulting in many broods with mixed parentage.

A close relative of the chickadees, a member of the same genus, is the European great tit *(Parus major)*. Boxes for these birds to nest in have been constructed on the island of Gotland in southeast Sweden. In a 1985–89 experiment, eggs were removed from one pair's nest and transferred to another pair in order to observe the effect on divorce. Pairs whose eggs were removed divorced more often, presumably because they were not able to raise as many young as the population baseline. Conversely, pairs given extra eggs divorced less often, presumably because they did raise more young. Success at raising a family thus seems to be a factor in whether birds decide to divorce.[15]

One survey of marriage fidelity among birds shows annual divorce rates as low as 2.4 percent in the barnacle goose *(Branta leucopsis)* of northern Europe, 2.5 percent in the silvereye *(Zosterops lateralis)*, a berry-eating forest bird from Australia, New Zealand, and Fiji, and 2.7 percent in Cory's shearwater *(Calonectris diomedea)* of Long Island to Nova Scotia.[16] The highs were 36 percent in the European shag *(Phalacrocorax aristotelis)*, which is similar to a cormorant but more marine, and 30.6 percent in *Parus major,* the European woodland songbird mentioned above.

Divorce rate correlates with mortality. In birds where the annual survival rate is only 40 to 80 percent, the divorce rate is high, and in birds where the annual survival rate is 90 percent or more, the divorce rate is low.[17] Lots of eligible widows and widowers make for a hot singles scene. And when the singles scene is hot, the action doesn't stay confined to singles. When divorce rates are high, lots of mating also takes place outside of the nest. The data for birds show a positive statistical relation between rates of divorce and EPCs.

Thus monogamy among birds seems to be an economically beneficial social institution, with divorce and some out-of-wedlock matings a regular part of the picture too. Bird females seem to have lots to say about their own lives, choosing partners and initiating divorce when advantageous. When we turn to mammals, though, we have to face the fact that monogamy seems rare. Why?

One explanation for why birds are more often monogamous than mammals is that flight endows female birds with more opportunity to choose their mates than female mammals have.[18] Female birds can check out prospective husbands by flying from party to party around town, whereas a female mammal is stuck walking to the nearest block party. With so much choice around her, a female bird can demand a husband who is faithful and helps with the dishes, while a female mammal can't. However, this theory assumes that a male generally doesn't want to stick around and help with the young, can't stand doing dishes, and must be manipulated to do so by a female's threat of turning to someone else if he doesn't. I don't accept this logic. I feel the male's perspective should be stated differently. He has two directions in which he can invest social effort. Within-sex effort involves competing with other males and/or building coalitions with them to access females. Between-sex effort involves "coalition-building" with a female to raise offspring together. Whether a male winds up with more offspring overall from within-sex or between-sex coalition-building depends on circumstances. This is the animal equivalent of balancing career and family.

Monogamy then emerges when (a) building relationships with a female is more advantageous to a male's reproductive success than building relationships with other males, *and* (b) building a relationship with a male is more advantageous to a female's reproductive success than raising young by herself or in conjunction with other females. In general, dif-

ferent mating systems emerge from different optimal allocations of social effort to between-sex and within-sex relationships.

Although not as commonly as in birds, monogamy among mammals
does happen. It occurs in 15 percent of primate species and is common
among wild canines, among others. In most monogamous species, the
husband contributes to parental care by building a den, burrow, or
lodge, defending the family's feeding territory, feeding his wife when
she's nursing, and carrying the young around (driving the kids to soccer
after school). In the monogamous prairie vole, *Microtus ochrogaster,*
when a female produces a bigger-than-normal litter, a second nest is
built, the young are divided up, and the male cares for one nest and the
female for the other. Thus monogamy occurs among mammals, although
not as commonly as in birds.

But why is monogamy rare in mammals? Mammalian females have
internalized embryonic development in a uterus or pouch, whereas avian
females leave the developing embryos as eggs in the external environment. This difference affects who can control the offspring. A mammalian male who wants to control offspring must somehow control the
female herself, whereas an avian male can directly control the eggs in the
nest. A mammalian female knows the embryo developing in her body is
hers alone, not an egg deposited there by some other female. In birds, a
female may derive from a monogamous marriage both male provisioning of the young and male protection of the nest, not only from predators but from "dumpers"—other females who deposit foster eggs in the
nest.[19] The male gains the female's initial investment in eggs, plus her additional provisioning. Neither male nor female mammals benefit from
marriage as much as birds do.

EXTENDED FAMILIES

Let's now take a look at two-gender families larger than two individuals: extended families. The groove-billed anis *(Crotophaga sulcirostris)*
is an insectivorous black bird with a large, deeply grooved bill. It lives in
marshes and open pastures in Central America and is related to the
cuckoo. Family organizations of the anis may consist of twosomes with
one female and one male, foursomes with two females and two males,

sixsomes, and even eightsomes.[20] The foursomes are not sixties-style communes of free love. An anis foursome is two couples cohabiting a one-bedroom flat with one crib. Nests are built in thorny trees or vines. Each male guards one of the females. A female lays an egg every one to two days. The time from egg-laying to fledging is three weeks.

The two females in a foursome wind up with four eggs total. A female in a twosome can also produce four eggs by herself. Thus the number of eggs laid per female is lower in a foursome than in a twosome. In a foursome, the females start laying eggs at different times. The starter lays bigger eggs and has a longer time between successive eggs than the follower. Both females stop laying eggs at about the same time. Each female "tosses" out of the nest some of the eggs already laid by the other, with the follower winding up mothering on average 63 percent of the eggs and the starter only 37 percent of the eggs. After all the tossing is over, four eggs are left in the nest.

Even though the follower mothers more of the eggs, she doesn't necessarily successfully raise the most offspring. The starter lays larger eggs, which hatch earlier, so these chicks have a better chance of surviving than chicks from the follower. All four birds in a foursome work together to provide for and protect the eggs. The males divide the day into unequal shifts. The oldest male incubates at night and most of the daylight hours. As a result, he also incurs the most hazard and the highest death rate. Yet he also fathers the most young.

But why should a female, who can lay four eggs in a twosome without worrying about a nest mate tossing her eggs out, bother living in a foursome? The answer is that the larger group provides protection against egg predators. Once loss of eggs to predators is taken into account, the starter in a foursome produces the most young, a female in a twosome produces an intermediate number of young, and the follower in a foursome the fewest. For the anis, the benefit from predator protection of living in extended families of two couples outweighs the disadvantages of a rancorous life at home.

The family life of tamarins offers a pleasant contrast to that of anis. The saddle-backed tamarin is a tiny monkey that lives in the tropical rainforest of southeastern Peru, including the Manu National Park.[21] Among tamarin families, 22 percent consist of one female with one male in a monogamous relationship, 61 percent of one female with multiple males, 14 percent of multiple females with multiple males, and 3 percent of males only. In the families with one female and multiple males, the female mates with all the males. The matings take place in view of the other males without any sign of aggression. The males in this species not only help take care of the young, but they cooperate with one another when doing so. The females usually give birth to twins, and the males carry the babies with them through the treetops. The males and females give the babies fruits and large insects to eat.

The twin babies are 20 percent of their mother's weight, and are 50 percent of her weight by the time they can walk and climb on their own. Just one female and one male aren't enough to raise the twins; three adults seem to be the minimum to do the job. Even the families consisting of one adult male and one adult female are accompanied by older children who help out. This family organization is called cooperative polyandry.

Other mammals with cooperative polyandry include African hunting dogs *(Lycaon pictus)* and dwarf mongooses.[22] Cooperative polyandry also occurs in birds, including the Australian white-browed scrubwren *(Sericornis frontalis)* that lives near Canberra, the Tasmanian native hen *(Tribonyx mortierii)*, the Galápagos hawk *(Buteo galapagoensis)*, the English dunnock *(Prunella modularis)*, the New Zealand pukeko *(Porphyrio porphyrio)*, and the Venezuelan striped-back wren.[23] No biological reason prevents these guys from cooperating with each other and helping around the house.

Lions, in contrast, seem to take the idea of a war between the sexes very seriously. We may be misled by the common picture of lions as cooperative hunters. Although lions seem to work together, not only at hunting but also in rearing cubs and roaring in a unified chorus, the truth makes human political infighting seem benign.[24] Lion family organization is multimale polygyny—gangs of males guarding groups of females, called prides. Males form lifelong alliances among one to eight lions. Most members of an alliance are brothers or cousins, but others are unrelated. Once mature, these coalitions take charge of a pride of females and father all the offspring born in the pride for a period of two to three years. After that, a rival coalition moves in and evicts them. The males work together more effectively when battling with a rival gang than in any other situation. How well a male does depends on how well his coalition does—gang warfare in the extreme. A male lion is nowhere if not in the right gang.

Victorious lions don't make ideal boyfriends. A cub matures in two years, and a female isn't interested in mating during that time. However, if a cub dies, a female mates again in as little as two days. In their hurry to become fathers, an invading gang of males may kill off more than one quarter of the cubs, which quickly brings their mothers into reproductive condition. The invading males don't share fatherhood equally. One or two father almost the entire pride's litter.

To counter the male danger to the cubs, the females band together to raise the young. The females enjoy a reproductive life span of about ten years, during which five gangs may come and go. The females give birth in secrecy and keep their litters hidden in a riverbed or rocky outcrop until the cubs can move on their own. Then they bring the cubs to a place where they are nursed together in what is called a "crèche," a word meaning a public nursery for infants of working women.

The lionesses nurse their own cubs and those of others as well. This shared lactation is not entirely altruistic. The lioness gives milk primarily to her own cubs and rejects the advances of others. She needs her sleep, though, and while she is asleep, cubs who are not her children are able to nurse. Although a lioness prefers her own cubs, the strength of this preference depends on how closely related the other cubs are to her. If a pride consists mostly of close relatives, a lioness is more generous with cubs who are not her own than if she is in a pride of comparatively unrelated females.[25] In sum, female lions raise their young in crèches to defend against infanticidal males rather than to provide nutritional benefits from shared nursing. Ironically, house mice have the same family dynamics as lions,[26] as though a lion were no more than a mouse that roars.

THE BIG CITY

Some animal species live in what might be called cities. In these cases, family life shows many of the sophisticated intricacies of human urban living. Consider the vampire bat. Your first impulse may be to shudder at the terrifying vision of vampire bats, in the dead of night, swooping down to drink the blood of an unsuspecting victim. Yet vampires have a wonderful story to tell of social cooperation.[27]

The vampire bat *(Desmodus rotundus)* is a rather small bat, not much bigger than a plum. A vampire can hang by its feet from the hair of a horse's mane and bite the horse's neck. It doesn't suck the blood; instead, it removes a small patch of flesh with its razor-sharp incisors and laps up the blood flowing from the wound. A vampire's saliva has an anticoagulant to keep the blood from clotting. After one bat has drunk its fill, another continues at the same spot. Horses can dislodge feeding bats by tossing their heads, swishing their tails, or rubbing against trees.

Life as a vampire is hard. Bats are warm-blooded and, without feathers or fur, lose lots of heat. Their requirements for energy are huge. A vampire bat consumes 50 to 100 percent of its weight in each meal. Yet up to one-third of the bats may not obtain a meal on any given night. Going without a meal is dangerous. A vampire dies after sixty hours without food because by then its weight has dropped 25 percent, and it can no longer maintain its critical body temperature. To survive, vampire bats have developed an elaborate buddy system for sharing meals. The sharing takes place between mother and pup, as well as between adults.

One study of vampires on a ranch in Costa Rica focused on a population divided into three groups of a dozen females. The members of a group often stay together for a long time, twelve years in some cases, and get to know one another very well. The group of a dozen adult bats is a family unit from a vampire's standpoint. Most of the group consists of females, each of whom usually cares for one pup. A female pup stays in the group as she matures, whereas a male pup leaves. The females in a group span several generations. Group membership is not entirely static, however. A new female joins the group every two years, so at any time the females in the group belong to several lineages, called matrilines.

The bats live in the hollows of trees. Imagine a hollow tree with an opening at its base and a long vertical chamber reaching up into the tree trunk. The females congregate at the top of the chamber. About three males hang out, so to speak, in the tree hollow. One male assumes a position near the top of the chamber, nearest to the females, and defends this location against aggressive encounters from other males. This dominant male fathers about half of the group's young. Subordinate males take up stations near the base of the tree by the entrance. Other males are out of luck, roosting alone or in small male-only groups rarely visited by females.

The pattern of food-sharing is especially interesting. The food is transferred by one bat regurgitating into the mouth of another. (You wouldn't want to be a bat, would you?) Most (70 percent) of the food transfers are from a mother to her pup. This food-sharing supplements the mother's lactation. The other 30 percent involves adult females feeding young other than their own, adult females feeding other adult females, and on rare occasions, adult males feeding offspring.

Some adult females have a "special friendship" with females who are not their kin (males also have same-sex relations; see p. 141). This bond is brought about in part by social grooming. The bats spend 5 percent of each day grooming and licking one another. Some of this grooming is between special friends, and the remaining among kin. A hungry bat grooms one who has recently fed to invite a donation of food. To solicit food, a hungry bat licks a donor on her wing and then licks her lips. The donor may then offer food.

The mutual assistance is significant. If they didn't help each other, the annual mortality of vampires would be about 80 percent, based on the chance of missing a meal two nights in a row. Instead, the annual mortality is around 25 percent because food-sharing tides bats through their bad nights.

Biologists assume that animal species don't readily cooperate with each other. If natural selection is the survival of the fittest, shouldn't natural selection reward selfishness and discourage cooperation? Biologists suggest two forms of cooperation that can evolve by natural selection. The first is cooperation restricted to helping kin, and the second is cooperation restricted to helping special buddies—those who regularly reciprocate the cooperative acts.[28] Vampire bats help not only kin as do many species, but also unrelated friends—which is what makes vampires so interesting. This mutual helping, called "reciprocal altruism," takes place primarily between animals who have lived together and gotten to know one another. Each helps the other at various times, and each instance of helping benefits the recipient much more than it costs the donor.

Critics of the idea of reciprocal altruism have argued that natural selection favors the "cheater" who takes food without reciprocating. If cheaters are evolutionarily more successful than food-sharers, the altruism eventually disappears and all the animals wind up being selfish. The vampires solve the problem of cheaters by developing special friendships through what might be considered same-sex courtship. This involves continual mutual grooming and food solicitation using the bat equivalent of kissing, all of which reinforces the pair-bond and promotes long-term survival.

Other species have different tactics to exclude or retaliate against the selfish.[29] For instance, Rhesus macaques who find food sources and don't give food calls telling everyone else about it are subsequently targets of aggression.

Little is known about whether animals acquire a "reputation" that others use to decide whether to include them in cooperative activities. The idea of reciprocal altruism invites thinking in terms of pairs. Yet in my field studies of lizards, whenever I've seen two animals interacting with each other using head bobs, pushups, and color changes, all the other lizards in the vicinity were watching too. Do they remember what they've just seen? Probably. The lizards can probably remember who won or lost in a showdown over territory, and they can probably remember who cheated and who reciprocated in an instance of cooperation. Animals may talk about each other as well, indulging in animal gossip.[30] Animal interactions, from mating to territorial spats, to grooming and food-sharing, are often done out in the open, so that everyone can see and later discuss what happened. Animals with "nice" reputations may be included in cooperative activities and "meanies" left out. Reputations may provide a way for an animal to know whether another is likely to reciprocate, without having to learn the hard way.

Similarly, little is known about animal "generosity." A social system effective at excluding cheaters promotes evolution of the desire to share. Generosity depends on society's promise that what goes around comes around. If vampires someday prove to be among nature's most generous creatures, future children's comic books may feature vampires as friends rather than foes.

The gold medal for cooperation between mammals is held by small, almost hairless rodents that live underground in parts of Kenya, Ethiopia, and Somalia. Their subterranean families consist of certain individuals specialized for reproduction and others who routinely groom, feed, and protect the offspring. If this society sounds like a colony of bees, with a reproductive queen surrounded by the workers, you're right. These mammals, called naked mole-rats *(Heterocephalus glaber)* because of their exposed smooth skin, are the vertebrate counterparts of the social insects.[31]

A family of naked mole-rats typically consists of about one hundred individuals. Naked mole-rats are underground all the time. Their only aboveground signs are volcano-shaped mounds about one foot high, created by ejecting loose soil from their burrows. Naked mole-rats make these mounds when excavating tunnels, primarily at dawn and dusk, and during the winter rainy season. To find food, the naked mole-rats dig until they bump into a juicy root. The rats can't see or smell through the dirt, so finding a root is like a miner striking a vein of gold. Because naked mole-rats are so difficult to study in the field, most observations are based on captive colonies in the lab.

Naked mole-rat families are really close—more than 80 percent of the matings occur between brothers and sisters or between parents and offspring. Typically one female and one to three males do the breeding. The breeding female is aggressively dominant over other females. A breeding female gives birth every two to three months, producing a litter of about ten pups. A female produces thirty to sixty offspring per year. Nonbreeding naked mole-rats are not sterile. If a breeding male or female dies or is removed, a nonbreeding mole-rat of the same sex steps up to take his or her place.

Breeding is a demanding occupation. Although the breeding female typically remains the largest and heaviest animal in the extended family during her tenure as chief breeder, the breeding males lose weight after they become breeders, quickly shedding 17 to 30 percent of their weight, and appear emaciated after several years. Meanwhile the breeding female becomes not only heavier but also longer, adding vertebrae to her spinal column.

Who becomes a breeder (either male or female) seems to be determined by conflict among the aspiring females. Upon the death of the female breeder, the would-be successors not only attack one another but also target specific males, shoving and biting them. Of seven fights started by females against males, five fights led to the death of the male.[32]

The males attacked were either mates of the previous breeding female or were pair-bonded to rival females, as indicated by courtship activities such as frequent anogenital nuzzling.

The nonbreeding males and females provide parental care to the offspring of the breeders. From shortly before the pups are born until they're weaned, the nonbreeders huddle with the pups in the communal nest to provide a stable thermal environment, a warm nursery. The nonbreeders regularly nudge, handle, and groom the young; retrieve pups that fall out of the nest; transport pups when the family moves to a new nest site; and evacuate pups from the nest during a disturbance. The nonbreeders also provide food to the pups in the form of caecotrophies, partly digested fecal pellets. The pups routinely solicit and obtain these morsels of candy from the nonbreeders of both sexes, but not from the breeders themselves. After the pups are fully weaned, they are able to eat food that has not been preprocessed. Nonbreeders also defend, maintain, and extend the family's system of tunnels. They collect and transport food through the tunnels back to the nest, where they feed other family members, including the breeders.

The distribution of reproductive activity throughout a group is called its reproductive skew. A social group where everyone reproduces has low skew. High skew occurs in the naked mole-rats because only two to four individuals in the entire group of one hundred or more reproduce. The reproductive skew in an animal society is the most fundamental attribute a society has from an evolutionary standpoint—the index of a society's reproductive equity. Little is known about what determines a society's reproductive skew to begin with, but once in place, the skew sets a baseline for how each individual in the society structures a life plan for reproductive success.

Cliff swallows *(Hirundo pyrrhonota)* are perhaps our closest cousins when it comes to life in the big city.[33] Swallows live as monogamous pairs in colonies. Their nests, which look like small pitchers, are arranged side by side. Up to five thousand birds form what amounts to a city of mud huts. Not all swallows live in the big city, though. Some live in small villages of twenty or so nests, and some males hang out outside of town without any nests.

Cliff swallow life includes many features of our own city life—a hot real estate market, trespassing, robbery, hanky-panky with the neighbors, plus presumably some compensations. Nonetheless, most observers of cliff swallow life emphasize the problems, perhaps because the birds live in the countryside, where the virtues of city living are underappreciated. The nests are packed in so densely that occasionally a bird is trapped inside and dies when a neighbor's construction project blocks its entrance hole. Even worse, the droppings from a nest above may clog up and bury a nest below—swallows seriously need expertise in civil engineering. Swallows also have a major public health crisis. The colony density promotes the growth of bugs that harm the chicks and adults.

Birds occasionally "trespass" into one another's nests by either barging in on the owner unannounced or following the owner in before he or she can turn around and block the entrance. Seventy-five percent of the trespassers are males. Of nest entries considered "successful," 14 percent of the time, the trespasser stole grass used to line the nest; 9 percent of the time, a male trespasser forced himself on his neighbor's wife; 7 percent of the time, the trespasser stole some still-wet mud before it had dried; 3 percent of the time, a female laid an egg, or transferred an egg, into her neighbor's nest; 1 percent of the time, the trespasser tossed one of the neighbor's eggs or chicks out the window; and in 0.3 percent of the cases, the trespasser evicted the owner. Enough intrigue for a new TV drama, "Cliff Swallow Vice."

Females congregate in flocks while gathering mud and grass to make nests. "Surplus" males hang around mud holes waiting for females to alight. The males circle above and then pounce, "forcing" copulation as the pair flails about in the mud. Nonetheless, some males may be innocent of evil intent, traveling to a mud hole for regular ol' mud, where they encounter a female who "elicits a forced copulation." Also, "females did not always appear to struggle with the males attempting to copulate with them at a mud hole. Some females clearly allowed successful cloacal contact."[34] Indeed, 86 percent of the extra-pair copulations (EPCs) at the mud hole appeared to "achieve" cloacal contact.

What do husbands back at the nest do about all this? They are "suspected of dealing with the threat of cuckoldry by frequently copulating with their mates." Indeed, "the male copulated with his mate virtually each time she returned to the nest," leading to conjugal love "dozens" of time in a single morning.

Biologists have observed which males are "perpetrators" of illicit romance. Among thirty-eight male birds observed in EPCs, one male "committed" twelve copulations, another eleven, the next eight, and so forth. Thirty percent of the EPCs involved the top three, with the numbers trailing off to those who had only one dalliance apiece. Thus only a few males commonly "engaged in this behavior," whereas most "did it" casually or not at all.

EPCs lead to extra-pair paternity, or EPP—that is, to eggs in the nest fathered by a male other than the male tending the nest. Females place eggs in one another's nest, leading to varied egg maternity in the nest too. Extra-pair maternity, or EPM, refers to eggs mothered by females other than the female tending the nest.

Females either lay eggs in their neighbor's nest or transfer eggs laid in their own nest to other nests by carrying them in their bills. Females transfer eggs primarily to nests nearby, within five nests of their own. The transfer typically happens when the recipient nest is left unattended. In several cases, though, a male nest owner allowed a neighboring female to enter his nest and lay an egg there while he was present. The female does not toss an egg already there to make room for hers; she simply adds her own egg to those already there. Around 15 percent of the nests wind up with one or more eggs with extra-pair maternity.

Biologists call transferring eggs "brood parasitism"; the bird owning the recipient nest is called a "host"; and the bird delivering the egg, a "parasite." Host birds lay 71 percent fewer eggs than parasite birds, implying that parasites are somehow taking advantage of the hosts. However, the parasites are themselves often parasitized, as they leave their own nest unguarded.

Just as some males are more likely to "perpetrate" an EPC, some females are more likely to be brood parasites. In one study 29 percent of the females labeled as parasites laid eggs in two or more nests, whereas other females laid eggs only in their own nests. Females did not broodparasitize to get out of housework. The females labeled as parasites contributed just as much parental care and raised as many offspring in their own nests as did their hosts. The advantage to brood parasites is simply leaving more eggs in the nests of other females, not in lowering the size of their own nest. Just as certain females were more likely to lay eggs in another's nest, some females were more likely to receive the eggs of others. Also, females sometimes transfer baby nestlings who have already hatched.

When EPCs and brood parasitism are taken into account, 43 percent of the nests are estimated to contain an egg unrelated to one or both birds tending the nest. Clearly, cliff swallows have decoupled economic monogamy from reproductive monogamy.

Both EPPs and EPMs lead to eggs in the nest unrelated to one or both of the paired birds at the nest. EPPs and EPMs are not symmetric, however, because an EPP implies that one gamete was transferred, whereas an EPM implies the transfer of two gametes, one from the mother and one from the father. In fact, a female transferring an egg may not be intending to get a free ride, but rather may be transferring the egg to the father's nest. So-called brood parasitism hasn't been demonstrated to be competitive at all.

The appetite for seeing theft and deceit everywhere has blinded biologists to other interpretations of what's going on. Swallows apparently have a distributed system for raising young. Throughout the colony the parenting workload is essentially parceled out to work teams of two adults apiece, which amounts to economic monogamy, even though the egg and gamete trading implies an absence of reproductive monogamy. Each team winds up tending about the same number of eggs and nestlings in its nest, and each clutch of nestlings contains offspring from the neighborhood. (For more on a system of distributed paternity in a closely related species, the tree swallow, see chapter 7.)

After fledging, the juvenile birds gather in flocks called crèches. The adults continue to feed their young for a few days after they have flown the nest by searching them out in these crèches and giving them food there. They can recognize their own offspring by listening for an individually distinctive signature in their calls. Some juveniles do not join the crèches, returning instead to the nests, much as an eighth grader might return to kindergarten for cookies. Called kleptoparasites, the juveniles block the nest entrance and intercept food destined for the baby nestlings inside. The parents "willingly" feed these juveniles. The adults never evict them as they routinely do other adults who trespass. Why? The parents are supposedly unable to recognize the juveniles as thieves and are duped into disgorging their food to someone other than their nestlings.

Why isn't the food given the juveniles considered a voluntary donation by the adults? Calling juveniles at the nest kleptoparasites criminalizes these birds and implies that the adults are incapable of knowing their own best interests. Indeed, any swallow who doesn't do what they're supposed to is criminalized. The males in EPCs are "perpetrators" of copulations, the females are "elicitors" of copulations, and females who place eggs in other nests are "parasites." Birds haven't been corrupted by sex and violence in the movies; shouldn't they be better behaved?

The plot thickens. Males sometimes copulate with other males at mud holes, described as "fights in the mud." In an experiment in which stuffed models of birds were placed near a mud hole, 70 percent of the copulation attempts were directed by males to the male models. The interpretation was that the males were "mistaken"; they were unable to distinguish the sex of the stuffed birds. Hmm . . .

Cliff swallows have a complex dynamic going. City life in these birds taxes the ability of biologists even to describe what's happening. But the pejorative language biologists have used so far undercuts the attempt to understand cliff swallow society in greater depth.

A DIFFERENT MODEL

Do animals really own anything? The assumption that they do naturalizes human property rights. Because of this assumption, animals can be described as stealing. Biologists are willing to impute ownership to animals, as though animals cared about property as much as people do. A biologist interprets a bird feeding another as wasting food. The bird is said to be incapable, too dumb, to know it's been duped into giving up hard-earned wages. Yet the same bird is acknowledged to be smart in so many other ways. Why is a bird smart in some ways and dumb in others? Time and again, biologists assume that ownership is well defined, and explain away the failure to be selfish as a limitation of ability, rather than as falsifying the assumption that selfishness is adaptive.

Close genetic relationship among individuals undercuts natural property. Imagine living in a place where you're closely related to everyone else. In a colony where every individual is related 50 percent to another, do you own half your things or half-own all your things? I don't know. The clarity of who owns what becomes fuzzy here. Reciprocal and distributed altruism also undercuts natural property. I live in San Francisco among the homeless. I often see people of limited means give to the poor on the streets. Hasn't someone told them to be selfish? Perhaps they, or their genes, have been there too. What goes around comes around.

Twenty years ago, Sandra Vehrencamp, an evolutionary biologist, introduced the theory that a society's reproductive skew was connected to what might today be called a "labor market" for cooperative effort within the society. Her focus was on insect colonies with multiple queens. The basic idea is that an animal helps another in exchange for access to reproductive opportunity.[35] Some individuals, the privileged, are envisioned to have control of reproductive opportunity, and to pay out some of this opportunity to others who do not have similar access. In return for this paycheck, the underprivileged contribute labor to assist the privileged in their reproduction.

The inequality of reproductive opportunity initially available to different individuals is called a "distributional inequity" by economists. Distributional inequity may reflect territories that vary in exposure to predators or availability of food, water, and a mix of sunny and shady spots. An animal's political connections may also give it control of resources. Distributional inequity may develop because of inheritance, age, abilities, and luck.

Exchanges of labor for reproduction are especially profitable between relatives, leading to the formation of an extended family wherein the individual who does the breeding is a parent of helpers who remain at the nest. The value to a helper of assisting its parent's reproduction depends on its genetic relationship to the parent's offspring. The highest value accrues to offspring who are full brothers or sisters—in this case a helper may not bother with reproducing at all but let the parent do all the work, called kin selection.[36]

An exchange of help for reproductive opportunity is possible even in the absence of a genetic relationship if the amount of access the helper is paid exceeds the reproductive opportunity the helper would have in the absence of supplying any labor. The reproductive opportunity granted by a privileged individual who employs an underprivileged individual is called a "staying incentive" because this payment leads the underprivileged individual to stay at the nest as a helper instead of leaving to start a new nest. Overall, the theory envisions the animal society as a political economy held together by transactions in the currency of reproductive opportunity.

Extended families form depending on supply and demand within the labor market. If demand for labor is tight—no jobs for youngsters outside of home—then even a small staying incentive will induce them to stay at home and join an extended family. With lots of opportunity outside of the home, even a large staying incentive will not persuade the youngsters from striking out on their own.

According to this thinking, family structure is fluid, changing when opportunities outside the nest vary and when the breeder changes mates (which devalues the genetic paycheck helpers receive for their labor). Great distributional inequity causes reproduction to concentrate in a few individuals by mutual agreement between breeders and helpers, resulting in a high reproductive skew that may amplify the initial inequity. If resources are evenly distributed, almost everyone breeds for himself or herself, and the social system has a low reproductive skew. Sandra Vehrencamp termed these extremes "despotic" and "egalitarian" societies.[37]

Let's see how this theory works in an actual case. White-browed scrubwrens of Canberra, Australia, fit the labor-market theory of family dynamics.[38] The social groups consist of a breeding female and one or more males (polyandry). In this case, the young males have to decide whether to stay and help mom and dad or to leave and set up a new home. Females build their nest alone and incubate the eggs alone. The males feed the female while she is fertile before the eggs are laid, and while she is incubating the eggs. The males and the females both feed the chicks while they are in the nest and up to eight weeks after they leave the nest. The males fight it out, leading to a dominance hierarchy, with the alpha male on top and the beta male as a subordinate. The question is what the alpha male should allow the beta male to do so that the beta male remains as a subordinate and doesn't strike out on his own. How does the alpha male hire the beta male as a helper? Four types of multi-male families occur:

1. If the beta male is related to both the alpha male and the female, the beta male helps at the nest and doesn't father any of the nestlings himself. The alpha male is the sole male parent. Even

though the beta male is not the father, the eggs are very worthwhile to him because the eggs are his full siblings. When combined with limited outside opportunity, the beta male finds it reproductively profitable to stay, even without the incentive of sharing in some of the matings. This family has high reproductive skew.

2. If the beta male is related to the female but not the alpha male, the beta male again helps at the nest without mating, even though the eggs are not as valuable to him as they are when both the female and the alpha male are his relatives. Again the alpha male provides no staying incentive. This family too has high reproductive skew.

3. If the beta male is related to the alpha male but not the female, the value of the eggs is even less because the paternity of the eggs is uncertain. About 15 percent of the eggs are fathered by extra-group matings. From the beta male's perspective, any eggs the alpha male doesn't father have no value. As a staying incentive, the alpha male allows the beta male to sire about 20 percent of the brood by mating with the female. This family has moderate reproductive skew.

4. If the beta male is not related to either the alpha male or the female, the beta male sires about 50 percent of the brood himself. The alpha male has to share half of all matings, the maximal possible incentive, to keep the beta male as a helper. This family has zero reproductive skew.

Thus the alpha male can be viewed as allotting the beta male access to reproductive opportunity within the family group in whatever amount is necessary to induce him to stay as a helper. In situations 1 and 2, the beta male doesn't need any staying incentive at all. In situation 3, the alpha male allows the beta male to sire 20 percent of the brood, and in situation 4, to sire half of the brood.

Do you think these families are happy groups of individuals sticking together by mutual consent to fashion productive lives for all? Some biologists are critical, raising three objections.[39] The breeder, for one, may find the price of the staying incentives too high and not agree to pay. The helper would then abandon the nest and set forth alone. But the breeder might coerce the helper to stay anyway. However, the breeder might not be able to completely control the helper. The helper would therefore breed surreptitiously to whatever extent he could. This is not a peaceful home, but a family at war.

A second objection concerns whether the net effect of the "helper" is actually to help or to hurt. Does hanging around the nest and bringing in some food now and then yield a net benefit to the breeder? In naked mole-rats, the breeding female aggressively prods "lazy" workers, suggesting a tension between employer and employee.[40]

A third objection is that mutual consent is beyond what animals are capable of doing. Perhaps animals can't really negotiate labor contracts among one another when people can't even do this very well. But I believe that animals can do anything—I'm a hopeless animal chauvinist.

These objections are perhaps merely growing pains in the early stages of a new theory. The approach of looking at labor relations among family members seems promising to me. I'd like to see this theory become more dynamic and interactive. As it is, the breeder is assumed to know what the helpers are willing to accept and then offer that price. The breeder is what economists call a perfectly discriminating monopolist, a sole seller who has perfect information about what buyers are willing to pay. Economic theory also allows for price negotiation between seller and buyer, and competition among sellers and among buyers. Extending biological labor relations theory to include ongoing negotiations between breeders and helpers may solve the present limitations, and allow us to predict when societies will become peaceful or violent.

The take-home message from this theory is that reproductive inequity emerges from distributional inequity combined with genetic relationships. In animals, we have the counterpart of human democracies and dictatorships. We see in these biological theories the same issues that political scientists deal with in human societies. We see an even distribution of resources lead to widespread participation in breeding and a concentrated distribution of resources lead to power hierarchies, family feuds, and labor strife. We see economic markets for transactions of reproductive opportunity.

As we now move to social systems featuring multiple genders, the language biologists use to describe how animals behave becomes particularly loaded. The language always lauds the individuals who hold territories and possess mates, as though each male were biologically entitled to a castle of his own, complete with princess. Words like stealing, parasitism, deceit, and mimicry dominate the discussion and distort the sophisticated reality of what really happens in societies that contain a biological diversity of participants. Instead, the theory of transactions of reproductive opportunity seems to extend nicely to family organizations with multiple genders.

6

Multiple-Gender Families

The social roles of multiply gendered animals are indicated by their bodies. Males or females in a species may come in two or more sizes or colors. The morphological differences are the tip of the iceberg. The two morphs approach courtship differently, have different numbers of mates, have different arrangements of between-sex and same-sex relationships, live different life spans, prefer different types of real estate for their homes, exercise different degrees of parental care, and so on. Because body shape, color, and posture—the important modes of communication in fish and lizards—are so easily visible to biologists, multigender societies are better described in these groups than in groups that communicate more by sound and scent.

Biologists have struggled with naming these within-sex polymorphisms. They may be called alternative mating strategies to emphasize the different approach of each morph to courtship. Or they may be called alternative life histories to emphasize the different life path each morph takes. Previously, I suggested that we call these different expressions of how to live "genders." I feel this word best captures the totality of the differences between the morphs, extending from mating, through lifestyle, to length of life.

Phrases like "alternative" mating strategy or "alternative" life history are especially poor names for multiple genders because the "alternative" strategy is usually the most common strategy. Singling out a minority strategy as "normal" and labeling the rest "alternative" is simply prejudice. Societies with multiple genders are not easy to describe because we're not prepared to find what we actually see. But let's wade in.

TWO MALE, ONE FEMALE

The first step beyond one-male one-female two-gender societies are those with two male genders and one female gender, making three altogether. Here's a sample.

Bullfrogs *(Rana catesbeiana)* have two male genders: large males who call at night, giving bullfrogs their name, and small males who are silent.[1] Both are reproductively competent, and females mate with both. Silent males turn into calling males as they grow older. Male frogs in other species[2] and males in many vertebrate groups also have to decide when to begin breeding—whether to wait until established enough to flaunt wealth and power, or to begin sooner with fewer resources but lots of charm. Perhaps silent males should not be considered a different gender from calling males, but rather an early developmental stage of the same gender. Compare this case with others, though, and you may agree that it makes more sense to view males who mature from a silent stage into a calling stage as changing genders.

Did you know a fish could sing? There is a fish in the bays and estuaries of the Pacific coast, including San Francisco Bay, called the plainfin midshipman *(Porichthys notatus)* because of its bioluminescent spots, which resemble rows of buttons on a Navy uniform. These fish are also known as the California singing fish or canary-bird fish. The species has two male genders who behave somewhat like bullfrogs. The large male gender consists of fish who defend territories and guard the eggs laid in them. To signal readiness to mate, a large male emits a low humming sound for as long as fifteen minutes, and a female may respond by entering the territory and laying eggs there. Females lay only one batch of eggs. A large male guards a big collection of eggs laid by five or six females. The small male gender consists of fish that mature at a younger age and are silent, like the silent bullfrogs. They don't defend territories. Instead, they mate by darting in to fertilize eggs being laid in a large male's territory.

This fish is but one of hundreds of known species where males come in two or more genders. This species is unusual because its habit of singing has allowed biologists to determine how deeply the anatomy of the two male genders differs. For biology wonks, here are the technical details: The large male gender relative to the small male gender has a relative sixfold advantage in the mass of sound-producing muscles, a three- to fivefold increase in the number and diameter of sound-producing muscle fibers, a cellular ultrastructure with enlarged zones densely filled with mitochondria, a more branched endoplastic reticulum, larger sarcomeres, and Z-lines that are twenty times wider. Motoneurons and pacemaker neurons are also three times larger, as well as sonic axons with terminal boutons two to three times larger. And so on. Even without being a biology wonk, the large male and small male genders clearly represent deeply different developmental programs, involving different expressions of entire suites of genes.[3]

Thus, both the bullfrog and plainfin midshipman have a calling large male gender and a silent small male gender. A bullfrog male changes from the small gender to the large gender as he ages, whereas a male plainfin midshipman is locked into one of these genders for life.

In the Pacific, coho salmon *(Oncorhynchus kisutch)* have two types of males. A "jack" spends two years in the ocean before returning to streams to breed, and a "hooknose" spends three years in the ocean before returning to breed. The female spends three years in the ocean too. All three types die after breeding. A jack is small and cryptically colored, and a hooknose is big, has a pronounced snout (hence its name), and is brightly colored.

The females excavate a nest in the gravel in which they lay their eggs. When they do, the closest male fertilizes the most eggs. Hooknose males are better than jacks at fighting for position near a female and wind up with the most fertilizations. The jacks obtain some fertilizations by darting in under the female while she is laying eggs. The benefit for the jack of being able to breed one year earlier and avoiding the hazard of living another year at sea compensates for its relative disadvantage in fertilizations compared with the hooknose. Jack and hooknose coho salmon appear to have equally successful strategies of life.[4]

In the Atlantic, salmon *(Salmo salar)* have some males that migrate from rivers to the sea as smolts and return after about five years as large anadromous males about 75 centimeters in length. Other males, called parr, don't bother journeying to the sea. They remain in the streams and mature in about three years at near 50 centimeters. The females also migrate to the sea and return. At spawning, a large anadromous male defends access to a female, while the parr hang out downstream. When the anadromous male and female are mating, the parr dart in and obtain some fertilizations. These two male life strategies work out to have about the same overall success when survival and mating access are factored in.

Male red deer *(Cervus elaphus,* also called elk) who don't have antlers probably are counterparts of the silent male frogs.[5] Called hummels (or notts), these deer are in better physical condition than males with antlers, and may at times be more successful at mating.[6]

THREE MALE, ONE FEMALE—SUNFISH

Now to species with three male genders. The females here have one gender, making four genders in total. A good example is the sunfish, a deep-bodied fish averaging about 10 centimeters in length and exceedingly common in North American lakes. When I was in high school in New Jersey, I remember seeing sunfish underwater. Every time I went snorkeling in one of the nearby lakes, I would see them through my face mask. If I went fishing, all I would catch were sunfish. Everybody took these fish for granted and hoped to catch a perch or other rarer fish. I would never have guessed that these everyday freshwater fish from the United States and Canada would someday challenge the foundations of gender and sexuality.

One sunfish species, the bluegill sunfish (*Lepomis macrochirus*), has been studied in detail at Lake Opinicon, Ontario, Canada, and at Lake Cazenovia, in upstate New York.[7] Spawning males consist of three distinct size/color classes, and together with females, fall into four morphological categories, corresponding to four distinct genders:

1. *Large males* are about 17 centimeters long and eight years old. Their gonads constitute 1 percent of body weight, and they have a light body color with a yellow-orange breast.

2. *Medium males* are about 10 centimeters long and four years old. Their gonads are 3 percent of body weight, and they have a dark body color marked with dark vertical bars.

3. *Small males* are about 7 centimeters long and three years old. Their gonads are 5 percent of body weight, and they have a uniformly light body color with neither a yellow-orange breast nor dark vertical bars. The testes may occupy most of the body cavity, crowding the stomach and displacing the intestines.

4. *Females* are about 12 centimeters long and six years old, making them larger than the medium males by about 2 centimeters and older by about two years. At breeding, females bulge somewhat with eggs. Females have a dark body color with vertical bars, like the medium males. The medium males somewhat resemble small young females because of the similarity of color pattern.

The yearly spawning episode lasts only one day. In preparation, large males aggressively stake out territories next to one another in aggregations of a hundred or more, called leks, along the bottom of the lake at a depth of 1 meter. Large males are called on to defend their space against neighbors about once every three minutes. Large males make nests for eggs in their territories by scooping out a depression in the mud with their tails. Females aggregate at the locales with many males and do not visit isolated or peripheral nests. Females prefer nests belonging to large aggregations because the presence of many males affords more protection from egg predators.

The large males are not Mr. Nice Guys. Their acts of aggression include biting, opercular spreading, lateral displays, tail beating, and chasing. Although primarily directed at intruding males, aggression sometimes is directed at a female in the territory—domestic violence, sunfish style. The male apparently tries to control the speed and timing at which a female lays eggs. Females simply leave if harassed too much in this way.

The females arrive in a school, and one by one they enter the territories of the large males. When a female arrives, a large male begins to swim in tight circles, with the female following. Every few seconds as the pair turns, the female rotates on her side, presses her genital pore against that of the large male, and releases eggs that the large male fertilizes. The egg release is visible as a horizontal dipping motion.

A female may spawn in many nests. A large male accumulates up to thirty thousand eggs from various females during the one-day spawning episode. A female lays about twelve eggs at a time with her dipping motion, so this total egg accumulation involves some female laying in the nest about once every thirty seconds. The scene is fast. Still, large males somehow find the time to enter the nests of neighbors, and about 9 percent of the fertilizations in a nest are by a neighboring large male.

Meanwhile, the small males are active. They stay at the borders between territories of large males and in the periphery, often close to rocks or in vegetation. Eggs remain viable in lake water for about an hour and sperm for only a minute. When the female releases eggs, the small males dart in quickly to release sperm over the eggs and carry out their own fertilizations. The large males try to repel the small males from their territories, but the small males are more numerous than the large males—about seven to one in shallow-water colonies. Chasing all these small males, as well as neighboring large males and the occasional predator, takes a large male away from fertilizing the eggs being laid in his territory. In these circumstances, the females spawn readily with small males while the large male is busy with all his chasing.

There are more small males in shallow-water colonies than deep ones because there is more vegetation for cover. It is important to hide because predators—large-mouth bass, small-mouth bass, and pike—lurk in the lake. Thus the ratio of small to large males depends on the surrounding environmental context. All in all, the small males seem to be the gender counterpart of silent bullfrogs, silent singing fish, jack and parr salmon, and antlerless male deer.

The medium males—the third male gender—are really surprising. No one knows where the medium males live most of the time, but they may school with the females. A medium male approaches the territory of a large male from above in the water and descends without aggression or hesitation into the large male's territory. The two males then begin a courtship turning that continues for as long as ten minutes. In the end, the medium male joins the large male, sharing the territory that the large male originally made and defends.

Although the medium male sometimes joins the large male before a female has arrived, more often the medium male joins after a female is already present. The large male makes little if any attempt to drive away the medium male, in contrast to the way the large male drives away small males that dart into the territory. When a female and two males are present, the three of them jointly carry out the courtship turning and mating. Typically, the medium male, who is smaller than the female, is sandwiched between the large male and the female while the turning takes place. As the female releases eggs, both males fertilize them.

Occasionally, two females may be within a large male's territory at the same time. Although the large male mates with both females, the three do not participate in any common ritual similar to the three-way interaction of the female with a large and a medium male.

After the day's excitement is over, each large male remains in his territory for eight to ten days to guard the eggs. The large male repels nest predators. During this period he never leaves the nest to forage and loses body weight.

In all, 85 percent of spawning males are either small or medium, with the remaining 15 percent large males. Though in the minority, large males take part in most of the matings. Among the large males, the reproductive skew is high and only some of the large males apparently survive the mutual aggression that is necessary to acquire a successful territory. The small and medium males obtain about 14 percent of the spawnings. Overall, 85 percent of the territories in which spawning occurs consist of one male with one female, 11 percent of two or more males and one female—usually a large male accompanied by a medium male—and 4 percent of one male and two females.

Developmentally, the small and medium males are one genotype, and the large males another. Individuals of the small male genotype transition from the small male gender into the medium male gender as they age, whereas individuals of the large male genotype are not reproductively active until they have attained the size and age of the large male gender.

Explaining the medium male gender has caused big-time confusion among biologists. Three theories have emerged.

DECEIT

The most popular theory is that by sharing some female coloration and participating in courtship turning, a medium male deceives a large male into thinking he's a female.[8] This female-disguised male then steals some of the fertilizations that rightfully belong to the large male.

I don't find this theory plausible. Sunfish, which prey on tiny shrimp in the water during the day, have great eyesight. As predators, they estimate size and distance very well, constantly using these abilities to decide which prey to catch and which to ignore. Lots of visual cues distinguish a medium male from a female, including simply size and shape. A fish that can detect the difference between a 1- and 2-millimeter shrimp 1 meter away can surely detect 2 centimeters of difference between two fish right next to him. In addition, the large male has lots of time to check out a medium male as they turn together in courtship for up to ten minutes. During this time, the large male can see that the medium male isn't laying eggs.

If the large male were being deceived, he should occasionally break off the three-way mating ceremony. For a large male and a female to swim in formation with the medium male squeezed in between, precision is needed. If the large male were being deceived, he would maneuver to exclude the medium male once the medium male was discovered to be producing sperm rather than eggs, just as the large male actively chases off small males. Furthermore, when two actual females are in the territory with a large male, the three don't swim together in a three-way mating ceremony. If the large male believed the medium male was a female, then he shouldn't carry on a three-way mating ceremony with him.

The explanation for why the large male doesn't chase off the medium male once the deceit is discovered is supposed to be his worry about losing fertilizations while the female is actively laying eggs. But the large male does chase small males in spite of this same cost. To counter this, it's further argued that the medium male's female coloration suppresses the large male's tendency toward aggression.

All in all, this theory is another instance of biologists explaining away something surprising by assuming the animals are somehow incapacitated. The large male should chase away the medium male, shouldn't he? And if the large male doesn't—well then, for some reason he just can't. A large male is too dumb to tell a medium male from a female. A large male can't turn his aggressiveness on when he needs it.

COMMON UNGENDERED SIGNAL

An alternative theory is that the medium male is helping the large male, that they are working together as a team.[9] How? One possibility is that the two males together are more successful at attracting a female than one male is by himself. Females prefer to lay eggs in territories in the midst of many males because collectively many males protect against predators better than an isolated male does. Therefore, a territorial male might enhance his chances of attracting a female if he teamed up with a medium male. Teaming up with a medium male might seem more appealing from his point of view than teaming up with another large male because the medium male, with smaller gonads, produces fewer sperm than a large male would. By teaming up with a medium male, he obtains a helper at the least cost.

According to this theory, the fertilizations obtained by the medium male are not stolen from the large male, but actually offered to him by the large male as an incentive to stay, a transaction based on reproductive opportunity. The courtship that precedes the medium male joining the large male's territory amounts to a job interview. Medium males have female color patterns, this theory claims, to act as a white flag, an invitation to cease hostility and aggression. The medium male resembles a female by coincidence, because both are sending the same signal, flying a white flag. By this account, the medium male is a bona fide male known to all as male, even though he happens to be flying the same colors as a female.

Although this theory does seem plausible to me, I'm still suspicious. It seems a bit too male-centered, suggesting that the more males the better—more males mean more protection, more aggregate masculinity means more attractiveness.

Well, maybe. But what about the function of female and medium male coloration? Are these only ungendered white flags? Or does the medium male really intend to be a *feminine* male, and is the large male specifically employing a *feminine* male as a helper, rather than a smaller version of himself?

MALE FEMININITY

I suggest a third interpretation, that females view a large territorial male as dangerous, to be approached with caution. A female might wonder if she will suffer domestic violence from this male who's trying to look big and powerful. She sees the large male chase neighbors and small males.

All she sees is violence. Where's the evidence that this good-looking guy with a great territory is safe to be with? Conversely, how is a large male to say that he's gentle after all—that all the tough stuff is reserved for male colleagues?

Perhaps the courtship between the large male and the medium male offers the female a chance to see how the large male behaves with a feminine-looking fish who is slightly smaller than she is. She can watch how the large male does his courtship turning with the medium male. She can watch whether the large male is aggressive toward the medium male. Of course, a large male who is kind to a medium male is not guaranteed to be kind to her too, but at least watching how the large male behaves with the medium male supplies evidence, which is better than just going on a hunch.

Furthermore, once the medium male is sandwiched between the large male and the female during their combined courtship turns, he may somehow facilitate the mating process by synchronizing the release of egg and sperm. The medium male may protect the female from spawning harassment through his position between her and the large male. Also, the medium male may have developed a relationship with the females while schooling with them, and thus be able to vouch that the large male is safe.

In my interpretation, the medium male's femininity as such has a genuine, nondeceptive role. I suggest that the feminine male is a "marriage broker" who helps initiate mating, and perhaps a "relationship counselor" who facilitates the mating process once the female has entered the large male's territory. This service is purchased by the large male from the small male with the currency of access to reproductive opportunity.

Thus, the second and third theories both view the medium male as working in tandem with the large male, rather than as stealing from him, and extend the concept of a helper. Sharing fertilization represents an incentive to stay, not theft. To coin some new biology jargon, we might say that a medium male is a prezygotic helper, in contrast to the postzygotic helper, who assists in caring for offspring that have already been born. Nothing prevents animals from cooperating in bringing about a mating, as well as in caring for young after a mating—the animal counterparts of a dating service selling prezygotic help, and a pediatric clinic selling postzygotic help (see the discussion of ruffs in chapter 7).

In view of the roles played by the three male genders, let's agree to call the large male a "controller," the small male an "end-runner," and the medium male a "cooperator."

THREE MALE, ONE FEMALE—OTHER CASES

The services of the third male gender, typically the intermediate body size type, are purchased by controller males with the currency of access to reproductive opportunity. The services that are most valuable to the controller vary with the circumstances.

The wrasses of Europe are as interesting as those on coral reefs. These species don't do any sex changing, but they do have multiple male genders. The two-male species have a controller morph that is colorful, territorial, and guards eggs, plus an end-runner morph that is smaller and plain-colored. The three-male species add a medium-sized male, offering a useful comparison to the North American sunfish.

Take the spotted European wrasse *(Symphodus ocellatus),* which lives in the shallow water of the Mediterranean along rocky shores. Biologists have observed these wrasses while scuba diving in Revellata Bay, west of Calvi on Corsica.[10] These rather small fish live for up to three years, breed during the summer, and have a maximum length of 8 centimeters. Of the three morphs, the medium male is again the most interesting. This medium male does not look or act like a feminine male. It's a bit bigger than a female, has its own distinctive coloration, and can be aggressive. Nonetheless, the medium male is enticed by the large male into his territory and fertilizes some of the eggs laid in the territory. Why? The large male has apparently hired the medium male as a security guard. The medium male chases away small males that the large male would otherwise have to chase away himself. More interestingly, this species happens to have more females than it does males with territories. During spawning, several females arrive at one nest, and because only one female can spawn at a time while the large male is there, the others wait at the rim of the nest. But females may try to crowd in and interrupt the female who is laying eggs. In this situation, the medium male expels the excess females so the spawning can continue. The medium male then stays with the nest for three days, compared with the full week that the large male sticks around.

The spotted European wrasse has three developmental pathways. One type skips early reproduction and matures into a controller. The second starts as an end-runner and transitions into a cooperator. The third starts as a cooperator and transitions into a controller.[11] Thus, comparing wrasses to sunfish, the operational sex ratio may determine whether the medium male is feminine or masculine. If females are scarce, as in sunfish, the large male will need help attracting them, and a feminine male can assist. If females are common, as in wrasses, the large male will need help keeping order at home. In this case, the controller employs a bouncer instead of a marriage broker.

Another particularly graphic case is found among the cichlids, a family of colorful perchlike fishes found in the tropical freshwaters of Africa, South and Central America, India, Sri Lanka, and Madagascar. About 1,500 species are known, or 5 percent of all vertebrate species! Most species occur in the Great Rift lakes of eastern Africa—Lake Malawi, Lake Tanganyika, and Lake Victoria. Cichlids are most closely related to the saltwater damselfishes, wrasses, and parrotfishes, among others.[12] They are the freshwater equivalent of the colorful and diverse coral reef fish.

Oreochromis mossambicus, a kind of cichlid, from the Incomati River in Mozambique were studied in an aquarium in Portugal.[13] These fish are rather small, around 6 centimeters, and males come in three genders. The controller recruits the cooperator through a courtship that includes remarkable same-sex sexuality.

The black territory-controlling males form dense aggregations, leks, in the sand or mud during the breeding season. A male digs a pit to attract a female, and after courtship, she lays eggs there. The male then quivers and releases spawn over the eggs. The female inhales the mixture of eggs and spawn into her mouth where the actual fertilizations take place. The female then broods the eggs in her mouth, continuing brooding even after the young fry have hatched, for a total of three weeks. The young are "born" when they swim out of the female's mouth.

The second male gender is the familiar end-runner, who darts into a controlling male's territory during spawning and adds some of his own spawn to the mix inhaled by the female. The controlling male aggressively repels the end-runners. The third male gender is once again the most puzzling one. These males have a neutral light color and are actively courted by controller males using the full courtship repertoire used for females, including tilting, signaling the nest, circling, and quivering.

Of six hundred courtships observed, two hundred were directed to these light-colored males and the remaining four hundred to females. In three of the male-male courtships, the light-colored male placed his mouth on the genital papillae of the dark territorial male, then the territorial male quivered and released spawn, whereupon the light-colored male moved his mouth as a female does when she inhales the sperm/egg mix. The end-runners did not intrude into these male-male courtships, although they did dart into male-female courtships, indicating that everyone around knew what was going on.

In most groups of these fish, males courted females more than males. In one group, though, males courted males more than females. The authors concluded that "further experimental work is needed." In particular, the benefits to the controller presumably provided by the third male gender need to be described.

Male genders may also range from territorial stay-at-homers to nonterritorial travelers. Although many of the three-male species are organized according to the template of controller, end-runner, and cooperator, not all are. Among tree lizards *(Urosaurus ornatus)* living in the American Southwest,[14] the males come in various colors—nine are known for the species. Some populations have only one color, others only two colors, and still others as many as five.

At one site upstream of the Verde River in Arizona, two colors each account for 45 percent of the total males. The orange-blue form is a punk rocker's delight—an orange chin with a big blue spot in the middle, a throat fan with orange near the body and a blue band at the tip, and blue on the stomach. In contrast, the orange form is solid orange on the chin, throat fan, and stomach. The orange-blue males are the most aggressive, and their body proportions are short and stocky. They are the controllers, defending territories large enough to overlap the territories of three to four females. The orange males are end-runners, but they come in two subvarieties, nomadic and sedentary. Unlike most end-runners, these males are the same weight or heavier than the controllers, although longer and leaner. These males are not aggressive and defer to controllers when challenged.

In a typical dry year, orange males are nomadic, spending only a day or two at a site before moving on. In a rainy year, orange males settle down for the season, becoming sedentary and occupying relatively small territories, the size of a female home range. The controllers and end-runners are fixed for life, although an end-runner can transition back and forth between nomadic and sedentary styles in successive years.

In this species, the hormonal dimension of gender expression has been worked out. Progesterone determines whether a male matures into a controller or an end-runner. A single injection of progesterone given to a tiny hatchling on the day he hatches from the egg will ensure that he matures into an orange-blue controller. In contrast, males with low progesterone develop into orange end-runners.[15] No intermediates occur. Presumably genes produce high or low progesterone levels on the day of hatching, thereby determining whether a male develops into a controller or an end-runner.

As the season progresses and the lizards wait for rain, the oranges listen to the lizard version of Emmylou Harris's song "Born to Run." How does an orange male get to feelin' like it's time to hit the road rather than settlin' down for a spell? Orange males are sensitive. When conditions are dry, orange males show high levels of the hormone corticosterone, an indicator of stress. In orange males, corticosterone causes testosterone to decline. This drop in testosterone in turn causes orange males to hit the road and become nomadic.[16] Meanwhile, orange-blue males are indifferent to weather conditions—they tough it out no matter what.

The tree lizards illustrate what are called the "organizing" effects of hormones (irreversible effects that occur early in development) and also the "activating" effects (reversible effects, which usually occur later in life). Progesterone on the day of hatching *organizes* the male body to mature into an orange-blue controller. Corticosterone from the stress of going thirsty during a drought *activates* the orange end-runner to turn nomadic rather than remaining sedentary.

TWO MALE, TWO FEMALE

When multiple genders occur in both males and females, we may wonder whether some gender combinations don't mesh together especially well. Would a feminine male paired with a masculine female be just as successful as macho male with a femme female? What about other pairings too?

White-throated sparrows *(Zonotrichia albicollis)* of Ontario, Canada, have four genders, two male and two female:

1. A *male with a white stripe* is the most aggressive, calls often, and is the most territorial.

2. A *male with a tan stripe* is less aggressive and unable to defend a territory from the white-striped male.

3. A *female with a white stripe* is aggressive, calls spontaneously, and defends a territory.

4. A *female with a tan stripe* is the most accommodating of all. When challenged with a territorial intrusion, she continues foraging.[17]

So, in both males and females, the white-striped individuals are more aggressive than the tan-striped individuals. Ninety percent of the breeding pairs involve either a white-striped male with a tan-striped female or a tan-striped male with a white-striped female—attraction between opposites.

The white-striped male appears to have everything going for him. A female who chooses to pair with a tan-striped male has to settle for an inferior territory. Yet some females do prefer tan-striped males. In studies in both 1988 and 1989, more tan-striped males found mates, and found them sooner, than did the white-striped males. So why is the macho white-striped male doing so poorly at attracting mates?

A tan-striped male and a white-striped female work as a team to defend the territory. When the performance of a pair is considered, the team of the tan-striped male and white-striped female is just as effective at repelling intrusions as that of the white-striped male and tan-striped female. The aggressive potential of both teams is the same. The tan-striped males don't acquire territories until the females arrive to help them out.

Still, why doesn't a white-striped female pair with a white-striped male? That way she'd form a team with the most aggression of all, and together they would get the best territory of all. Well, tan-striped males provide more parental care than do white-striped males, so when nesting survival is factored in, the white-striped female is better off with a more domestic partner who leaves her to do more of the fighting. Conversely, the tan-striped females provide more parental care than their white-striped counterparts, so a team composed of a white-striped male and a tan-striped female provides the same total parental care as the other type of team.

White-throated sparrows are a neat case of gender meshing. Two kinds of teams provide the same total amount of protection and parental care, but divide the labor differently. These genders represent a genetic polymorphism, in which the body differences are not limited to colored stripes. The brain architecture of the morphs differs. Just as with the morphs of the singing fish, the plainfin midshipman, the differences among the genders extend deep into the body (see also p. 224).

THREE MALE, TWO FEMALE

So far, the most genders that have been described in one species is five: three male and two female genders. The present medal-holder—the side-blotched lizard *(Uta stansburiana),* from the American Southwest and West—has both males and females of multiple colors, signifying different genders in both sexes. Mortality is high, and the population turns over annually. Three male and two female color morphs occur at a grassland site in Los Baños Grandes in central California:

1. *Orange-throated males* are controllers. These "very aggressive, ultradominant, high-testosterone" males defend territories large enough to overlap the home ranges of several females.
2. *Blue-throated males* are less aggressive and juiced with less testosterone. They defend territories small enough to contain only one female, whom they "guard."

3. *Yellow-throated males* don't defend territories. Instead, they cluster around the territories of the orange males, "sneak" copulations, and masquerade as "female mimics."

4. *Orange-throated females* lay many small eggs, 5.9 eggs per batch. Orange-throated females, like their male counterparts, are very territorial and, as a result, must distance themselves from one another, achieving a maximum density of only one female per 1.54 square meters.

5. *Yellow-throated females* lay fewer but bigger eggs, 5.6 eggs per batch. Yellow-throated females, like their male counterparts, are more tolerant of one another and can achieve a maximum density of one female per 0.8 square meter.[18]

Females lay up to five batches of eggs at monthly intervals during a season. The ratio of males in each gender cycles over time. In one four-year period, blue males predominated in 1991, orange in 1992, yellow in 1993–94, and blue again in 1995. The ratio of females in each morph also cycles, but over a two-year period. The total abundance also fluctuates in a two-year cycle. The female cycle synchronizes with the two-year cycle of total abundance.

Concerning the male genders, an explanation for the male cycling was proposed based on the child's game of rock-paper-scissors: rock beats scissors, paper beats rock, and scissors beats paper, leading to a never-ending cycle of who's winning: "Trespassing yellows can fool oranges with their female mimicry. However, trespassing yellows are hunted down by blue males and attacked. Although oranges with their high testosterone and high stamina can handily defeat blues, they are susceptible to the charms of yellows."[19] And so on, in an ecological perpetual motion machine.

Does this theory seem too cute to be true? The problem, as usual, lies with how the nonaggressive male gender is interpreted. First, all the males "sneak," not only the yellow ones. Thus, the yellow-throated male isn't distinguished correctly as a sneaker. More important, the yellow male isn't a female mimic after all. What is supposedly feminine about the yellow-throated male? Early studies indicated that all females had yellow throats. Therefore, the yellow male was thought to resemble a female in throat color. Later study revealed that the orange females had been at a low point of their population cycle during the original study. Once the orange females peaked, the yellow-throated and orange-throated males both resembled corresponding morphs in the females. The loss of throat color as a criterion of femininity left only one other trait: "The most intriguing display that males make, which is restricted to yellow-throated males, is an imitation of the female rejection display. This rejection display is characteristic of post-receptive females and consists of a series of rapid head vibrations [called buzzing]. The male extends his yellow throat, assumes a humped back, and comes in and nips the dominant male on the tail. The parallel between the yellow male . . . and an actual rejection by a bona fide female is extraordinary."[20]

Why would performing this one behavior be sufficient to fool an orange male into thinking the yellow male is a female? The blue males aren't fooled, why only the orange males? This question seems to bother the investigators a bit too. A revealing passage on their website entitled "Are You Blind?" states, "The orange male is somewhat blind and can't recognize the yellow male in front of him as a male." A blind lizard that makes its living as a visual predator catching insects? Impossible— the orange males would starve if they were blind. The possibility that the orange male knows what he's doing and actually wants the yellow male around is never remotely considered.

As for the females, the synchrony between population size and morph ratio in the females suggests that alternating low and high crowding drives the alternating ratios of orange to yellow females. The orange female is more valuable when crowding is low and growth is at a premium. The yellow female is more useful when crowding is high and the ability to pack into a small space is at a premium.[21] Thus the polymorphism between orange and yellow females is theorized as a polymorphism between a genotype adapted to low-density conditions and a genotype adapted to crowded conditions.

Still, the alternation of high and low crowding can't be the whole story. Helping takes place too, not just the negative effects of crowding: "When orange females had more orange neighbors their fitness was reduced, but fitness increased with more yellow neighbors. Yellow female fitness was not affected by the density of either morph."[22] Also, one wonders how the two-year female orange/yellow cycle connects with the four-year male orange/yellow/blue cycle. All in all, this social system with multiple genders would benefit from rethinking and further study.

FEMININE MALES—THE DECEIT MYTH

From what we've seen, the notion of a universal male or female template is clearly false. Let's focus specifically on the males who would seem to most clearly violate the universal male template: the feminine males. The third male gender in bluegill sunfish consists of males that look like females. Are such cross-dressing animals rare? Members of one sex often dress in the clothes of the other. Feminine males especially provoke biologists to froth at the mouth. Why would any self-respecting male want to appear feminine? Well, maybe it's okay if the purpose is deception. Hey, it's war out there—a guy does what he has to, even wearing a dress, in order to win.

Let's look into cases of male-to-female cross-dressing to see if biologists have really demonstrated that the function is deception. If not—banish the thought—we might have to consider that being a feminine male might be adaptive in itself.

The pied flycatcher *(Ficedula hypoleuca)* is a common insectivorous European bird. Males vary in plumage from a striking black and white to brown, and the colors are inherited. Females are also brown. Some biologists have suggested that the brown males are female mimics, even though brown males have a darker tail and more white on their wings than females. So a human observer can tell the sexes apart, but the birds somehow can't.[23] Do you think it likely that biologists are more observant than birds?

In woods near Oslo, Norway, male flycatchers who had set up territories were presented with individual caged birds to see if they could distinguish the sexes.[24] A territory-holding male who hadn't already attracted a female reacted to a female by showing off the entrance to his nest hole and calling enticingly. When the same male was presented with a macho black-and-white male, he was not so hospitable, and jumped on the cage of the visitor, pecking at it, trying to attack, and not bothering with any welcoming calls. When a feminine brown male was presented, the male again showed off his nest hole and called invitingly. Is the territory-holding male making a mistake, believing the feminine brown male is a female?

Later in the season, after these territory-holding males had attracted a female, they were again presented with the feminine male. This time, about half of the territory holders did not court the feminine male visitor, but instead reacted aggressively. The investigators concluded that the territory holders had now acquired enough "experience" with females to tell the difference between a female and a feminine male, so they were no longer deceived.

But some of these very same territory-holding males had bred the preceding year. Didn't they become "experienced" at that time? The investigators concluded that the territory-holding males forgot over the year how to tell a feminine male from a female, and needed "recent sexual experience for correct sex recognition," which had to be "refreshed each year." Could territory-holding male birds be this dumb?

At the beginning of the breeding season, the feminine males tend to arrive late, and they must find space for their territories amid the territories of birds that arrived earlier. The feminine males are allowed to settle closer to the macho males than the macho males can settle next to each other. If you were a macho male, wouldn't you allow a friendly neighbor to settle closer to you than an aggressive neighbor? Not that the feminine males are necessarily wimps. When forced to compete with a macho male for a nest box in experimental aviaries, the feminine male attacked first and won, provided he fought at all. Twenty percent of the time the feminine male didn't bother fighting and simply let the macho male have the nest box.[25]

Males with territories who had not yet attracted a female have been observed in the wild advertising to feminine males. The territorial male shows off his nest hole, gives enticing calls, and the feminine male joins him and they enter the nest cavity together. Does it seem plausible that the feminine male has used deceit to enter the home of a territorial male "to obtain information" about his "nest site quality"? Would the feminine male be that devious?

A simpler explanation is that territorial males who have not yet attracted a female are horny and invite romance with feminine males. Once the territorial males have attracted a female, they are no longer horny and no longer interested in courting a feminine male. A simpler explanation is that no one is deceived, no one forgets from year to year, and no one requires continual updating of his limited memory. A simpler explanation is that the two male birds who retire together into the nest hole are enjoying a romance. These birds may be neighbors building a cooperative relationship based on same-sex sexual attraction.

The problem with deceit theories of animal behavior is that not only must some animals be implausibly dumb, but others must be remarkably devious—there must be great asymmetry in cognitive ability. Imagine a bird sneaking into the nest of another to spy on it. What would a bird do with what it saw? Does a bird keep a file cabinet in its head full of dirty secrets about its neighbors? I don't think so, and scientists have not shown any such thing.

The European kestrel *(Falco tinnunculus)*, a dramatic bird of prey, offers another instance of cross-dressing. Males two years and older are blue-gray with brick-red on the back and spots on the head. The females are mostly brown, with a barred pattern on the head. One-year-old males resemble females, so much so that observers find it difficult to tell young males from females. Biologists have therefore suggested that year-old males are female mimics who deceive older males into thinking they are females.

About thirty birds were housed in isolation at a field station in central Finland.[26] A bird was placed in a box with one-way glass on the sides. Birds were also placed on the sides; these birds couldn't see the central bird because of the one-way glass, but the central bird could see them. The biologists then noted which side bird the central bird paid most attention to and tried to associate with—called the "preferred bird."

When the central bird was a macho male and was offered a macho male on one side and a female on the other, he always preferred the female. When he was offered a feminine male on one side and a female on the other, he preferred either in a fifty-fifty ratio. The investigators claim they've shown that a macho male bird can't distinguish a feminine male from a female. Clearly, though, there is another possibility: a macho male may be quite able to tell the difference between a feminine male and a female, but he doesn't care which he sits next to.

When the central bird was a female and was offered a macho male on one side and a feminine male on the other, she always preferred the macho male. The investigators claim they've shown that females are better able to distinguish sexual identity than males are. The experiment doesn't speak to whether a female can distinguish a feminine male from a female. The female wasn't presented with a choice between a feminine male and a female; the female was offered only males.

The investigators go on to speculate that "the better sex recognition ability of females compared with males may have evolved because she is the 'choosy' sex. Males . . . do not need to be so good at sex recognition as females." Deceit theory is a trap. Deceit theory forces scientists to take sides on who is smarter—in this case, claiming that females are smarter than males.

A different kind of cross-dressing is found among red-sided garter snakes *(Thamnophis sirtalis parietalis)*.[27] One of seven distinct subspecies of the common garter snake, this nonpoisonous snake is predominantly black with yellow stripes and red bars, and eats invertebrates and small rodents. Females are 10 centimeters larger than males, averaging 55 and 45 centimeters, respectively.

The red-sided garter snake has made Manitoba's interlake region world-famous for snake-watching, and the town of Inwood has even created a monument in recognition of its large population of garter snakes. Snake dens (or hibernacula) can be found in tree roots, shale cliffs, rock piles, sewers, foundations, animal burrows, rock outcrops, and sinkholes. Twenty thousand garter snakes may gather in a single den during the winter. In spring the snakes' mass emergence creates an awesome natural spectacle. As far back as the 1880s fashionable picnics were held near Stony Mountain just to watch this phenomenon. But the snake gatherings have also provoked fear. In a labor strike, penitentiary construction workers once refused to work at Stony Mountain until the den was destroyed.

After the snakes emerge, they mate and then disperse to their summer homes in marshes and shallow lakes. Males hang around the dens longer than the females, so that the ratio of males to females near the den entrances is ten to one. Courtship takes place in small groups called "mating balls," in which one animal is courted by several others. These "suitors" align their bodies with the courted individual and vigorously work to position their tails base to base with the tail of the courted individual, resulting in a ball of writhing snakes.

Females have special lipid perfumes in their skin that turn males on.

In 1985 some male garter snakes were found with female perfumes in their skin. Of two hundred mating balls, about 15 percent consisted of a male, presumably with female perfumes in his skin, surrounded by courting males. The males with female perfumes were called—you guessed it—female mimics. The feminine males who joined female-centered mating balls in progress were thought to be distracting the males already there, thus giving themselves greater access to the female.

As with the other claims of female mimicry, the story had internal contradictions. In choice experiments, the males preferred a female to a feminine male, showing that males could tell the difference—they were not deceived. Moreover, in 2000 it was found that *all* male garter snakes have female perfumes when they emerge from the den in the spring and that *all* males court these perfumed males in addition to females. At this point, the investigators floated four deceit-based theories to explain why all males have female perfumes on emergence: (1) a perfumed male may confuse the other males while carrying out his own mating; (2) the perfumed male may avoid wasting energy in courting before he has fully awakened from hibernation; (3) a perfumed male may induce the other males to waste time and energy courting him while he gets ready to start his own courting; and (4) a perfumed male may distract other males from the females so he has more to himself when he does get going.

The investigators discarded what seems like an obvious hypothesis. Imagine you're a male garter snake just waking up in the spring. You're chilly and covered with cold dirt and mud. You see a spot of blue sky through the ceiling. You crawl out and are greeted by the smiling faces of ten horny males hoping you're a female. You're not. But they're hot and fast, and you're cold and slow. Wouldn't it be nice to roll around in the sun—why not use some perfume to signal your intentions? And if you're one of the ten males watching this face poke up through the ground, and you see he's not the female you were waiting for, why not welcome him into the sunlight and get acquainted? Better than attacking him, with nine of your buddies looking on ready to attack you too.

This explanation, that the female perfumes might protect a male from attack when he is emerging, was dismissed because garter snakes just aren't "agonistic." By nature they're friendly. Indeed, they'd better be, with twenty thousand of them in one spot! But why are they so "notoriously amorous"? And what makes them so friendly? Group sex.

These studies are disturbing because they attempt to sensationalize at the expense of transgendered people. An article in the scientific journal *Animal Behaviour* begins, "Female mimicry, whereby a male takes on a female's appearance, is a rare but widely publicized trait in human societies. Remarkably, parallels can be seen in other animal species." Feminine males and masculine females are not rare among humans. Nor are transgendered people comparable to snakes. *All* male garter snakes wear female perfume and participate in same-sex copulation every year. No human society has ever enjoyed such a rite of spring! An article in another scientific journal, the *Canadian Journal of Zoology*, refers to female mimicry as "bizarre."[28] The problem with female mimicry is not that it is bizarre; the problem is that female mimicry is a myth.

Both articles refer to feminine males in text and figure captions as "she-males." This language, derived from pornography, is derogatory. A she-male is a woman with a penis. The transgender community has better words to describe transgendered bodies. In the biological literature expressions like "gynomorphic male" and "andromorphic female" are preferred when describing a feminine male or a masculine female.[29]

The articles derogate not only transgendered people but also their partners. The title of one article includes the phrase "transvestite serpent," and another claims to be about the "behavioural tactics of 'she-males' and the males that court them." This writing not only stigmatizes transgendered people with certain body types, but also transfers the stigma to their friends. I hope future work on these animals is carried out with more professionalism, and that future publications on this subject receive better editorial oversight.

The examples just cited involve males whose appearance is somewhat feminine. A different kind of example, involving a male bighorn sheep that is morphologically indistinguishable from other males but quite different behaviorally (called an "effeminate male"), might also have been discussed here (see chapter 8). What seems common among all these feminine males is a lessening of hostilities. The cessation of hostility may be temporary, as in garter snakes, or last a year or so, as in young birds whose juvenile color matches the color of females, or be permanent, as in the sunfish cooperator morph whose occupation is to assist controller males in their courtship. The feminine males may exhibit a distinctly feminine signal, such as colored bars or stripes, or simply share with females the absence of the threatening colors of controller males. In either case, feminine imagery seems to be adopted by males to reduce hostility and promote friendship.

Overlooking the positive value of feminine males is part of a larger problem of overlooking cooperation among animals. Even apart from gender expressions, many forms of cooperation occur. Let's look at a few more.

FRIENDLY FISH

Books on fish behavior are the ponderous rivals of telephone directories. Without going into extensive detail, a few more examples are worth noting to fill out some missing colors in the rainbow of fish gendering.[30]

Some species don't bother with a controller morph. Instead, they spawn in large groups of two genders—just male and female. Simple. In surgeonfish, thousands of individuals aggregate for one giant love feast. Other species spawn both in pairs and in groups.[31]

In the bluehead chub *(Nocomis leptocephalus)*, males form partnerships to build nests together. Two large, individually recognizable males were observed together building five different nests in succession. Males of the northern greenside darter *(Etheostoma blennioides)* form partnerships to court females. Several species of temperate freshwater fish carry out joint courtship, including lake trout *(Salvelinus namaycush)*, the yellowfin shiner *(Notropis lutipinnis)*, and the sucker *(Moxostoma carinatum)*. In many species of suckers, spawning appears to occur *only* in trios. Two male spawning partners adjoin the female on either side and press against her flanks. This formation is aided by breeding tubercles, called "pearl organs," which roughen the body surface of males so that the three fish can hold position and not slip apart.

Males in some species, such as the Mediterranean peacock wrasse *(Symphodus tinca)* — "hot bed," that is — lay eggs in a nest while the nest owner is between spawning periods. For some reason, biologists term this nest sharing "piracy," again reflecting a preoccupation with theft. In the tesselated darter *(Etheostoma olmstedi)*, a male who spawns in a breeding hole abandons it to find another one. Another male takes over the nest and cares for the eggs, including cleaning and guarding them.[32]

Some species show cooperative brood care in extended families, just as in the extended families of birds with postzygotic helping at the nest: *Lamprologus brichardi,* a small fish from Lake Tanganyika, is one of six species there exercising cooperative brood care.[33] There is even cross-species brood care! All broods older than five weeks guarded by Midas cichlids *(Cichlasoma citrinellum)* in Lake Malawi contained young of *Neetroplus nematopus* in addition to their own, and conversely, some guarded broods of *Neetroplus nematopus* contained young of the Midas cichlid.[34]

All in all, fish show lots of cooperation. Although not my first choice, I could live with being a fish.

BIASED VOCABULARY

Silent bullfrogs, antlerless deer, and small, medium, and large male sunfish are happily ignorant of how they've been described by biologists. If they knew, they'd be mad.

The silent bullfrog has been termed a "sexual parasite" by the biologists who study it. Instead, the bullfrog who croaks all night long is the model bullfrog, what every young male frog should aspire to. Why is the noisy male so privileged? If I were a female frog, I'd certainly prefer a male who didn't keep me awake all night. I see no reason to admire a large, noisy male bullfrog as the masculine norm for frogdom while disparaging the silent bullfrog as a parasite.

Biologists call a small male fish who darts in to fertilize eggs a "sneaker," a medium male who resembles a small female a "female mimic," and a large aggressive territorial male a "parental," to place a positive spin on his egg guarding. Both the sneaker and the female mimic are "sexual parasites" of the parental male's "investment" in nest construction and territorial defense. The sneaker and the female mimic are said to express a gene for "cuckoldry," as though the parental male were married to a female in his territory and victimized by her unfaithfulness. In fact, a territorial male and the female who is temporarily in his territory have not pair-bonded. Scientists thus sneak gender stereotypes into the primary scientific literature and corrupt its objectivity. Are these descriptions only harmless words?

No. The words affect the view of nature that emerges from biology. Animals are not warrior robots—wind them up and all they do is lie, cheat, steal, and fight. The biology I know reveals sophisticated relationships among animals, relationships that involve honesty and cooperation as much as or more than deceit and competition.

Scientists are open about their predilection for seeing deceit everywhere. They write, "Natural selection favors those individuals that are able to increase their own fitness by manipulating the behavior of others," and "Cooperation might be seen as the opposite of competition . . . it is instead another form of selfish behavior."[35] These attitudes spin how animal behavior is interpreted and predetermine what data are taken.

The expression "female mimicry" prevents the study of gender variation. The words suggest a male deceptively impersonating a female. In biology, mimicry usually refers to such cases as an edible fly that looks like an inedible bee. "Looks like" here means *exactly like,*" not "approximately like." A fly that mimics a bee almost totally resembles a bee. A good magnifying glass and technical knowledge are needed to tell them apart. A bird flying quickly over the ground can't spot the difference.[36] So-called female mimics don't exactly resemble females, and all the players have a long time to examine each other. I doubt that female mimicry exists anywhere outside the imagination of biologists.

Thus biologists project scripts of their own prejudices and experiences with male-male competition onto animal bodies and use insulting language about animals. Far from being a sexual parasite, why not see the silent male bullfrog as nature's antidote to excess macho, preventing the controller from grabbing unlimited power? Far from being a cuckolder, why not picture the feminine male sunfish as nature's peacemaker? Biologists need to develop positive narratives about the diversity they're seeing. Then a new suite of hypotheses will emerge for testing, taking the place of the shallow, pejorative, and far-fetched ideas that deceit theory requires.

TRANSGENDER SPECIES

Some species have an appearance or behavior that invites the term "transgender." These species contain polymorphisms of feminine males, masculine females, masculine males, and feminine females all together, and/or gender-crossing behavior. One study offers comparative data on transgender morphology from museum specimens.

Hummingbirds, the world's tiniest birds, feed on nectar from flowers and on insects. Their name comes from the buzzing sound their wings make as they hover while feeding at a flower. About 340 species exist worldwide. Hummingbirds typically live three to five years. The smallest is the bee hummingbird of Cuba—a bird only 2.25 inches long.

Hummingbirds also have the smallest eggs of all birds—half the size of a jellybean. A female builds the nest and broods the young alone. Males only fertilize. A typical nest is tiny—about the size of a bubblegum ball. A female incubates two eggs for two to three weeks and feeds them for three more weeks. Because females do all the work, males would seem to be in great excess, allowing females to choose among them. Indeed, the males of many hummingbird species are spectacularly colored, which Darwin would argue is what female hummingbirds find handsome. Recently, though, hummingbirds have begun to emerge as the best documented example of transgender expression in birds.

Male sunangel hummingbirds *(Heliangelus)* of the Andes, from Venezuela through Colombia, Ecuador, Peru, to Bolivia, have colorful feathers on their throats called a gorget. The name comes from a biblike collar of metal armor used in fencing to avoid being pierced in the throat by a sword. In birds, a gorget is a broad band of distinctive color on the throat and upper chest.

Museum specimens revealed that eight of nine sunangel hummingbird species have some percentage of masculine females with gorgets just like those of males. A few instances of males with female coloration were also detected.[37] The investigation of masculine females and feminine males has now been extended to forty-two species of hummingbirds from five genera, yielding the first statistical information about transgender expression in birds.[38] Of the forty-two species, seven had both masculine females and feminine males, nine had masculine females and no feminine males, two had feminine males and no masculine females, and twenty-four had neither masculine females nor feminine males. Pooling the species with either masculine females, feminine males, or both, yields data on the total variation of gender expression in both sexes. The appearance of a bird's gorget was divided into four classes, from most femalelike to most malelike. Fifty-two percent of the adult females (288 of 548) were masculine, including 34 percent who were very masculine. In contrast, only 2 percent of the adult males (18 of 745) were feminine.

Not only the total gender variation but also its distribution varied between the sexes. The distribution of masculinity among females was gradual. That is, most of the females were feminine, and the percentage dropped off gradually from most feminine to most masculine. In contrast, the distribution of femininity among males was bimodal. The great majority were masculine, with no intermediates and a small second peak at the most feminine category. Thus the sexes are not symmetric in either the total amount or the distribution of transgender expression.

What does this variation in gender expression mean? Hard to say. The strength of this study on hummingbird museum specimens is its breadth of coverage. The study's weakness is the absence of field data on how the genders behave. Still, clues are provided by other traits besides the gorget.

Male hummingbirds tend to have shorter bills than females.[39] In hummingbirds, bill length is important in indicating the type of flower that is visited—short bills for short, squat flowers, long bills for long, tubular flowers. Overall, short flowers and short-flower users are more abundant than long flowers and long-flower users. Hence, short-billed birds wind up having to defend their flowers and are aggressively territorial compared to long-billed birds, who don't defend flowers. With its relatively short bill, then, a male typically also has a showy gorget and territorial behavior.

Masculine females have a shorter bill than feminine females, and presumably defend territories containing flowers. It is possible that a male might prefer to mate with a female whose offspring are guaranteed access to resources in the territory she defends. In fact, studies have recently shown that males prefer ornamented (masculine) females in seven bird species, including an auklet, pigeon, swallow, bluethroat, tit, and two finches.[40] Thus one guess is that masculine female hummingbirds represent a gender of female controllers maintained in part by male choice.

The feminine males have longer bills than masculine males, even longer than feminine females. Hence feminine males must be using different flowers than the masculine males. The feminine males also have smaller testes than the masculine males, indicating a lower allocation of energy to sperm production. Perhaps the feminine males are pair-bonded to masculine females, as another case of gender meshing. Alternatively, they may have a role in facilitating courtship. Also, mating in some of the species takes place in leks, suggesting a comparison to the feminine male sunfish and ruffs found in leks. Perhaps the feminine males have a role in facilitating courtship at the leks. In any case, the data on museum specimens show that transgender expression is widespread in humming-birds, inviting follow-up fieldwork.

For a case of transgendered behavior, let us turn to the opposite extreme in data collection, a single individual in the field. Hooded warblers *(Wilsonia cirtina)* live in woods of the mid-Atlantic United States. They are named for the black plumage that adult males have on their heads—a hood. Some females also have these black hoods, and can't be distinguished from males by birdwatchers.[41] Early on, variation in female plumage color was thought to represent age, but later work showed that the color is permanent, suggesting a genetic polymorphism for color. About 5 percent of the females very closely resemble males.

Of particular interest is one transgendered black-hooded warbler that was discovered in Maryland.[42] The bird was originally assumed to be a masculine female, but was later discovered to be gonadally male. Black-hooded warblers are monogamous, and the transgendered bird behaved as the female member of a monogamous pair, consistently showing female-typical behavior throughout two years, including nest building, incubating and brooding young, and not singing or engaging in territorial defense. The bodily appearance of the transgendered bird was typical of males, the behavior typical of females. This bird pair-bonded to a male who was also typically male in both appearance and behavior.[43] In this case, a male-bodied bird behaved in all respects like a female, except for laying eggs. Gender identity in this individual hooded warbler evidently crossed over from that typical of the sexed body.

In conclusion, families with multiple genders can be explained using the concepts developed for two-gender families. The idea of helping at the nest in return for reproductive access that was devised for social insects and applied to extended families of birds and mammals also works for how multiple genders are integrated into a social system. Extending kin selection theory now leads to a theory for a labor market that trades access to reproductive opportunity for service, with genetic relationship merely affecting the worth of a unit of reproductive access. The different genders represent different sectors within this economy. While some sectors, like the end-runners, clearly compete with the controllers, others (like the cooperators) are service providers working under contract. Understanding this complex and interesting social dynamic, an animal political economy, I believe is the next step for evolutionary social theory. The part of Darwin's theory of sexual selection that predicts universal male and female templates may be false, but an evolutionary approach to social behavior is alive and well.

7

Female Choice

As further evidence of the difficulties with sexual selection theory, let's consider how real-life female choice differs from female choice in Darwin's sexual selection theory. Darwin focuses on mating only. A female is supposed to select males according to their attractiveness and prowess. Males are supposed to compete among themselves for mating opportunities and to advertise their good looks to females. This peculiar emphasis on the mating act alone is simply not supported by actual female choices, which are more concerned with the totality of reproduction, including the growth and protection of the young.

"Darwinian fitness" is a technical term that refers to production of the young who will partake in the next generation's reproduction; in mathematical terms, fitness is the product of fertility and probability of survival. Evolution depends on this overall measure of reproductive success. Mating is one component of fitness, but a preoccupation with "mating success" has led to an emphasis on mating to the exclusion of other components of fitness. In reality, female choice considers the overall production of offspring, keeping mating in perspective. Darwin is incorrect in almost all details concerning female choice, although he must be credited with recognizing that female choice among animals exists in the first place.

What, then, are the preferences of female animals, and how do females vary in their preferences? What do females want from a male, how many times do females want to mate, how many males do females want to mate with, how does a female find Mr. Right, and how do females decide how many eggs to produce?

DEADBEATS NEED NOT APPLY

Is a male's true mettle tested in combat with other males? Does the best male surface as the winner and assume dominance over a hierarchy of wannabes? Shouldn't a female yearn to shack up with a proven winner? Shouldn't a female respect the winner of male-male competition as the best father for her baby, a stud with the best genes? Does mating with him guarantee the best and brightest child?

Let's see what female gobies think about male dominance. Sand gobies *(Pomatoschistus minutus)* are small fish (5 to 6 centimeters) common along European coasts. To see what a female goby wants in a male goby, specimens were collected from a shallow sandy bay near the Klubban Biological Station in Sweden and housed in seawater tanks for observation.[1] After the experiment, they were released back to the sea.

Sand gobies live for one or two years and experience one breeding season. Both males and females reproduce often during the breeding season, which is two months long (May and June). Males build nests under empty mussel shells by covering the shells with sand and excavating a cavity underneath. They attract females with a courtship display that includes showing their colorful fins. During spawning, a female attaches her eggs to the nest in a single layer.

In an experiment, two goby males were allowed to compete for a clay pottery fragment to use as a nest in order to determine the dominant male. The winner was usually slightly larger than the loser, although only by 3 millimeters. They were then placed in chambers at opposite ends of a tank. The tank was divided into thirds using transparent partitions. The middle chamber was left empty. The winner and loser were given new pottery fragments and allowed to build nests by themselves.

Next, a female was introduced into the middle chamber. The female could choose which of the males she preferred, indicated by the side of the chamber where she spent her time. After the female's preference was determined, she was placed with one of the males, either the one she preferred or not, by flipping a coin, and then the time needed for spawning to occur was noted. Another female was placed with the remaining male, and the time they took to spawn was also noted. Thus both males were able to spawn.

Finally, after spawning, the females were removed, as were the partitions separating the males, leaving two males, each with a nest containing eggs, at opposite ends of the tank. A small crab was introduced, which is an egg predator. Observers counted the number of eggs lost to the predator in order to determine how good the males were at protecting the eggs.

The results are striking. Whether a male was dominant in competition for nests did not correlate with whether he was a good father in protecting the eggs. Also, female preference didn't correlate with dominance in male-male competition. The females didn't care if the male they preferred won his fight with another male. The females did care whether the male would protect the eggs. Somehow females were able to predict who would or wouldn't be a good father, and decidedly preferred mating with males who later turned out to be good egg protectors. A female could somehow look a male in the eye and tell if he was a deadbeat.

Now, let's take a look at the peacock wrasse (Symphodus tinca) that lives off the coast of Corsica in the Mediterranean in a shallow rocky habitat.[2] The female peacock wrasses have a choice of whether to lay eggs in a male's nest or to broadcast their eggs over the sea floor. Which they do depends on how they assess the offer of male parental care.

Large controller males construct guarded areas of a meter in diameter and place pieces of algae in the middle, to which the eggs attach. Nest construction takes one or two days, followed by two or three days during which females visit the nests and deposit about fifty eggs at a time, leading to as many as fifty thousand eggs in a nest. Thereafter, the male may guard the egg mass until hatching, which varies from twelve days in the cold water of mid April to six days in the warm water of mid June.

Smaller males take on two roles. They may be "followers," who swim at a distance behind gravid females and fertilize eggs broadcast on the open sea floor. Or they may hang out as end-runners around the territories of controllers and fertilize eggs laid in the territory. During the first half of the reproductive season, however, small males are absent. The small males arrive only for the second half, presumably when the ability of the large male to shoo them away is constrained by the need to guard the eggs that have already been deposited.

Males defending eggs lose weight and appear to have a higher mortality during this period, so they abandon nests that haven't accumulated enough eggs to be worth their while. Abandoned eggs are hung out to dry, so to speak. Because abandoned eggs are concentrated in one spot, they quickly attract predators. Thus the best chance of an egg surviving is to be laid in a nest that is not abandoned, the next best chance is for an egg to be broadcast on the sea floor, and the worst is to be laid in a nest that is subsequently abandoned.

The males stay with only 20 percent of the nests early in the season, remain with 85 percent of the nests at midseason, and drop off to 20 percent again by the end of the season. Thus laying eggs in a male's nest is a good bet only in midseason. Indeed, only 15 percent of the females lay their eggs in nests at the beginning of the season, rising to 85 percent at midseason, and falling back to 15 percent as the season ends.

What does a female peacock wrasse want of a male? A male who isn't a deadbeat, who won't abandon her eggs. And she can tell. The investigators write, "If a female chooses to lay her eggs outside a nest, she tends to do so only after visiting several nests."

INVITING FEMALE COMMITMENT

How does a guy convince his gal that he isn't a deadbeat? Fish offer some advice on this ancient question too. Females know that males abandon nests that don't accumulate enough eggs to be worthwhile from the male standpoint. From a female's standpoint, adding eggs to an egg mass that's already large makes sense, because the male guarding it is more likely to stick around than if it was a small egg mass. So, how does an egg deposit get started? A female has to take a chance on a male or go it alone.

Various male fish have structures on their body that resemble eggs, a common example being the fantail darter *(Etheostoma flabellare)*. These small fish are found in freshwater streams in North America, including central Kentucky.[3] During the spring, males excavate nests beneath flat rocks, defend small territories, and mate with various females who deposit eggs in their nests. Males then guard the eggs until hatching.

Each of the seven to eight dorsal spines on the male's front fin is tipped with a fleshy knob. These knobs are smaller than the size of real eggs but, on the largest males, approach the size of actual eggs. Deceit theorists have suggested that a female is fooled by these structures into believing that a male is already tending eggs, so that adding her own to the collection is safe. This hypothesized deceit is called "egg mimicry."

Two facts compromise this interpretation: females also have these fleshy knobs, and the knobs on males are smaller than real eggs. Why do females have these knobs if their only function is for males to deceive females? Why would these fish, who are visual predators, be fooled by fleshy knobs that are smaller than real eggs?

Experiments suggest that the females prefer to lay eggs in the nests of males with fleshy knobs rather than the nests of males whose knobs have been snipped off with scissors. Although the study was preliminary, we can still ask what such a result would mean. Were the females fooled? The alternative is that the fleshy knobs are symbols of eggs, not mimics of eggs. When a male swims close to the underside of a rock, he might be showing where the eggs should be placed.

Female fish want male fish to live up to their promise of guarding the eggs. The male must communicate that he is serious about his willingness to provide for the young. The invitation to lay eggs in his nest must somehow show he knows how to handle this responsibility. The female carefully assesses the credibility of the promise; she seems unlikely to be deceived by a trick such as egg mimicry.

HOW MUCH SEX IS ENOUGH?

Newspapers are filled with advertisements for new toys and chemicals to help people have sex more often. Well, how often is enough? Birds illustrate how females may take the lead in determining how often matings happen and when.

Female alpine accentors (*Prunella collaris*) from the central Pyrenees of France like sex.[4] These females don't worry about male harassment. If anything, it's the reverse—females harassing poor males into sex all the time. What do these horny females want? They want the same thing female sand gobies and female peacock wrasses want: males who do their share of the housework.

Alpine accentors go in for eightsomes, as many as four males and four females. A female is fertile for about two weeks, from about a week before her first egg until the last egg is laid. After the eggs hatch, the males may help feed the young.

Fertile females actively solicit copulation. A female approaches a male, crouches with her breast touching the ground, lifts her tail to expose a bright red, swollen cloaca, quivers her tail from side to side, and shivers her wings. Just to be sure he's awake, she often jumps in front of him and presents her cloaca directly in his face. Hard to miss. A female solicits in this way once every 8.5 minutes. A full 93 percent of all solicitations are initiated by the female approaching the male, the other 7 percent by him approaching her.

The males in the group form a dominance hierarchy. The alpha male follows behind any fertile female, limiting but not entirely preventing lower-ranking males from approaching her. Moreover, the males play hard to get, ignoring 68 percent of the solicitations. Still, they do a lot of mating anyway. In fact, a female copulates 250 times per clutch of eggs, although a single insemination provides enough sperm to fertilize all the eggs. So much for believing that the sole purpose of copulation is to conceive!

What's going on here? An alpha male doesn't stick around to help at the nest unless he's sufficiently occupied at home. He can easily visit nearby nests, so to keep him at home, the female invests more time in mating with him than with the lower-ranking males. The lower-ranking males don't have as many opportunities to shop around outside the nest, but if they are to remain as helpers at the nest, they require some minimum share of the action. Therefore, a female actively displays to the subordinate males as well, making sure that they have some share of the copulations and therefore of the paternity.

Alpine accentors provide an example of females preferring the alpha male, because most of the copulations are with him and are initiated by the female. This preference might seem to suggest that the alpha male offers some benefit to the female, such as his "great genes," and that female preference for alpha males is an endorsement of their superior quality. The data show, however, that the quality of the chicks sired by an alpha male is no better than that of chicks sired by the subordinate males, judging by the weight of the chicks at the time they leave the nest. In fact, the only reason the female appears to prefer to copulate more with the alpha male is that the greater availability to him of opportunities outside the nest makes it more of a challenge to keep him at the nest. From the female perspective, copulation provides the shared paternity needed as a "staying incentive," which is allocated to males of various dominance status according to what is required to keep them involved at the nest.

Do monogamous females mate only during the brief period when the eggs are ready for fertilization? Or do monogamous females like fun too? In fact, monogamous females may be even more sexually active than females in other types of families.[5] In birds such as the mallard duck and common guillemot, mating starts before the female is ready to produce eggs, and before the male is ready to produce sperm. Why should all this mating occur when it is apparently not needed? The obvious answer, one might have thought, is that mating maintains the bond between male and female. Regular mating keeps the pair in touch with one another, so to speak. By mating, they enjoy sexual pleasure with one another. One might theorize that the pleasure of mating evolved in such species in order to provide an ongoing motivation for the members of the pair to stay together.

But in the minds of deceit theorists, "excess" mating between members of a pair has nothing to do with building relationships; rather, it represents females using sexuality to manipulate males into giving them free food—a dinner date followed by sex. According to one model for the evolution of "female sexuality" in monogamous birds, males keep buying dinner because they can't "risk leaving." As a result, "females benefit from the presence of males in such a way that males get nothing in return."[6]

For the record, biology provides no evidence whatsoever that the function of sexuality in monogamous relationships is deceit. Instead, theories of male/female cooperation should have been considered as a rationale for sexuality in the monogamous family.

WHEN FEMALES LOOK LIKE MALES

What does female-to-male cross-dressing tell us about the role of female choice? Reports on feminine males are marked by deceit rhetoric and sensationalism. Reports of masculine females are scanty, suggesting un-derreporting. What emerges is that some females signal receptiveness with colors that coincidentally resemble male colors, whereas other fe-males modify their attractiveness to males to control how often males so-licit them.

At the northwestern tip of the Iberian peninsula lies the seaport city of A Coruña, Spain, where Bocage's wall lizard *(Podarcis bocagei)* lives. This lizard is the only vertebrate animal species so far in which females have been reported to imitate males, but the case isn't convincing.[7] Males have an intense green color on their back. Female wall lizards are usu-ally brown, but when they have fertilized eggs already in their oviducts or have recently laid an egg, they turn green to signal that they won't ac-cept courtship. Is being green masculine and therefore romantically un-appealing to other males, as some scientists have speculated? Whereas feminine males are cast as deceivers, masculine females are cast as unat-tractive. Or could green simply be a gender-neutral signal telling males not to bother courting?

The green color seems to be a gender-neutral signal rather than a mas-culine presentation that males find unattractive, because males do occa-sionally try to mate with green females and are rebuffed. These males are presumably learning what green means. If males found green females unattractive, they wouldn't court them to begin with.

Interestingly, the phrase "male mimicry" is not introduced. Females are not seen as deceiving males. If this was a case of male mimicry, the males who do try to mate with green females would have to be mistak-ing the females as males and soliciting a same-sex courtship, something not (yet?) described in this species.

A comparable lizard species in western Ecuador, *Microlophus occip-italis,* also has females that display a special color when unreceptive to courtship.[8] Hatchlings of both sexes have red throats and chins for about a month. Then males lose the red pigment, while females retain some of the red in skin folds on the side of the neck. The males develop black markings on their back and grow larger than the females.

During the reproductive season, some females develop bright red pig-mentation covering the throat and chin similar to that of juveniles. Imag-ine painting Texas-red on your chin and neck, all the way down to your breastbone: you'll get noticed. Females wear red on their chin and neck when carrying undeveloped eggs in their oviducts or after laying eggs. Males were found to make more courtship approaches to nonred females and pursue the courtship with them more ardently. Conversely, red females rejected courtship advances more than nonred females did. Out of thirty-eight matings observed during three years of study, thirty-three involved either unpigmented females or ones with but a small trace of red, whereas only five involved fully red-throated females.

Thus females in both Spanish and Ecuadorian lizards signal when they are not receptive. In the Spanish species, the signal (green on the back) is a color that males coincidentally also have on their backs, whereas in the Ecuadorian species the signal (red on the chin and neck) is distinctive from the color that males have on their chins and necks. Bright colors have been described in the females of more than thirty species of lizards so far, and in eighteen of these, the bright colors are expressed when the females are carrying oviductal eggs.[9] Thus females using color to signal to males to back off is apparently quite general in lizards.

To find cases of genuinely masculine females, we visit the insects. Since the 1800s, naturalists have known that in many species, female damselflies appear in two color morphs, one distinctively female and the other resembling a male.[10] A species of damselfly from ponds in central Florida, *Ischnura ramburi,* has colorful males with green spots on the head, green on the thorax, and a black abdomen.[11] Feminine females have orange spots on their head, orange on the thorax, and a green-black metallic abdomen. (Gucci, are you listening?) The masculine females are green like the males but can still be identified by their female external genitalia and a bit of feminine color on the wings. What are we to make of these masculine females?

Male damselflies don't mate-guard. Instead, male and female damselflies take their sweet time in the mating itself. Copulation ranges from over one hour to over six hours, averaging three hours. While a long copulation might seem like great fun, this can waste a whole day and be too much of a good thing, especially if carried out day after day over a life span that is only a few days long.[12] Indeed, from a female's perspective, copulations beyond the first would be redundant, because one copulation supplies enough sperm. Extra copulations simply increase the risk of falling prey to some hazard.

The masculine females average half as many copulations as the feminine females. The behavior of a masculine female approached by a male resembles the behavior of a male to another male—a face-to-face stand-off, like a baseball coach getting in the face of an umpire. Still, the masculine females definitely do mate, and therefore the males presumably do know what's happening.

A follow-up study on another species of damselflies, *Ischnura elegans,* shows that the advantage to a female of looking masculine depends on how many males are around. At high densities, masculine females benefited by avoiding sexual harassment from males and having freer access to water, where they could lay eggs with less disturbance compared with feminine females. But in sparse populations, masculine females were courted less by males and more often remained unmated compared with feminine females.[13] I find these damselfly cases convincing. Masculine females have a higher survival rate because of diminished harassment from males, but they can incur a lower chance of being mated.

Still, you never know what turns guys on. Although masculine females are in the minority (about 30 percent in most damselflies), in *Enallagma boreale Selys* the masculine females constitute about 60 to 80 percent of all females. In this species, males are actually attracted to the masculine females.[14] Thus what happens when a female looks masculine depends partly on male tastes.[15]

Some insect species have females that synthesize male perfumes, reversing what we saw in garter snakes. Females use these perfumes to keep males away—like a woman wearing after-shave lotion. To unload a guy, wear Jade East on your next date! During mating, a male fruitfly *(Drosophila melanogaster)* transfers an "antiaphrodisiac" to the female. Although most evaporates four to six hours after the first mating, females later synthesize this compound themselves during courtships, making them less attractive to males.[16] Butterflies also use antiaphrodisiacs.[17]

Well, at this point you might conclude that vertebrates offer no examples of masculine females because the colors in female lizards that signal an unwillingness to mate are only occasionally the same as male colors, hence the overlap is probably coincidental. True, the most extensive studies of masculine females come from insects. But recall the Andean hummingbirds and the hooded warbler of the eastern United States, both cited in chapter 6 as examples of species with transgender expression. In fact, female hummingbirds and female hooded warblers introduce the phenomenon of "female ornaments." Here's where the underreporting of gender variation in females is taking place.

Female ornaments in birds are brightly colored feathers, skin flaps, beaks, and crests that are found in males and also expressed in a few females. Darwin suggested that female ornaments were male traits being "accidentally" expressed in females because the genetic system in females wasn't up to the task of shutting off their development during embryogenesis. Today, interest increasingly focuses on how females benefit from these traits. Other birds with ornaments causing some females to resemble males include the crested auklet, feral pigeon, barn swallow, bluethroat, blue tit, house finch, and zebra finch.[18]

Among wattled starlings *(Creatophora cinerea),* grassland birds of eastern and southern Africa, most males develop a special appearance during breeding season consisting of two hanging skin flaps (wattles) on each side of the beak, loss of feathers from the head to expose yellow or black skin underneath, and fleshy combs on the forehead. (The feather loss has been compared to male pattern baldness in humans because both are brought about by male hormones.) About 5 percent of the females also develop these fleshy folds and feather loss, qualifying them as masculine females. Not much else is known.[19]

Female deer with antlers, usually a trait limited to males, might be thought of as masculine females. In white-tailed deer *(Odocoileus virginianus),* 1 percent of the females have antlers, and antlers are reported in some female black-tailed deer *(Odocoeleus hemionus)* as well.[20] In reindeer *(Rangifer tarandus),* females usually have antlers, as do the males, but not every female—the frequency is anywhere from 8 to 95 percent, depending on the population.[21] Thus many deer species offer the possibility of looking further into why females might adopt a masculine appearance.

FINDING MR. RIGHT

Sometimes it's hard to get enough information about prospective mates. Some male genders appear to help bring couples together, like the medium-sized male sunfish (see chapter 6). Here's a similar case involving birds who mate in leks, just like sunfish do. A lek is a male red-light district in which males congregate, each defending a personal space within this patch, called his "court." Females come to the lek, and each male tries to attract a female onto his court so they can mate there. From a female's point of view, what basis is there for choice? How to find Mr. Right?

Ruffs *(Philomachus pugnax)* are sandpipers—shorebirds that breed during the summers in northern Europe from England to Siberia.[22] Ruffs owe their name to a ring of feathers around the neck of the male that is reminiscent of the large collars, or ruffs, worn in the Renaissance. Male ruffs occur in at least two genders. One gender has a dark ruff, accessorized with dark feathers on the head to make a tuft, while the other has white feathers in both ruff and tuft. These genders are genetic, with about 20 percent white-ruffed, and the remaining 80 percent black-ruffed.

Ruffs mate in leks, but not exclusively so. Males also follow females as they forage, displaying to them while they are feeding. If the resources are so spread out that the female density is thin, males stop following females and instead congregate in a lek, letting the females come to them.[23] At a site in Finland, 12 percent of the males participated in a lek, and 90 percent of the displays to females took place outside of leks.[24] Males off the lek spend 75 percent of their time feeding and the rest trying to attract a female. On a lek, males have a mating rate five times higher than in the fields, despite all the effort spent displaying to females while off lek. A female off lek is busy feeding too, and her mind is on other matters. Females who go to a lek have the same thing on their mind that males do—sex, sex, sex.

What differentiates the ruff from other lekking birds is its two male genders. The dark-ruffed males are controllers who defend small courts of about 1.5 square meters against other dark-ruffed males on the lek. White-ruffed males are solicited to join as assistants. When a white-ruffed male is nearby and a dark-ruffed male is alone on a court, the dark-ruffed male does a half–knee bend and bows his bill downward. This invites the white-ruffed male to join him on the court.[25] Females who arrive at the lek prefer a dark/white team rather than a single dark-ruffed male. Both males jointly court and then mate with the female. While mating, a dark-ruffed male may try to limit the white-ruffed male's access to the female, short of actually evicting him. A dark-ruffed male obtains more matings when a white-ruffed male is present than when alone, even though the matings are shared. Thus the two male genders act as controller and cooperator. Somehow the cooperator assists the controller in providing a more attractive mating court, and in return is paid a staying incentive of shared matings. The two genders exist specifically to address the demands of female choice.

I haven't located any theories about why a female finds a court with a dark/white team more attractive than a court with a single dark-ruffed male. Most investigators seem to assume that a female automatically finds two males better than one—the more masculinity the better. If more total masculinity is so important, then two dark-ruffed males could simply team up with each other. Why two genders? My hunch is that a white-ruffed male builds relationships with the females while he is with them off the lek. While the dark-ruffed male is defending a court from other dark-ruffed males, the white-ruffed male is flying with females in the field and presumably getting to know them. Perhaps the white-ruffed male can, so to speak, make introductions when the females arrive at the lek. He can facilitate a mating by knowing the females after having spent time with them, and also by knowing the dark-ruffed male after their initial courtship together. He can act as a go-between, a marriage broker.

FAMILY SIZE

Who determines the size of a family? From an evolutionary standpoint, family size is ultimately controlled by a female determining the size and number of eggs she lays. A female chooses an egg size and number based on the parental investment she expects to provide plus a discounted expectation of what the male will contribute to their combined investment pool. In mammals, a female may also be coerced by a controlling male to produce more young than she would if allowed reproductive freedom. Little is known about female choice of family size among vertebrates. More attention has been focused on female choice of mates and frequency of mating.

Females of the side-blotched lizard *(Uta)* have two color morphs, yellow and orange, which differ in egg size and egg number, as mentioned in chapter 6. In salmon, the largest egg can be two to three times the size of the smallest egg. Since they start with the same amount of material to put into eggs, this egg-size variation translates into some families being up to three times bigger than other families.[26] A large variation in egg size has also been noted in some bird species.[27]

Family size is one aspect of reproductive choice. Do females control their reproductive destiny? In biology, the traditional assumption has been that a female sets the number and size of the eggs she produces, and the males fight it out among themselves to acquire paternity of those eggs. An alternative theory is that female choice of mating partners allows a male to cooperate with her in jointly setting the family size. If a male promises to assist with parental care and doesn't defect, the female can elect to have more offspring than she would have if she were raising them alone. This cooperative solution to family size would generally lead to higher egg production than a competitive solution would. The role of courtship may be more to establish mutual trust that a cooperative solution will be honored than for the male to advertise his qualities, power, and possessions.

NUMBER AND IDENTITY OF PARTNERS

Another aspect of female reproductive choice is the number and identity of mating partners. Female partner choice is yet one more area of biology showing severely biased language. Studies describe females who prefer one mate as "faithful" and females who prefer multiple mates as "promiscuous." A clutch of eggs with multiple paternity is said to contain "legitimate" and "illegitimate" offspring, and a male tending a clutch with multiple paternity is said to be "cuckolded." This overlay of moralizing obscures the facts.

Razorbills *(Alca torda)*, colonial seabirds of the North Atlantic, have been studied on Skomer Island off the coast of Wales.[28] Males and females have the same color and overall shape and live in pairs at nests in a colony. A pair provides joint parental care for one egg laid each year, which can be thought of as economic monogamy. Yet, as we saw in

chapter 5, an economically monogamous pair is not necessarily reproductively monogamous.

Openly visible areas, called arenas, are located near the colony. Most
mating occurs in the arenas, even that between bonded pairs. Approximately 75 percent of the within-pair matings take place in the arena,
even though the pair shares a nest in the colony, while 87 percent of the
extra-pair matings occur in the arenas. A goodly number of same-sex
matings occur there too (see p. 136).

One-third of the females accepted extra-pair matings, while two-
thirds did not. Over two consecutive years, the identities of the females
who did, or did not, accept extra-pair matings remained the same. Of the
females who did accept extra-pair matings, most accepted only one, and
the remaining accepted matings with two, three, or even seven other
males. The investigator concluded that two types of females exist: two-
thirds "faithful" and one-third "promiscuous."

All the males participated in extra-pair mating attempts. The males
who pair-bonded with promiscuous females were slightly more successful in obtaining extra-pair copulations (EPCs) themselves than males
paired with faithful females. The investigator concluded that all guys
normally play around, although playboys tend to pair-bond with play-
girls.

Why was this study done? To decipher the feminine mystique. The investigator writes, "The benefit of EPC's to males is clear; by fertilizing
additional females, males can increase their reproductive success at the
expense of other males. . . . While it is now clear that some female birds
pursue EPC's, the possible benefits accruing to females remain obscure."
The list of conjectured reasons for why a female might want to mate with
more than one male includes wanting great genes, wanting a variety of
genes, storing sperm in case one of the males turns out to be infertile, and
checking him out in view of switching later. These conjectures assume
that all a guy delivers is genes.

Let's think. Could mating involve more than the transfer of sperm? Indeed, why is the mating being done in open arenas? Even the within-pair
matings? So everyone can see, of course! Public matings have symbolic
significance. If a birdwatcher can see the matings, so can the birds. Not
only is the mating done in public—the mating act is often just for show.
Female razorbills can control whether sperm is transferred. Females have

long, stiff tails that they must lift for the male to make cloacal contact. Females can carry out copulatory behavior, including being mounted, while preventing sperm transfer. In over six hundred extra-pair copulations, observers never saw a male force a female to lift her tail and make cloacal contact. The investigators conclude that males "do not appear capable" of forcing a copulation. But perhaps males don't want to force a copulation. Perhaps the show of copulation is what's important, not the sperm transfer.

The focus on sperm transfer as the sole purpose of copulation leads to one difficulty after another. If a male is truly "cuckolded," he should abandon his unfaithful mate. Male razorbills do not abandon their mates, nor do they attack them when they accept an EPC. Nor do male razorbills reduce their parental care in proportion to their mate's promiscuity. Why not? Didn't they read the book? Are females getting away with something that males must grudgingly tolerate?

The story doesn't add up. I suggest instead that mating is as much about managing relationships as about the transfer of sperm. By mating in public arenas, both males and females are advertising the network of relationships they participate in. Two-thirds of the females apparently find it advantageous to concentrate the paternity of their eggs in one male, and one-third to distribute the probability of paternity across several males. Because this arrangement has been broadcast to the entire colony by mating in public, the alliances and power relationships that flow along this network of relationships are publicly known too.

But why might some females want to distribute the probability of paternity and others not? What are the implications of a network of power relationships for birds? Because the males do not prevent females from distributing the probability of paternity or retaliate against them, could they too be benefiting from the formation of a network of alliances?

Consider another colonial species. Like razorbills, female tree swallows *(Tachycineta bicolor)* from Ontario, Canada, are reported to have "two alternative copulation strategies."[29] These birds were also studied to decipher the feminine mystique. The investigators write, "Previously, much attention was focussed on benefits to males. . . . Later, it was realized . . . that females may not be just passive targets for EPC." Hello! The investigators continue, "The conflict between the extra-pair male and the pair male is obvious and straightforward; the extra-pair male will seek to enhance his reproductive success at a relatively low cost by parasitizing the parental care of the pair male, whereas the pair male will try to protect his paternity and avoid caring for unrelated offspring. The interest of the female seems more obscure." Again, the investigators admit they don't understand the females. They also disparage the EPCs as a theft of the parental care that rightfully belongs to the pair male, never considering that the pair male might be trading some of the probability of paternity that he apparently controls in return for some benefit. The investigators conclude, "In some species, females actively seek EPCs . . . in other species females are generally resistant toward copulation attempts by non-mates."

To find out why females stray from their marriage vows, the investigators first tried to determine if only certain females do. Surveys of paternity using DNA fingerprinting showed that 50 percent of nests contained one or more nestlings sired by an extra-pair male. Furthermore, a brood with extra-pair paternity (EPP) didn't contain just one "illegitimate" nestling; 65 percent of the nestlings were sired by extra-pair males. So half of the females lay clutches with no extra-pair eggs, and the other half of the females lay clutches with a majority of eggs fertilized by extra-pair matings.

In an experiment, ten females were allowed to lay an egg or two and then the pair male was removed (shot). The "widow" was then allowed to acquire a replacement male. The first two eggs would have been fathered by the original pair male if the female was a stay-at-homer, but by diverse males if she was a swinger. Would the replacement male father the remaining eggs? The stay-at-homer females declined to copulate with the replacement male and used stored sperm from the original pair male to fertilize subsequent eggs, so that 78 percent of the eggs laid after the first two were still fathered by the original pair male, even though he was now dead. The swinger females, however, readily copulated with the replacement male. But the eggs laid after the first two often were fertilized not by the replacement male but by other, neighboring males. The swinger stayed a swinger, and her brood continued to be fathered by multiple males, while the stay-at-homer remained "faithful" to her original pair male.

The investigators invite the possibility that "the two types of copulation behavior are obligate strategies, i.e., that some females are always faithful while others are always promiscuous." One-third of the extra-pair matings were solicited by a female who flew to the nest of another male and mated with him there, while the remaining two-thirds took place at the nest of the female, with a male who came to her. "The ability of females to effectively resist copulations may also explain why forced EPC attempts are rarely seen." Thus the responsibility for playing around rests with the females who volunteer to play.

How do the males react to "their" females playing around? The investigators say that mate guarding is not possible in colonial species because males must guard nests and can't guard the females themselves. Instead, males copulate frequently. Fifty copulations occur per clutch, when one is enough. Thus, according to this theory, females play around because the homebound males can't guard them. Instead, males mate extensively when the females return home after a night on the town, hoping their sperm will outnumber the sperm from any other males the female played around with. Still, the males don't copulate any differently when their mate is a swinger versus a stay-at-homer. For this reason, the theory claims that "the males cannot be sure whether or not their mates are faithful."

Keep in mind that the copulations take place in the open, where birds as well as birdwatchers can see them. I can't imagine any reason why the males are always unaware of the copulation history of their pair mate. Furthermore, recent copulation doesn't guarantee paternity, because females can fertilize eggs with stored sperm. Finally, a goodly number of the EPCs are actually same-sex matings between males (see p. 137).

Again we have a story that doesn't add up. The theory doesn't offer any reason why some females accept extra-pair matings and others don't. The theory doesn't explain why males should care more about defending the nest instead of guarding the female, nor why males should be seemingly indifferent to whether their pair mate is a stay-at-homer or a swinger. The overlooked clue is that replacement males are unhelpful, even dangerous. In an undisturbed nest, males make half of the trips to bring food to the nestlings, so male and female share this workload equally. In nests where the original male was removed after only one egg was laid, the replacement male defended the nest real estate, but only half of these males provided any food to the nestlings, and the remainder completely ignored the nestlings' need for food. In these cases, many nestlings died from starvation because the female couldn't fully compensate for the lack of male cooperation.

Even more dramatically, if the original male was removed after two or more eggs were laid, the replacement male actually killed the nestlings. Thus a male who takes over an undefended nest when it already contains a few eggs is certain to commit infanticide. The observation of male infanticide is not new. In a pioneering study twenty-five years ago, Sarah Hrdy showed that female langurs (an Indian monkey) distribute paternity to purchase protection against male infanticide.[30] I suggest that some female tree swallows also deliberately distribute the probability of paternity among the males most likely to take over the nest if the original male is lost. A female can allocate all the probability of paternity to the nest male if she feels he is not likely to die or be evicted and wants maximum parental assistance from him.[31] Or she can distribute the probability of paternity among the males likely to take over the nest if the risk of losing the nest male seems high, thereby ensuring some safety for her offspring.

Now, the nest male may even agree to "his" female distributing the probability of paternity to neighboring males. If his neighbors have some likelihood of paternity in his nest, there is less chance that they will wish to evict him, or kill his brood if he dies. Regardless of how often biologists claim that the only goal of a male is to fertilize as many eggs as possible, in fact the male also has an interest in whether the eggs successfully hatch. A male's parental care need not be limited to providing food for the nestlings, but can extend to ceding some probability of paternity in order to help ensure the survival of the nestlings he is helping to feed. The female's distribution of paternity among males may amount to a "peace incentive" to purchase protection for her brood, a household expense that the male approves of. Of course, the male may work to keep this expense as low as possible by mating extensively with the female when she comes home for the night, but monitoring a cash flow is different from trying to close an account.

We need not think of tree swallow females either as choir girls honoring their marriage vows or as loose women cheating on their husbands. Instead, females may be part of a social system for raising young, in which they allocate matings so as to balance the danger of male power with the benefit of male parental investment, all with the acquiescence of the males. The social system thereby decouples economic monogamy— male and female working together to feed the young at a nest— from reproductive monogamy (compare discussion in chapter 5 of decoupling in the closely related cliff swallows).

More generally, I'm suggesting that females publicly choose mating partners to manage the genetic relationships of their offspring. Females guarantee their offspring safety by buying membership in the old genes club and choose their extra-pair partners with the tacit consent of the pair male. Females choose not males with supposedly "great genes," but males with well-connected genes. In genetic lingo, females are concerned with genetic identity by descent, not genetic identity by state. When a female chooses a male with some special color on his tail, she is not following the dictates of some inexplicable taste for fashion, but rather endowing her offspring with a bodily marker of culturally inherited power, like the Tudor nose.

Thus Darwin was fundamentally on the wrong track in his conceptualization of female choice. Sand guppies and alpine accentors show that dominant males don't have any better genes than subordinate males, according to any known metric (such as the weight and vitality of the nestlings). Sand guppies and peacock wrasses demonstrate that females choose males not for their great genes but for the likelihood of actually delivering on their promise of parental care: females are looking to avoid deadbeats. Alpine accentors and tree swallows suggest that females may choose males to distribute the probability of paternity so as to balance the incentive for a male to provide parental care with the danger to her nest from other males.

Damselflies reveal that females may tune their gendered presentation to control the number of male advances. Female wattled starlings, hooded warblers, reindeer, and other females with male ornaments suggest that gendered symbolism may also be tuned among vertebrates to regulate the frequency of male advances.

The side-blotched lizard shows that females can vary family size by varying egg size, inviting the suggestion that family size is set to accommodate the discounted expectation of male parental care. Courtship is then not about a male advertising great genes to a female, but rather a negotiation over the degree of parental care the male will provide, together with the female's assessment of the credibility of the male's promise. To aid in this assessment, a female may require the services of a marriage broker to testify on behalf of a male. The cooperator morphs in bluegill sunfish and ruffs are apparently male genders that evolved to fill this need for a go-between, suggesting that female choice has contributed to the evolution of gender multiplicity among males.

This sophisticated constellation of decisions that females make about males goes far beyond the simplistic conceptualization that Darwin put forth that all a female is searching for is a hulk with great genes.

8

Same-Sex Sexuality

The final nail in the coffin of Darwin's sexual selection theory is the discovery of extensive same-sex sexuality in nature. According to Darwin, homosexuality is impossible because the purpose of mating is to transfer sperm with the intention of producing offspring, and a homosexual mating can't produce offspring. So, if homosexuality is discovered, and if one also wishes to retain sexual selection theory, some fancy footwork is needed. Typically, biologists quickly assert that homosexuality is an "error" or, if not an error, then some devious and unsavory trick. On the other hand, if matings serve as much to manage relationships as to transfer sperm, then mating need not be restricted to between-sex encounters. That is, mating isn't the same as breeding, and in fact these activities are often decoupled. But we're getting ahead of ourselves . . . Let's review same-sex sexuality among vertebrates and then return to considering implications.

SCIENTIFIC DENIAL

In 2000 a distinguished scholar summarized a career's work with a book on the biology of freshwater cichlid fish. "When animals have access to members of the opposite sex," he wrote, "homosexuality is virtually unknown in nature, with some rare exceptions among primates."[1] Yet we've already reviewed many cases of fish, including cichlids (see chapter 6), with same-sex courtship, including genital contact. Moreover, a year earlier, a young scholar, out as gay, had published a book two inches thick with 751 pages reviewing same-sex courtship, including genital contact, in over three hundred species of vertebrates.[2] All the cases were drawn from peer-reviewed scientific literature, and detailed references were provided.

How can one scientist assert that homosexuality in animals is virtually unknown, while another demonstrates that homosexuality is common? Perhaps this difference of opinion about the facts reflects a simple difference in experience. Some species do include homosexuality in their social lives, others don't, and perhaps the differing conclusions reflect a fundamental difference in study systems. Perhaps each scientist is innocently, although incorrectly, overgeneralizing. Still, the scientific silence on homosexuality in animals amounts to a cover-up, deliberate or not.

Why might scientists cover up homosexuality in animals? Perhaps some scientists are homophobic, refusing even to consider homosexuality, while others are embarrassed or fear they might be suspected of being gay themselves if they talked positively about homosexuality. Some may think that homosexuality is evolutionarily impossible and doubt their own eyes when they see homosexual behavior. A final reason is an absence of consensus that homosexuality is theoretically important. "Is that worth a paper? Would it get me a job?" In fact, homosexuality in animals is exceptionally important and challenges basic premises of evolutionary biology.

To counter this suspicion of cover-up, scientists must start to teach the truth, and organizations who present nature to the public must start offering an accurate picture. Scientists are professionally responsible for refuting claims that homosexuality is unnatural. The dereliction of this responsibility has caused homosexual people to suffer persecution as a result of a false premise of "unnaturalness," and to suffer low self-worth and personal dignity. Suppressing the full story of gender and sexuality denies diverse people their right to feel at one with nature and relegates conservation to a niche movement—the politics of a privileged identity.

What stories should museums exhibit or nature films depict regarding same-sex courtship? The stories could begin with the courtship we've already seen between animals who are members of visibly different genders. These courtships are homosexual, yet heterogenderal. Does courtship, including genital contact, also occur between animals who belong to the same sex *and* the same gender? These animals might be thought of as "gay"—that is, both homosexual and homogenderal. Yes, let's visit such a species.

LESBIAN LIZARDS

Recall the all-female species of whiptail lizards from the American Southwest and Hawaii (chapter 1). You'd think a clonally reproducing species wouldn't bother with courtship. After all, what could be easier than laying an egg that doesn't need to be fertilized? No worries about getting a male for his sperm, no heartache of relationships, and no putting up with someone else grabbing the remote. Yet females in the all-female lizard species of the American Southwest do go through an elaborate courtship, including genital contact, prior to laying eggs.

In ordinary sexual species of American whiptail lizards, such as *Cnemidophorus inornatus,* the male begins courtship by grabbing the female with his mouth on the hind leg or tail. If she doesn't reject him, he transfers his bite to the nape of the neck, mounts, and curls his tail under hers. Male lizards have a Y-shaped penis so the animal can mount from the right or the left. The male wraps his body around the female, assuming a "donut" posture, and everts one of his hemipenes into the female's cloaca, whereupon intromission occurs.

Courtship in an asexual species is almost exactly the same.[3] One of the females copies the male role down to the last detail. One mounting female was even seen everting her cloacal region to contact the cloacal area of the mounted female. Courtship between female whiptail lizards is not a sloppy parody of male-female courtship left over from its sexual ancestry, but an intricate and finely honed sexual ritual.

When two females are housed together, they quickly wind up with alternating hormonal cycles.[4] As one female cycles into high estradiol, her eggs mature and she assumes the female role in courtship. At the same time, the other cycles into high progesterone (not testosterone) and assumes the male role. Then they switch roles a few weeks later as their hormone cycles switch.

Why do asexual female whiptail lizards bother with this elaborate ritual? One suggestion is that they're evolutionarily stuck with it. A female is said to "depend" on males for stimulation. Without any males, one of the females has to step up and take over. The asexual species haven't had time to evolve to the point of losing an undesirable dependency.[5] Yet it is clear that asexual females benefit from the copulations with one another. In nature, asexual females lay an average of 2.3 batches of eggs each season. If a female is housed alone, she lays only about 0.9 batches. If housed with a female whose hormonal state leads to male behavior, she lays 2.6 batches during the season.

Copulation between asexual females was originally reported for three whiptail species, the one from Arizona *(Cnemidophorus inornatus)* and two from Colorado *(Cnemidophorus velox* and *Cnemidophorus tesselatus),* although more of the females engaged in same-sex courtship in the Arizona species than in the Colorado species. The original report was attacked, claiming "no evidence that homosexual activities normally are involved."[6] But the attack actually confirmed the presence of same-sex mounting in four asexual species, and further revealed that females mount females and males mount males in sexual species too. An additional attack claimed that same-sex courtship happened only in the laboratory, and not in the field. But data showed that asexual females in the field have V-shaped marks on their sides identical to the copulation marks on sexual females, which come from the male's biting behavior during courtship.[7] Evidently, asexual lizards do participate in same-sex matings in the field as well as in the laboratory.

Apparently, then, asexual female lizards enjoy a form of same-sex courtship and benefit from it reproductively. Does this benefit express a physiological dependency they're stuck with because of evolution? Is their same-sex courtship the behavioral equivalent of our own useless appendix? Consider another possibility . . .

Asexual female lizards show social behaviors and live in a social system; they are not clonal robots, living solitary lives and laying eggs in solitude. Perhaps they form pair-bonds. At the very least, two females who are cycling in alternation have a thing going. Reports suggest other forms of cooperation as well.

Asexual whiptail lizards from New Mexico and Colorado, along with sexual lizards from New Mexico, were collected and raised in outdoor metal enclosures in Oklahoma.[8] During the night, the lizards live in burrows in the ground, coming out during the day to forage. Whiptail lizards are not territorial. Constantly on the move in search of food, they are called "searchers" to differentiate them from the other culinary style of lizard life—"sit-and-wait foragers"—who feed only within the confines of the territories they defend. While looking for food during the day, the sexual lizards showed four times more aggression directed at one another than the asexual lizards did. The sexual species had more fights, more chasing as they tried to steal one another's food, and a stronger dominance hierarchy.

An interesting experiment split this higher aggression in sexual species into two possible components. The asexual lizards are more closely related to one another than sexual lizards are because of their clonal reproduction, and so might be expected to cooperate with one another because of shared kinship. Furthermore, the asexual lizards do not include males. Maybe the presence of males in sexual species causes higher levels of aggression there. So is the willingness of asexual lizards to get along with one another a result of being close kin (kinship effect) or of having no males around to mess things up (male effect)? The study showed that both are factors.

But is the peaceful life of asexual whiptail lizards due only to close kinship and the absence of males? Or does peace also result from females forming pair-bonds through same-sex courtship? An intriguing clue that pair-bonding occurs comes from the observation that the asexual females often share burrows, sleeping together in the same hole in the ground. Such cohabitation was observed only once in sexual whiptails, and then between a male-female pair. Thus the female asexual lizards may have an even more cooperative relationship than can be explained by kinship and the absence of males, an extra cooperation resulting specifically from pair-bonding brought about by same-sex courtship.[9]

The extensive study of same-sex courtship in all-female lizard species can easily overshadow the same-sex courtship that occurs in two-sex lizard species. Though much less is known, lizard species for which same-sex mounting is described include the common ameiva *(Ameiva chrysolaema)*; six species of anoles *(Anolis carolinesis, A. cybotes,*

A. garmani, A. inaguae, A. porcatus, and *A. sagrei)* from Cuba, the Dominican Republic, and Jamaica; the blue-belly fence lizard *(Sceloporus undulatus)*; the side-blotched lizard *(Uta stansburiana)*; plus skinks, geckoes, and a curly-tailed lizard.[10]

AVIAN ARRANGEMENTS

Information about same-sex courtship is hard to come by because of the chilly attitude toward research on this topic in the United States, where most of the world's science is done. The U.S. Congress has meddled directly. In 1992 a study of the sexual habits of Americans to help prevent the spread of AIDS was halted by Senator Jesse Helms of North Carolina and Representative William Dannemeyer of California. Helms saw the study as a "gay agenda" and insisted that the money be used instead to promote sexual abstinence outside of marriage. The research was eventually carried out anyway, with private funding.[11] The censorship didn't ultimately succeed, of course, although information to stop the spread of AIDS in straight as well as gay couples was delayed. Two studies of same-sex courtship in birds, which we are about to review, were done in settings where homosexuality is not so stigmatized. The chilly attitude in America toward research on same-sex courtship means Americans are the last to know what's going on.

There is a lovely large bird, called the purple swamp hen *(Porphyrio porphyrio)* in the United States and the pukeko in New Zealand. A deep, almost iridescent indigo blue, it has a large, chunky, scarlet bill and orange-red legs. In the United States, it lives in Louisiana, but it can also be found in southern Europe, Africa, India, Southeast Asia, New Guinea, Melanesia, western Polynesia, Australia, and New Zealand. Pukekos are abundant in New Zealand, where they are thought to have been introduced a thousand years ago. They eat swamp and pasture vegetation, as well as insects, frogs, small birds, and eggs, and are quite omnivorous overall. They are now frequently killed by automobiles after feeding at roadside ditches.

The pukeko enjoys same-sex as well as between-sex courtship and matings. A study conducted north of Auckland, New Zealand, documented the context for the same-sex courtships.[12] A male-female mating typically involves three steps: (1) a male approaches a female in an upright posture and gives a loud humming call, (2) the female assumes a hunched position and allows the male to step onto her back, and (3) the female raises her tail as the male brings his cloaca into contact with hers and sperm are transferred.[13] A male-male mating or a female-female mating is identical to the male-female mating in all three steps, the only difference being the sex of the birds. Observation over three years revealed 555 between-sex matings, 29 female-female matings, and 12 male-male matings.[14] Thus about 10 percent of the matings are same-sex.

How did the female-female matings take place? Within a breeding group of two females and two or three males, the females and males maintain separate dominance hierarchies. In seventeen instances, an alpha female approached a beta female while both were alone, and they mated. In nine instances, the alpha female approached a beta female who was courting with males at the time. The alpha female ran in front of the males and mounted the beta female before the males could. In three instances, all involving the same female, a female evaded males who were courting her by running to another female and initiating courtship with her instead. The males then transferred their attention to this other female.

What about the male-male matings? Dominant males initiated all twelve instances. However, in only two of these did the dominant male approach the beta male in an upright position. In the remaining ten instances, the dominant male ran in front of the subordinate male and assumed a hunched position. In these cases, the subordinate male was in the process of courting females. He then stopped and mounted the alpha male instead.

Only some of the birds participated in same-sex matings. Three of nine alpha males initiated almost all of the male-male matings, and three females initiated most of the female-female matings.

What do these same-sex matings mean? Consider the circumstance where an alpha male intercedes during a beta male's between-sex mating and mates with the beta male. A competition advocate would hastily conclude that the alpha male is using same-sex mating as a tactic to prevent the beta male's mating with a female, thereby ensuring that his own sperm are used instead. The investigators, however, caution against a competitionist theory, noting that dominant males often totally ignore the matings of subordinate males without trying to interrupt them. Moreover, the hunched posture used by the dominant male to solicit the mating is also used outside of courtship as a signal to reduce aggression after a territorial dispute. Thus, the solicitation of the mating could have a social significance in reducing overall hostility, and not have anything to do with competition for fertilizations.

Consider the female-female matings. Aggressive interactions between pukeko females are rare, and females don't attack each other's eggs, as do some other communal nesting birds. The investigators raise the possibility that the alpha female is communicating to the beta female how many eggs to lay so that the total number laid is manageable by the group. The female-female matings occurred only when eggs were about to be laid, whereas the male-female matings were taking place for almost two months before this. Thus the female-female matings take place at a time when the total brood size is being decided and communication would make a difference.

Finally, the investigators note the lack of overt aggression among breeding females and males in the highly developed social system of pukekos, and same-sex matings clearly occupy a place in this social system.

The second study looks at the Eurasian oystercatcher *(Haematopus ostralegus)*, a conspicuous marine coastal bird, a wader, with a black-and-white body, a long orange bill, rose legs, and red eyes that is common on mudflats. Most oystercatchers breed in economically monogamous male-female pairs, with some distributed paternity coming from divorce and extra-pair copulations. The female lays enough eggs that both members must care for the young, and if one doesn't do his or her job, the nestlings suffer.[15]

However, at a site in Holland studied intensively from 1983 to 1997, about 2.5 percent of the breeding groups consisted of one male and two females. By itself, this grouping isn't especially noteworthy—after all, polygyny is one of the most common mating systems in nature. What is extraordinary is that these threesomes occur in two different forms: aggressive and cooperative. In an aggressive threesome, the females compete. Each defends her own nest, and the male defends a territory that encompasses both females. In a cooperative threesome, the two females share one nest, laying eggs in it together, and all three together defend the nest and raise the young.[16]

About 60 percent of the threesomes are aggressive. The females show some aggressive interaction with each other about every two minutes during the time of breeding. The females lay eggs about two weeks apart. The male contributes most of his parental care to the nest of the female who laid her eggs first. The net result is that few fledglings are produced. The male's attention is divided between two nests, so one remains un- guarded for long periods, and neither gets enough incubation time to keep the egg temperature right for proper development.

The other 40 percent of the threesomes are cooperative. The females mate with each other about once every six to seven hours and show none of the constant aggressive behavior that continually takes place in the aggressive threesomes. The alpha female mates with the male every three hours or so, and the beta female every five hours or so. Thus the females mate with each other only slightly less often than they do with the male. Females switch back and forth between being mounted or doing the mounting, so neither could be identified as hav- ing a male or female "role." They also sit and preen their feathers to- gether. They lay eggs about one day apart, placing them in their com- mon nest. Although cooperative threesomes may seem advantageous, the arrangement is still somewhat undesirable overall. The birds are only so big and can sit on no more than four eggs, so three or four of the eggs fail to develop for lack of warmth. The net effect is that co- operative threesomes produce more nestlings than aggressive three- somes, but still less than an economically monogamous male/female couple does.

The investigators conclude that the threesomes form because females are positioning themselves to obtain a territory in a subsequent year. Be- cause of the high abundance of oystercatchers, real estate is at a pre- mium. A female who was a member of a threesome had a higher chance of moving into a monogamous relationship the following year than a fe- male who was not part of any breeding group. So being in a threesome is better than being in no reproductive group at all, and a monogamous pair is best of all. Still, if a female is in a threesome, then being coopera- tive is more effective than being aggressive.

The role of the same-sex copulations between the females in the cooperative threesomes is to promote cooperation, to help cement and continually reinforce the bond between them. Same-sex copulation may also be the mechanism that leads to females laying their eggs at the same time rather than two weeks apart. Thus same-sex copulation is part of the normal social repertoire of oystercatchers, not used often, but sometimes invaluable.

A recent survey of same-sex matings in animals found ninety-four descriptions for bird species.[17] Let's round out the descriptions so far with some other examples. Geese *(Anser anser),* for example, are well known as the avian example of the human social ideal of a lifelong marriage. Geese may live for twenty years, and the pair-bond lasts more than a decade. Gay geese marriages are stable too. About 15 percent of pairs are male-male, and some couples have been documented to stay together over fifteen years. A male is reported to show "grief" after his partner dies, becoming despondent and defenseless, just as between-sex partners do when one dies. Geese sometimes form threesomes that are the reverse of oystercatchers: a male pair is joined by a female and the trio raise a family together.[18]

Swans *(Cygnus atratus)* also form stable male-male pairs that last for many years. Gay swans may even raise offspring together as a couple. A female may temporarily associate with a male-male pair, mate with them, and leave her eggs with them. The male couple then parents the eggs and is reported to be more successful than a male-female couple because together they access better nesting sites and territories, sharing the workload more equally than between-sex couples. A full 80 percent of the gay couples successfully fledge their young, compared with 30 percent for straight couples.[19]

The "mystery" of why monogamous razorbill *(Alca torda)* females participate in extra-pair copulations has already been mentioned (see chapter 7). The "mystery" deepens because males solicit EPCs with one another, not only with females. Of all out-of-pair-bond matings, 41 percent are between males, which constitute 18 percent of *all* mountings, including the within-pair male/female matings. Nearly two-thirds of all males mount other males (an average of five partners apiece, and as many as sixteen), and more than 90 percent of the males receive mounts from other males. Thus extra-pair copulation by males is distributed across both males and females.[20]

The little blue heron *(Egretta caerulea)* and the cattle egret *(Bubulcus ibis)* also have same-sex EPCs—about 5 percent of the EPCs are male-male matings.[21] The ring-billed gull, common gull, western gull, kitti-wake, silver gull, herring gull, black-headed gull, laughing gull, ivory gull, caspian tern, and roseate tern offer a large variety of mixed same-sex and between-sex pair-bonding, all facilitated with same-sex and between-sex copulation.[22] In some species, up to 20 percent of all pairs consist of two females, while in other species, up to 20 percent of all pairs consist of two males. Trios with two pair-bonded females and one male sometimes occur; the male's presence may be temporary, allowing a lesbian couple to raise the chicks on their own. Conversely, in other species, trios of two pair-bonded males with one female may culminate in the departure of the female, allowing the gay male couple to raise the chicks by themselves.[23]

Tree swallows *(Tachycineta bicolor)* were also discussed earlier (see chapter 7) in relation to the "mystery" of why females participate in exta-pair copulation. Some of the EPCs in this species are between males.[24] In the blue tit *(Parus caeruleus),* pairs occasionally consist of two females. Both females incubate eggs that only one has laid, and usually the eggs are infertile. In this species, same-sex pair-bonding occurs only sporadically.[25] In the orange-fronted parakeet *(Aratinga canicularis),* pairs appear to be about 50 percent same-sex for both males and females.[26]

WE, LIKE SHEEP

By 1984 male homosexual behavior had been reported in sixty-three mammalian species.[27] A 1999 review featured detailed descriptions of male and female homosexual behavior in over one hundred mammalian species.[28] From the many examples now available, I've selected sheep to begin with because both behavior and physiology have been studied in the field and in the lab.

Bighorn sheep are card-carrying members of the charismatic megafauna high on people's conservation priority list. Living on rugged slopes of the Rocky Mountains, bighorn sheep inspire visitors to the Banff and Kootenay National Parks in Canada, and the National Bison Range in

Montana. The males (rams), with large thick horns that curl back from above the eye to behind the ear, weigh up to 300 pounds. Their macho appearance has become a symbol for many male athletic teams. The females (ewes) live separately from the males. The sexes associate only during the breeding season, called the rutting season, which extends from mid fall to early winter. A female is receptive for about three days, and will not allow herself to be mounted outside of these three days.[29]

The males have been described as "homosexual societies." Almost all males participate in homosexual courting and copulation. Male-male courtship begins with a stylized approach, followed by genital licking and nuzzling, and often leads to anal intercourse in which one male, usually the larger, rears up on his hind legs and mounts the other. The mounted male arches his back, a posture known as lordosis, which is identical to how a female arches her back during heterosexual mating. The mounting male has an erect penis, makes anal penetration, and performs pelvic thrusts leading to ejaculation.

The few males who do not participate in homosexual activity have been labeled "effeminate" males. These males are identical in appearance to other males but behave quite differently.[30] They differ from "normal males" by living with the ewes rather than joining all-male groups. These males do not dominate females, are less aggressive overall, and adopt a crouched, female urination posture. These males refuse mounting by other males. These nonhomosexual males are considered "aberrant," with speculation that some hormone deficiency must underlie the effeminate behavior. Even though in physical appearance, including body size and horn development, these males are indistinguishable from other males, scientists urge further study of their "endocrinological profile."

This case turns the meanings of normal and aberrant upside down. The "normal" macho bighorn has full-fledged anal sex with other males. The "aberrant" ram is the one who is straight—the lack of interest in homosexuality is considered pathological. Now, why would being straight be a pathology, requiring a hormone checkup? According to the researchers, what's aberrant is that a macho-looking bighorn ram acts feminine! He pees like a female—even worse than being gay!

This Alice-in-Wonderland mixing of what's normal and what's pathological continues in laboratory studies of homosexuality in domesticated sheep. These studies, funded by the U.S. Department of Agriculture, were carried out at the government's U.S. Sheep Experimental Center in Dubois, Idaho. The project escaped the congressional knife because it was camouflaged with the goal of improving the economic success of sheep farmers.[31]

Homosexual behavior in male domesticated sheep has been documented since the 1970s.[32] Two investigators write, "It is commonly accepted that male-male mounting of prepubertal animals is important in the development of normal rear orientation in mount interactions," and they cite a study on homosexual cattle.[33] This argument has been a familiar escape over the years from dealing with the reality of gay animals—gayness in youth is necessary for straight life later on. Really, though, let's be honest. Homosexual sheep and cattle are actually gay, not playing make-believe.

Investigators coined the term "dud stud" for a ram attracted only to other rams. Preference, not ability, is the issue: "No matter how many bullets are in the clip, nothing happens until firing commences."[34] Now, consider that rams cost from $350 up to $4,000. A sheep farmer has a ratio of one ram to 30 to 50 ewes, and a "high-performance" stud can service 100 to 125 ewes. Having a dud stud, then, deprives a sheep farmer of profit because more studs must be purchased for each lamb produced. The investigators developed tests to determine whether a ram has bullets in his clip ("servicing test") and whether he knows where to aim them ("preference test"). The long-term goal was to determine the biological and genetic basis of homosexual behavior, so that duds could be weeded out of domestic sheep, enhancing the economics of sheep raising.[35]

To separate gay sheep from straight sheep, rams were exposed to receptive ewes for various periods. If the rams didn't mount the females, they were candidates to be considered homosexual. Next a candidate gay ram was strapped into a stanchion, a big crate with holes on each side to keep sheep in a fixed position. By arranging sheep of both sexes around the stanchion, the candidate gay ram was allowed to respond to a variety of females and males. The candidate gay male could indicate his preference for a female, or another male, within the stanchion.

A candidate wasn't offered just any male. He was specifically provided with males previously identified as "receivers" based on their willingness to be mounted by other males in the pen where they lived. If a candidate male chose a receiver rather than a female for courtship, he was classified as homosexual. If a candidate failed the gay preference test, he was given a second chance. If he wouldn't mount one of the receiver males, but would mount some other male in the home pen, he was then provided with that special other male. Eight out of ninety-four males representing Rambouillet, Targhee, Colombia, Polypay, and Finnish Landrace breeds tested as gay. Interestingly, six of these males would mount receiver males. Two, though, would only mount each other and none else, suggesting that they were somehow pair-bonded.

After all of this preparation, the hormonal response of the gay males was determined. The investigators hypothesized that gay rams would respond hormonally to receiver males in the same way straight males respond to receptive females. They were wrong. Gay males don't think another male is a female, and don't respond as if they did.

Gayness is expected in domesticated sheep because wild sheep are gay. The social structure in which being gay makes sense in nature is undoubtedly also present to some degree in the pens where the animals live. The removal of gayness from rams to increase sheep-farming profits would also produce a change in their social system in the pens. I bet any economic gain from breeding out duds will be offset by lower survival rates among the remaining studs in an increasingly dysfunctional domestic social system.

Many other creatures with hair have been documented as engaging in same-sex mating.[36] White-tailed deer, black-tailed deer, red deer (also called elk), reindeer, moose, giraffes, pronghorns, kobs, waterbucks, blackbucks, Thomson's gazelles, Grant's gazelles, musk oxen, mountain goats, American bison, mountain zebras, plains zebras, warthogs, collared peccaries, vicuñas (a llama), African elephants, and Asiatic elephants have all been documented in scientific reports as engaging in some degree of same-sex mating.[37] In some species, same-sex mating is sporadic; in others, very common, comprising over half of all copulations. In some, males engage in most of the same-sex matings; in others, mostly females do it; and in still others, both sexes participate. Same-sex mating is common among female red deer, male giraffes, female kobs, male blackbucks, male and female mountain goats, male American bison, and male African and Asiatic elephants.

To continue, lions, cheetahs, red foxes, wolves, grizzly bears, black bears, and spotted hyenas have been documented as engaging in same-sex mating. Again, the frequency varies from sporadic to common, with either or both sexes involved, depending on the species.[38] The gray kangaroo, red-necked wallaby, whiptail wallaby, rat kangaroo, Doria's kangaroo, Matschie's kangaroo, koala, dunnart, and quoll all enjoy same-sex mating too, although at relatively low frequency.[39]

The red squirrel, gray squirrel, least chipmunk, olympic marmot, hoary marmot, dwarf cavy, yellow-toothed cavy, wild cavy, long-eared hedgehog, gray-headed flying fox, Livingstone's fruit bat, and vampire bat show various degrees of same-sex mating.[40] For example, female red squirrels occasionally form a bond, with sexual and affectionate activities leading to joint parenting. The female squirrels take turns mounting each other, and raise a single litter of young. Although only one member of the pair is the mother, both nurse the young. Only females form such pair-bonds; male and female red squirrels don't form pair-bonds. Among male red squirrels, 18 percent of the mounts are homosexual. Concerning vampire bats, recall that females form special long-lasting friendships with affectionate gestures, including grooming and kissing (see chapter 5). No genital-genital contact has been reported among female vampire bats, but male vampire bats hang belly to belly licking one another, both with an erect penis.

The bottlenose dolphin, spinner dolphin, Amazon river dolphin, killer whale, gray whale, bowhead whale, right whale, gray seal, elephant seal, harbor seal, Australian sea lion, New Zealand sea lion, northern fur seal, walrus, and West Indian manatee are exceedingly active in same-sex genital behavior.[41] Nearly everyone has marveled at the playful personality of dolphins, often featured in children's movies—lots of makin' whoopee going on in all directions. Male bottlenose dolphins are especially well studied. A male places its erect penis into another male's genital slit, nasal aperture, or anus. They nuzzle each other's genital slit with their beak, and they can interact sexually in threesomes and foursomes. In mixed-sex groups, homosexual activity occurs as much or more than heterosexual activity. The same-sex courtship is part of forming and maintaining lifelong pair-bonds between male dolphins of the same age. They bond as adolescents, becoming constant companions and often traveling widely. Paired males may take turns watching out while their partner rests, and they protect one another against sharks and predators. On the death of a partner, the widower must search for a new companion, usually failing unless he encounters another widower.

Sometimes one finds lifelong bonds among a trio of dolphins rather than a pair. And to complete this picture of facile sexuality, male same-sex matings occur *between* species too! Male bottlenose dolphins mate with male Atlantic spotted dolphins *(Stenella frontalis)* and may band together for interspecies cooperation. All is not sweetness and light, though. These pair-bonds are part of a system of "nested alliances." Teams composed of pairs and trios of the Indian Ocean bottlenose dolphin *(Tursiops aduncus)* fight other such teams in contests over females.[42] Thus, a huge story remains to be told about same-sex matings among mammals under the sea.

OUR CLOSEST RELATIVES

We come at last to our closest relatives in the animal world—the primates. By now you might expect that primates, like other mammals, would show a good deal of same-sex courtship and mating. You won't be disappointed.

Japanese macaques *(Macaca fuscata)* are one of the best-known of the old world monkeys and the northernmost of all primates other than humans. Both sexes have gray-brown fur, a noticeably red face, often red genitals as well, and a small tail. Their maximum height is about 2 feet, and they can weight up to 40 pounds. Japanese macaques are mostly vegetarian, eating fruits, seeds, leaves, and bark, with some snails, crayfish, and bird eggs thrown in. Japanese macaques become adults at about five years and can live up to thirty years. A free-living group (the Arashiyama West troop) was introduced to Texas in 1972 and has thrived there, even inventing a special alarm call for rattlesnakes.

Their social structure consists of mixed groups of around fifty to two hundred females and males in an area of 100 to 500 hectares. The females stay put, and a group consists of several matrilines, or female descendants of an elder female. The ratio of females to males in a group is typically around four females to one male. The males migrate between groups every two to four years. The social system revolves primarily around females and the interactions among them.

The females are described as having a rigid dominance hierarchy, and all the females can be ordered along a line from top to bottom. Here's how the hierarchy works. Suppose the three elder females are ranked A, B, and C from top to bottom. All female offspring inherit the rank of their mother. So, if the daughter of A is A1, the daughter of B is B1, and the daughter of C is C1, then the overall dominance hierarchy goes A, A1, B, B1, C, C1. And if A1 has a daughter too, say A11, and similarly for B11, C11, and so on, the overall dominance hierarchy spanning three generations would be A, A1, A11, B, B1, B11, C, C1, C11. However, matters are a bit more fluid than this picture suggests. This hierarchy is produced because elders come to the defense of their daughters and granddaughters. At birth, A11 is subordinate to the C matriline, but after interventions on her behalf by A1 and A, the status of A11 is raised to that of the A matriline. Dominance testing goes on continually. No one just accepts her place unquestioningly.

Against this backdrop, same-sex courtship and copulation are also going on. Same-sex relationships happen *not* between close kin but between distantly related individuals. These same-sex copulations produce bonds that go beyond the straight-line lineage-based dominance hierarchy, building cross-cutting links and suggesting a network structure in the social system. Same-sex courtship and copulation take place in what are called female-female consorts. These are short-term relationships (STRs) that last for less than an hour up to four days. During this time, the two females mount each other frequently, engaging in genital-genital contact. When not having sex together, they huddle, sleep, and forage together, groom each other, and defend each other from challenges. For the duration of their STR, a pair is monogamous. After a few days, though, they form new STRs. During the STR the mountings are bidirectional and mutually pleasurable, and there is no sign of dominance or submissiveness within the relationship.[43] Indeed, the presence of any aggressiveness in the relationship destroys the "mood," and forecasts the dissolution of the STR.

The males have a hard time while all this lesbian love is going on. When a guy approaches to mate with a female, her partner usually shoos him away.[44] Females back each other up while in an STR. The lower-ranking member of an STR increases in rank temporarily because of her partner's support.[45] This temporary increase in rank ends when the STR dissolves. Although the lower-ranking member of a consort temporarily rises in rank, a low-ranking female shows no tendency to pair up preferentially with a high-ranking female, as would be expected if she were interested in finding a more powerful ally. Instead, the higher-ranking female is responsible for starting the relationship. Neither is forming a consort a means for two females to share parental care. The consort partners do not help raise each other's young. A partner does not support the other's young in conflicts, nor does she groom the other's young. If anything, partners tend to be aggressive toward each other's young.

A subgroup in which females outnumbered males by eleven to one was formed experimentally to see if the females would show increased competition for the male's favor, or if they would instead increase the number of same-sex relationships. Instead of competing more intensely for the one male, they formed more female-female relationships. Nonetheless, the majority of females who rejected a sexual solicitation from the one lucky guy in favor of a same-sex partner did later form a heterosexual consortship with him, showing him to be an acceptable mate.[46]

As these studies show, research on homosexuality in primates has advanced beyond the "Gee, do they really do it?" stage of other vertebrate groups. Homosexuality is so conspicuous among primates, so in-your-face, that it cannot be ignored, resulting in a relatively extensive literature going back to the 1970s.[47] Yet, if the fact of extensive same-sex sexuality in primates is well established, the reasons for this homosexuality are open to debate. The macaques don't participate in homosexual relationships because heterosexual partners aren't available, or as an expression of dominance and submission, or to form alliances, or to acquire help in raising their young. So why do female macaques spend so much of their time in same-sex courtship and copulation?

With no ready explanation for female homosexuality in Japanese macaques, investigators have wondered "whether our line of questioning is not faulty."[48] Does homosexuality really have to further overall lifetime reproduction, albeit in some indirect way? An alternative view—what I call the neutralist position—is that homosexuality is a neutral byproduct of the evolution of other traits.[49] Homosexuality, it is argued, doesn't disappear during evolution because homosexuality is harmless. Female macaques have lots of offspring, and they participate in heterosexual matings whenever they need some sperm. Homosexuality doesn't interfere with their reproduction, so why would natural selection remove this harmless behavior? By chance, it is suggested, over the course of evolution homosexuality has drifted into prominence in some species, while remaining nearly absent in others. Or maybe in some species chance has genetically linked homosexuality to important genes, and homosexuality has "hitchhiked" into prominence on the coattails of those genes. Either way, it is contended, this behavior serves no evolutionary purpose.

Since lots of traits are neutral, does some causal mechanism determine which of these traits actually evolves? One suggestion is pleasure. Why do macaques participate in same-sex copulation instead of, say, the monkey equivalent of reading Kant in their spare time? Neither increases reproductive success, we may suppose. Yet sexual stimulation is pleasurable, and reading Kant isn't. Therefore sexual activity, and not reading Kant, evolves to take place when not carrying out reproductively important activities. Although "sexual pleasure was selected for because it motivates individuals to engage in fertile sex, . . . sexual pleasure is not specific to reproductive sex but can be satisfied by many non-reproductive sexual outlets as well."[50] In this view, evolutionary theory is incomplete because it applies only to traits that affect reproduction and survival. This pleasure principle extends evolutionary theory to explain which among the set of selectively neutral traits evolve and which don't. Homosexuality is viewed as existing under the radar screen of natural selection, subject only to the winds of passion.

I don't buy this neutralist position, at least not yet. I'm from the other school, the adaptationist school, which holds that nearly all behaviors and traits benefit organisms, and our task is to figure out how. Here's where I'm coming from. In my experience, animals don't have lots of free time for hanging out. The lizards I work with are busy all day. When not actively eating, mating, or displaying to one another, lizards are looking around intently for food or keeping tabs on their neighbors. Sure, lizards sleep now and then, and they stay in bed on a cold day, when they're too slow to catch prey anyway. Still, lizards use their time wisely and don't waste it. We're always underestimating animals. Long ago we thought lizards were mindless little robots. We find now that throughout the day they make complex decisions of the sort that top business executives pay expensive consultants to solve.[51] From this fieldwork, one becomes wary of claims that a trait is useless.

What about macaques? Is every day another day in paradise, an endless party filled with evolutionarily meaningless play? Why should macaques be so lucky, while the rest of us poor sods have to work for a living? To my mind, homosexual interaction occupies far too much time in the lives of female macaques to be evolutionarily incidental.

My hunch is that the social system has been oversimplified. Traditionally, if A attacks B, and B backs down, then A is said to be dominant over B. A is assumed to get whatever is wanted as long as A remains dominant. However, this assumption isn't always true. I've sat on a rock in the woods watching a juvenile lizard, about 1 inch long, take territorial space from a male lizard five times as big. Here's how. The juvenile goes to the edge of the big lizard's territory. The big lizard sees this and rushes over. The juvenile scurries away—he lost. So the big lizard goes back to his perch, and the juvenile lizard tiptoes back. The big lizard then runs over again and chases the juvenile away. The juvenile loses again. And so on. After five of these chases, the big lizard has had enough and doesn't bother to chase the juvenile anymore. The sliver of land the juvenile wants isn't worth the fuss from the big lizard's standpoint.[52] Overall, the juvenile lost every single pairwise interaction but still won a slice of the dominant male's territory. He lost every battle but won the war. Thus saying the big lizard is dominant over the juvenile just because the juvenile always backs down in pairwise contests doesn't capture the true power relationship. Furthermore, even pairwise interactions don't take place in a vacuum. A lizard lives on every tree. All these lizards are watching, and they remember what they see. So the meaning of the interaction extends beyond the two lizards in the pair to all the lizards who were watching. A social system isn't the sum of isolated pairwise disputes.

For macaques, the ranking scheme is assembled from records of many pairwise interactions, and may miss much of the true social organization. The question comes down to whether female macaques are participating in sex for fun only, or whether they're networking for future profit. If the homosexuality is only for fun, why are STRs not between close kin, but only between matrilines? In heterosexual mating, an "incest taboo" prevents inbreeding. In homosexual matings, no offspring are conceived, so there is no reason for STRs to be avoided between relatives. I suggest that STRs between close kin have no strategic value because close kin already share a bond, the kinship bond itself. STRs are needed to build bonds beyond kinship. Furthermore, if the homosexuality is only for fun, why should high-ranking females be more responsible for maintaining the STRs than low-ranking ones; shouldn't rank be immaterial and the STRs be formed solely on the basis of who is the most enjoyable sexual partner?

Let's look at it another way. Calling this social structure a dominance hierarchy may exaggerate the power of a high-ranking female. Is the alpha female given her status by the consent of the governed? Does she have to solicit their support, ask for their votes, to continue as their alpha? Perhaps high-ranking females need support from low-ranking females, and forming STRs secures their friendship. If the social system is based on a network of power rather than a dominance hierarchy, then female choice of homosexual partners may be adaptive not for climbing a dominance ladder but for navigating a political network.

So a horse race is shaping up. Neutralists have often squared off against adaptationists in evolutionary biology. The dispute may be over quickly, though. Evolutionary theory includes an acid test for neutralist and adaptationist theories.[53] For homosexuality to be evolutionarily neutral, females who do, and who don't, participate in short-term homosexual relationships need to have nearly identical average reproduction over a lifetime.

I doubt that homosexuality is selectively neutral in Japanese macaques. I suspect that the fate is bleak for a female who doesn't participate in homosexual STRs, as she is likely kicked out of the group and left to die. Being kicked out of the group would greatly reduce her average lifelong reproduction because she wouldn't live long enough to reproduce, and any offspring she did have would not have access to group resources or protection. If participating in STRs is necessary for inclusion in female social groups, then female same-sex sexuality in this species is what I call a social-inclusionary trait (see chapter 9). Not to participate in the STRs bound together with same-sex sexuality would be lethal.

Consider now another primate with extensively documented same-sex sexuality. The bonobo *(Pan paniscus)* is not just any primate. Bonobos, or pygmy chimpanzees, are our closest relatives, along with the common chimpanzee *(Pan troglodytes)*. *Pan* and human lineages split apart about eight million years ago. The *Pan* line went on to divide into pygmy and common chimpanzees, while the *Homo* line divided into various forms of early humans. Because *Homo sapiens* is the only remaining species from the human lineage, our closest living relatives are these chimpanzee species. Chimpanzees[*] have become well known for their male-male power games, whereas bonobos illustrate female-female relationships and the social uses of sexuality. Bonobos are less well known but just as relevant, if not more so, to the way people actually live.

Bonobos, which live in the tropical lowland rainforests of Zaire, in central Africa, grow to about 2 to 3 feet in height, weigh about 70 to 85 pounds, and have black hair. They live about forty years, beginning to breed after thirteen years or more. Bonobos eat fruit, insects, and small mammals, and are more nearly vegetarian than chimpanzees. Chimpanzees actually hunt monkeys for food, whereas bonobos don't. In captive situations bonobos use tools skillfully, but in the wild they show less tool use than chimpanzees. Bonobos are as intelligent as chimpanzees, and more sensitive. During World War II bombing in Germany, bonobos in a zoo near Hellabrun died of fright from the noise, while the chimpanzees were unaffected.

Bonobos live in mixed-sex, mixed-age groups of about sixty individuals. In bonobo females, a pink swelling around the genitals signals readiness to mate. Bonobo females are receptive almost continuously, whereas female chimpanzees are receptive for only a few days during their cycle. In bonobo between-sex matings, one-third take place face to face, the remaining two-thirds taking place front to back, with the male mounting the female. By contrast, all chimpanzee between-sex matings are front to back, with the male mounting.

In bonobo female same-sex encounters, the two females face each other. One clings with arms and legs to her partner, who lifts her off the ground. The females rub their genital swellings side to side, then grin and squeal during orgasm, a form of mating called genito-genital rubbing (GG-rubbing). In bonobo male same-sex encounters, the two males rub humps: standing back to back, one male rubs his scrotum against the buttocks of the other. Another position, penis-fencing, involves two

---

[*]The word "chimpanzee" will hereafter refer to the common chimpanzee unless otherwise qualified, and "bonobo" to the pygmy chimpanzee.

males hanging face to face from a branch while rubbing their erect penises together. Bonobos don't have anal intercourse, but they do have sporadic oral sex, hand massages of the genitals, and lots of intense French kissing. With all these choices of sexual activity, bonobos have even developed a set of hand signals to tell each other what they'd like. These signals are used in both between-sex and same-sex sexual encounters.

Bonobo life isn't a continuous orgy, as the extensive menu of sexual activity might suggest. Daily life consists of numerous brief episodes scattered throughout the day. Each female participates in GG-rubbing once every two hours or so. A sexual encounter lasts about ten to fifteen seconds, so in total, sex doesn't take a lot of time. At least six situations lead to sex:

1. Sex facilitates sharing. When a zoo caretaker approaches bonobos with food, male bonobos develop erections. Before the food is tossed in, males invite females, females invite males, and females invite each other for sex. After sex, the meal begins. In the field, at the Lomako Forest in Zaire, bonobos engage in sex after they enter trees loaded with figs, or after one has captured a prey animal. When five to ten minutes of sexual contact have passed, they all settle down to dine together. Sex facilitates sharing not only food but anything in demand. When a cardboard box used as a toy is given to bonobos in a zoo, they mount each other before beginning to play with the box. Most primate species would squabble and fight over it instead.

2. Sex is used for reconciliation after a dispute, such as arguing over who has the right of way when walking along a branch.

3. Sex helps integrate a new arrival into the group. When females migrate to a new group, the new arrivals establish relationships with the established matriarchs through frequent GG-rubbing and grooming.

4. Sex helps form coalitions. Female bonobos bond through GG-rubbing to form coalitions against males who would otherwise be dominant. When food is given to chimpanzees, the males eat their fill before the females are allowed their turn, whereas in bonobos, the females eat when they want, regardless of male presence. Females together chase off harassing males.

5. Sex is candy. In return for sex, a female may take a bundle of branches and leaves, or sugarcane, from a male.

6. Oh, I almost forgot—sex is used for reproduction.[54]

Why has homosexuality evolved between female bonobos? Females maintain strong friendships with unrelated females, females control access to food, females share food with one another more often than with males, and females form alliances in which they cooperatively attack and even injure males. Their increased control over food and the lessened threat from males allows bonobo females to reproduce starting at an earlier age compared to chimpanzee females, who don't form such friendships. An earlier age of first reproduction in turn leads to higher lifetime reproductive success.[55] A female who doesn't participate in this social system, including its same-sex sexuality, will not share in these group benefits. For a female bonobo, not being lesbian is hazardous to your fitness.

For these reasons, female same-sex sexuality in bonobos is what I call a social-inclusionary trait (see chapter 9). The evolution of female homosexuality is driven by the need to be included in the social group that controls resources, and not belonging is near-lethal. The selection in favor of participating in same-sex sexuality, given that this mode of bonding is already in place, is exceedingly strong.

Bonobos and common chimpanzees offer an interesting contrast in social organization. Physically, they are much the same, the main difference being that common chimpanzees have grayish hair under their chins and bonobos don't. Yet chimpanzees have a male-dominated society and bonobos a female-centered society, and as we've seen, bonobos have a love life that chimpanzees can only envy.

In case you're wondering about other primates, here are some in which same-sex courtship and mating are documented. Lemurs, such as Verreaux's sifaka *(Propithecus verreauxi)* from Madagascar, have limited same-sex mating between males—up to 14 percent of all matings in one study.[56] New world monkeys, like the squirrel monkey *(Saimiri sciureus)* and the white-faced capuchin *(Cebus capucinus)* from South America, have same-sex genital interactions.[57] Female squirrel monkeys form short-term sexual relationships, and also have close female "friends" with whom they travel and rest and occasionally coparent.

Males too have same-sex genital displays. Mounts between females occurred once every forty minutes during a week out of each month, and 40 percent of all genital displays were same-sex, one-quarter between females. In white-faced capuchins, more than half of the mountings were same-sex and included specialized courtship gestures and vocalizations.[58]

Old world monkeys have an extensive record of same-sex sexual encounters. In addition to the Japanese macaque *(Macaca fuscata)* already discussed, the rhesus macaque *(Macaca mulatta)* and stumptail macaque *(Macaca arctoides)* engage in same-sex sexuality as a regular part of life. In rhesus macaques, about a third of all mountings are same-sex, 80 percent of which are between males. Females form short- to medium-term relationships, as do male/female pairs.[59] Male-male genital contact in stumptail macaques includes anal intercourse, with penetration and ejaculation, and mutual oral sex. Homosexual activity accounts for 25–40 percent of all sexual encounters.[60]

The savanna baboon *(Papio cynocephalus)*, hamadryas baboon *(Papio hamadryas)*, and gelada baboon *(Theropithecus gelada)* from Africa have extensive same-sex genital relationships. Savanna baboon males have numerous same-sex genital contacts, including "diddling," in which males fondle each other's genitals, called "greeting" behaviors. Some males form long-lasting coalitions with mutual exchange of sexual favors. Savanna baboons protect and help one another, and their associations can last for many years, constituting long-term relationships (LTRs). Approximately 20 percent of the mountings are between males, and 9 percent between females.

Baboon life is marked by violence. Male savanna baboons coerce matings with females, often seriously injuring them. When an outside male takes over a troop, he may attack mothers and infants, injuring females, causing miscarriages, and killing infants. Males often "kidnap" infants, and the youngsters may be injured. In this social system, same-sex courtship is used specifically for coalition-building. For baboons, coalitions are threatening to those who exercise power and dominance.[61] Attempts by powerful males to break up threatening coalitions among subordinates emerge as a form of homophobia.

Hanuman langurs *(Presbytis entellus)* are medium-sized monkeys from India known for their exceptional violence toward infants and juveniles.[62] Attacks by adult males may account for half of all infant deaths. The stress of this violence causes females to abort fetuses spontaneously. Females may also induce abortion by pressing their bellies on the ground or allowing other females to jump on them. Females are not necessarily award-winning mothers, either. Maternal mistreatment includes abandonment; dangling, dropping, and dragging the baby; shoving it against the ground; biting or kicking it; and throwing the infant out of trees. Females from one group may kidnap a baby from a neighboring group, keeping it for three days before allowing the mother to retrieve it. The presence of this behavior was controversial when first reported.[63]

The same-sex sexual activity of langurs is rather ordinary. All females mount each other. Interestingly, females do mount close relatives (27 percent of all lesbian mountings are between half-sisters) and show no homosexual incest taboo, whereas an incest taboo does govern heterosexual mountings. Males also mount each other. The mountee initiates the mounting with a special head-shaking display. Males sometimes form duos, a pair-bond lasting for about a month. Males who have bonded through homosexual mounting may cooperate in launching attacks against males in neighboring groups.

The white-handed gibbon *(Hylobates lar)* of Thailand and the siamang *(Hylobates syndactylus)* of the Malay peninsula and Sumatra, in welcome contrast to langurs and baboons, are primarily monogamous, although some divorce happens. In a study over six years of eleven male/female gibbon pairs, five split up and six remained intact. Females breed for four or five months every two to three years, and between-sex sexual behavior is largely limited to breeding periods. Nonabusive intrafamily same-sex behavior is common. A male parent and offspring have sexual contact—primarily penis-fencing leading to orgasm and ejaculation—about as often as between-sex mating occurs during the breeding period. Although gibbons are monogamous, about 10 percent of the heterosexual matings are extra-pair. The whole system seems quite similar to avian monogamy, with the addition of the same-sex father/offspring sexual activity.[64]

Gorillas *(Gorilla gorilla)* live in groups of one male with three to six adult females plus their offspring, as well as in all-male groups. Same-sex genital contact occurs in females and males. Females have favorite female partners in the mixed-sex groups. Most of the male-male homosexual behavior takes place in the all-male groups. Males have preferred partners too; some interact with only one partner, others with up to five partners. Males also commit infanticide, causing more than 40 percent of the infant deaths in one study.[65]

All in all, lots of same-sex courtship and mating takes place among primates. A look at the family tree of primates suggests a pattern. From its base near the ground, the tree trunk splits first into the prosimians on one side and the anthropoids on the other. The prosimian branch, including bush babies, lemurs, and tarsiers, appears to have only incidental same-sex mounting while in heat and no evidence of a major social role for same-sex courtship. The anthropoid branch splits into two sub-branches: the new world primates and the old world primates. The new world primates, including marmosets, tamarins, and monkeys with prehensile tails, such as spider monkeys, show some homosexual behavior. It is in the old world primates that homosexual courtship becomes prominent. The old world primates, including macaques, baboons, gibbons, orangutans, gorillas, chimpanzees, bonobos, and humans, contain the most sophisticated of the primate societies. In these societies, individuals form complex relationships, relationships clearly fostered through both same-sex and between-sex sexuality.[66] This pattern of occurrence across the primate family tree suggests that homosexuality in primates is an evolutionary innovation originating around fifty million years ago, when the major prosimian and anthropoid lineages began their divergence.

OBJECTIVITY AND HOMOSEXUALITY

My coverage of homosexuality in animals will be seen by some as a litmus test of objectivity. Am I simply going to assert that because homosexuality is common in animals, it is legitimate in people? I want to be clear about where I stand on the issues of how widespread animal homosexuality is and what relevance such information has for affirming human homosexuality.

I believe the moral assessment of human behavior is independent of what animals do, as I've mentioned before. Infanticide by males is common in animals, and female animals choose mates in part to manage this danger. The naturalness of male infanticide in animals is clearly no justification for infanticide in humans—human infanticide is wrong, period. By contrast, I believe that affirming homosexual expression is right for people, not because animals are often homosexual but because endorsing homosexual expression makes for a just and productive society. Further discussion of the moral aspects of homosexuality appears later in this book. Here, though, the morality of homosexuality is not at issue. The issue is whether homosexuality is in fact common among animals.

I began this chapter by citing two disparate views: one scientist saying homosexuality is rare to the point of being nearly nonexistent, another documenting over three hundred species of vertebrates in which homosexuality occurs. This gap in scientific opinion is huge. Scientists who think homosexuality is almost nonexistent will feel that I am trawling for scraps of data, making up a story where one doesn't exist. Scientists who think homosexuality is common will wonder why their colleagues have been silent, and suspect a cover-up. I have gradually moved from the first camp into the second.

Prior to researching this book, I took for granted that homosexuality was rare. Personally, I still have never seen a mating between lizards that I am certain is homosexual, even after thirty years of working with them. I can sympathize with those who think homosexuality is rare—this has been my experience too. Yet I now know this experience is misleading. Previously, I was aware of one published reference to a homosexual copulation in nonparthenogenic lizards, but I felt the situation was unusual enough to be an isolated instance. As a result, I never checked. Every few days during fieldwork I see lizards mating. Sometimes I'm sure of the sexes involved. Usually, though, I would assume the bigger lizard was male and the smaller female. If both were about the same size, I would still assume one was male and the other female. To find out, I would have had to catch them while they were in the act of mating, separate them, and inspect their sex by physically palpating them and examining their cloacal opening. This action would be intrusive, would be disturbing and possibly injurious to the animals, and would take time away from a research project that had other goals. For this reason, I don't know how much homosexual mating occurs even in the species I have worked with for many years. I'm sure that many other scientists are in the same situation—we have never really looked.

I was stunned to discover how many reports of homosexuality there are in the primary literature. The seven cases I've reviewed above—whiptail lizards, pukekos and oystercatchers, bighorn and domesticated sheep, Japanese macaques and bonobos—are well documented over many years by multiple investigators. In these, no doubt whatsoever exists of homosexuality. Of the, say, 293 others, for a total of approximately 300 species, some will fall by the wayside and others will be reconfirmed. Today, it's very hard to know how common homosexuality is in natural social systems. The data aren't collected, or the sexes aren't checked, and when data are available, they often aren't reported. My overall conclusion is now that the more complex and sophisticated a social system is, the more likely it is to have homosexuality intermixed with heterosexuality. Any animal in a complex society has to manage both within- and between-sex relationships. Both types of relationships are mediated through physical contact, including embracing, grooming, and genital contact, as well as through vocalizations, bodily symbolism, and behaviors like food-sharing and warning calls.

IS SAME-SEX SEXUALITY PROBLEMATIC?

Let us suppose, on the basis of the available evidence, that same-sex sexuality is now known to be natural and common. Would this discovery be a problem? Here are the questions I'm usually asked.

What's the function of homosexuality? Same-sex sexuality promotes friendship. Genitals have sensory neurons that provide pleasure. Activating one another's genital neurons sends a friendly message and builds relationships. The friendship's purpose depends on context. The purpose of friendship might be innocent or threatening. Friendly bats huddling together through the cold nights are innocent. Friendly baboons building coalitions to overturn an alpha male are threatening; the alpha male will try to prevent this friendship, which would be seen as homophobia.

Does a gay gene exist? The question doesn't really ask about DNA. The question asks whether homosexuality is inherited. Homosexuality in animals is obviously inherited in some way, but no single gay gene exists. Homosexuality is a complex social behavior. Complex traits are not caused by single genes. Reports of a gay gene in humans are erroneous, as will be discussed later.

Doesn't homosexuality contradict evolution? This question usually confuses same-sex sexuality with nonbreeding. Nonbreeding *is* an evolutionary problem; same-sex sexuality isn't. The two aren't necessarily connected. Nonbreeders always exist, heterosexual and homosexual. In most species, only some members breed, while the rest don't, reflecting the population's reproductive skew. Some nonbreeding homosexual animals are expected to exist simply because of the population's reproductive skew, and this can be explained by whatever causes that skew. Indeed, the main evolutionary issue is to explain where the reproductive skew comes from, not whether same-sex genital contact occurs.

Although some homosexual animals don't breed, most do. Homosexuals who breed can have either a lower or higher fertility than heterosexuals who breed, depending on circumstances. Breeding homosexual animals could have a lower fertility than breeding heterosexuals because of a tradeoff between fertility and survival. Natural selection favors traits that increase average total offspring production throughout life, which depends on both fertility *and* survival. A homosexual strategy could increase same-sex matings to obtain higher survival though friendships. This homosexual strategy might result in decreased fecundity because of fewer between-sex matings. However, the homosexual strategy might yield more offspring averaged over a life span than an exclusively heterosexual strategy because of the increased survival.

Alternatively, breeding homosexual animals might have an even higher fertility rate than breeding heterosexual animals because homosexually bonded friendships might access more resources than those available to exclusively heterosexual animals, yielding a fertility advantage. Homosexuality can be a social-inclusionary trait. In this situation, homosexuality might increase both fertility and survival, and be favored by natural selection even without taking into account a possible tradeoff between fertility and survival. Thus same-sex sexuality doesn't necessarily go against evolution, either in general or in particular.

Although homosexuality doesn't contradict evolution, widespread homosexuality among animals does open new perspectives on how we think about bodies and social relationships. Let's move on to some implications of this realization that homosexuality is a common ingredient in the social life of our vertebrate relatives. One implication is that traits presently interpreted solely in heterosexual terms now need to be reexamined.

GENITAL GEOMETRY

Colorful feathers and other traits that make an animal attractive are called secondary sex characteristics and are assumed to attract members of the opposite sex. Because same-sex sexuality may be as important as between-sex sexuality in some species, secondary sex characteristics may attract same-sex partners as much as between-sex partners. A special case of secondary sex characteristics is the color and geometry of the external genitals. Many people are too squeamish to discuss the optimal design of genitals, focusing instead on safe characteristics, such as colorful tail feathers, fur, or scales. Yet genitals also have symbolic importance, serving not only to transfer and receive sperm but also to build and maintain social relationships both within and between the sexes.

Why the clitoris with its sensitive pleasure neurons is located some distance from the vagina in humans has long been an enigma. In males, the main sensory spot is at the tip of the penis. As a result, a male is motivated to insert the penis into the vagina to obtain orgasm. By contrast, a female may not experience orgasm during penile insertion because the clitoris may not be activated. It is a puzzle why genital geometry doesn't guarantee as much pleasure for females in heterosexual intercourse as for males.

Because same-sex matings can be as common as between-sex matings, the geometry of the genitals may be shaped to promote same-sex contact as well as between-sex contact. In bonobos, females participate in same-sex sexuality by facing each other and rubbing their genitals side to side. In 1995 the distinguished primatologist Frans de Waal wrote, "The frontal orientation of the bonobo vulva and clitoris strongly suggest that the female genitalia are adapted for this [frontal] position."[67] More explicitly, the noted behavioral ecologist Marlene Zuk wrote in 2000 that the bonobo clitoris is "frontally placed, perhaps because selection favored a position maximizing stimulation during the genital-genital rubbing common among females."[68] Bonobos are unusual because between-sex mating also often takes place face to face rather than face to back. The frontal position may be how bonobo males adjust to the position of the female genitals, a genital geometry that must work for both same-sex and between-sex sexuality. From the standpoint of female reproduction, little is gained by placing the clitoral neurons near the vagina to further between-sex mating when males are well motivated for intercourse anyway. Instead, the pleasure neurons are shifted to a location that promotes same-sex mating and may yield more effective same-sex bonds, increasing overall Darwinian fitness at no reproductive cost. A subject for the future will be to account for genital design across the vertebrates in a unified treatment that takes into account all symbolic and other functions of genitals beyond the transfer of sperm.

9

The Theory of Evolution

Diversity in gender expression and sexuality undercuts Darwin's theory of sexual selection. Saying this, however, does not mean *all* of Darwin's writings are incorrect. Indeed, I feel we should not lose sight of his overwhelming contribution, even though I believe one of his theories is seriously mistaken.

Perhaps Darwin's most important discovery is that all species are related to all other species through shared descent from common ancestors. The most grand and most lowly share in the unity of life. Darwin came to this insight as a young man, during his travels as a naturalist on a sailing ship called the *Beagle*. In his diary, Darwin compared the animals of an archipelago, the Galápagos Islands, with those of South America, which he had previously visited.[1] He wrote, "We see that this archipelago, though standing in the Pacific Ocean, is zoologically part of America. If this character were owing merely to immigrants from America, there would be little remarkable in it; but we see that a vast majority of all the land animals, and that more than half the flowering plants, are aboriginal productions. It was most striking to be surrounded by new birds, new reptiles, new shells, new insects, new plants, and yet by innumerable trifling details of structure, and even by tone of voice and plumage of the birds, to have the temperate plains of Patagonia, or the hot dry deserts of Northern Chile, vividly brought before my eyes." Thus

Darwin observed that the species unique to the Galápagos, which he called aboriginal productions, are nonetheless related to South American species.

Darwin continues by comparing the animals on different islands within the Galápagos: "I never dreamed that islands, about fifty or sixty miles apart, and most of them in sight of each other, formed of precisely the same rocks, placed under a quite similar climate, rising to a nearly equal height, would have been differently tenanted . . . one is astonished at the amount of creative force, if such an expression may be used, displayed on these small, barren, and rocky islands; and still more so, at its diverse yet analagous action on points so near each other." Here Darwin further observes that species diverge even within a group of islands, not only between the group and the mainland.

Today, one could not improve on Darwin's formulation. Darwin perfectly expressed the idea of evolution through common descent. He focused on *populations* of plants or animals rather than on single individuals. Strictly speaking, Darwin might have concluded only that the species he personally saw were related to one another by descent from common ancestors. Although at present the possibility of more than one independent origin for life perhaps can't be ruled out after considering the enormous diversity of single-celled organisms, all the organisms people are generally familiar with do share descent from common ancestors.[2]

DARWIN'S NATURAL SELECTION

Darwin's next task was to understand what "creative force" produces the diversity of new species. Darwin identified a force he named "natural selection," which causes species to change over time. Darwin's theory of natural selection is correct overall, although our contemporary understanding of the process is somewhat different from the way Darwin wrote about it.

Living in an agricultural setting, Darwin was well aware of animal and plant breeding—cows, horses, and crops for yield and dogs, roosters, and flowers for show. Animal and plant breeding was done every day and could obviously change the properties of a stock. Animal and plant breeding is based on selecting certain individuals to reproduce and eliminating the remainder. This process is now called "artificial selection" to indicate that a farmer, rather than the natural environment, determines who gets to survive and/or to breed.

Darwin was also aware that a contemporary, Thomas Malthus, was developing scenarios about the consequences of population growth. Darwin wrote, "On the principle of geometrical increase . . . more individuals are produced than can possibly survive, there must in every case be a struggle for existence. . . . It is the doctrine of Malthus applied with manifold force to the whole animal and vegetable kingdoms."[3] Darwin realized that if only certain types of individuals survive in crowded conditions, then the population will consist of descendants of those survivors. Thus was born the idea of "natural selection," the process by which the natural environment determines who gets to survive and/or breed. Natural selection is nature's equivalent of artificial selection for yield. Furthermore, if nature selects for different types of individuals in different locations, then the populations in those locations will diverge over time, eventually accumulating enough differences to be distinguished as different species.

A technical difficulty in Darwin's original account concerns how diversity is maintained. Darwin had not heard of Mendelian genes and could not account for why variation persists in a natural population rather than simply dissolving. Fifty years later, population geneticists Ronald Fisher and J. B. S. Haldane in the United Kingdom and Sewall Wright in the United States rescued Darwin's theory of natural selection using mathematical equations that incorporated Mendelian inheritance. Today, evolutionary textbooks all triumphantly teach how early population genetics theory provided Darwin's natural selection with a rigorous mathematical basis.

Although scientists are perhaps justly proud of early population genetics, they rarely bother to mention that those equations also fundamentally change the interpretation of how natural selection works. In the Malthusian scenario, the "struggle for existence" emphasizes competition for scarce resources, making aggressive combat the theme of natural selection. Yet the equations for natural selection do not concern a struggle for limited resources at all. Instead, each genetic type is associated with a measure of net reproductive productivity—fecundity times probability of survival, the so-called Darwinian fitness. Natural selection is today best described as survival of the productive based on the progressive improvement of natural yield. Evolution by natural selection takes place even in a population unlimited by resources, because some genetic types are inherently more productive than others, regardless of the scarcity or abundance of resources. A genetic type may become more productive by being cooperative, forming friendships, being frugal or innovative, or any number of strategies having nothing to do with "struggle." I believe scientists have failed to publicize effectively that the notion of a tooth-and-claw struggle for existence was discarded over fifty years ago as the central metaphor of mathematical natural selection theory. What actually happens in nature is much kinder than people have been led to believe.

Thus Darwin's concept of natural selection has been modified and invested with new meaning, showing that he was on the right track. However, evolution by natural selection is not completely settled even to this day. The issue remains of where the variation among individuals comes from. This has been the most problematic area of evolutionary biology.

In the 1970s the distinguished biologist Lynn Margulis discovered that all the plants and animals above the level of bacteria—so-called eukaryotic organisms—are really partnerships at the cellular level.[4] I vividly remember when, as a teenager in biology class, I peeled an onion's skin, placed the thin sheet under a microscope, and saw cells for the first time. I was taught that cells are the elemental building blocks of organisms, and there I was, looking at the building blocks of an onion. Well, it's now clear that the cell is not a unitary building block after all, but rather a partnership of many subunits, some of which lived separately by themselves at some time in the past. The places within a plant cell where the green chlorophyll is located and photosynthesis occurs— the chloroplasts—were once bacteria that lived on their own. The places within a cell where our food is broken down and converted into energy (the mitochondria) were also once bacteria existing independently. The genes in an onion's cells, and in our cells too, are located not only in the nucleus, but also in other places that were once free-living cells. A cell is thus a partnership, and its overall genome is distributed across all the formerly independent partners and not solely contained in the nucleus.

Biologists have been reluctant to think through what this partnership implies. If every one of our cells is a symbiosis among formerly free-living bacterial elements, then we are but clusters of bacteria ourselves. We're not only descended from bacteria—we still *are* bacteria, a deeply humbling thought. And cellular function is not the simple story of a nucleus whose genes impose its wishes on the cytoplasm. Instead, some subcellular negotiation was required to form our cells to begin with, and may still take place. Perhaps the nucleus and mitochondria have an ongoing biochemical discussion, whose breakdown shows up as disease.

Most genes in our cells are in fact located in the nucleus, and those few residing in mitochondria and elsewhere are exceptional. This fact allows the narrative of nuclear genetic control to persist unchallenged. One wonders, though, how long this biological fiction can be sustained. Traditional population genetics views genetic variation as blindly popping up through random mutation of nuclear genes, and natural selection as operating on these new genes to fashion innovative adaptations. This view implicitly accepts the story of nuclear genetic control.

Suppose instead that genes arrive by negotiation with other organisms: one cell says to another, "I need some of your genes," and the other replies, "Sure, and I need a home to live in." Well, this collaboration is exactly what occurs in corals. The coral is an invertebrate animal like a hydra, capable of catching food with tiny tentacles. But corals also welcome single-celled algae called zooxanthellae into their bodies. At any time, the genes in a coral cell may include those in the coral nucleus plus those in the algae nucleus. However, the zooxanthellae of a coral are still quite capable of leaving the coral and surviving alone, unlike the chloroplasts of land plants. The coral-zooxanthellae relationship breaks down in low light, where corals rely on what they can catch with their tentacles. No one knows what zooxanthellae do when living on their own. Many strains of zooxanthellae exist, and the total genetic composition within a coral cell varies as different strains of zooxanthellae shuffle in and out. The genetic variation in a coral cell thus does not depend on the blind mutation of single genes, as envisioned by traditional population genetics theory. Instead, a cell can adaptively negotiate its genome with other cells.

Evolutionary biology is nowhere close to engaging the implications of a genome whose composition originates by negotiation with other genomes instead of by blind mutation. I feel the discovery of the partnership basis of cells is as important as the discovery of DNA. But the DNA story has been relatively easy for people to absorb, a refinement of the narrative of genetic control we've been taught since grade school. The partnership theory of cellular function is wholly unexpected, and scientists haven't known what to do with the finding. The situation is analogous to gender and sexuality, where also no one was prepared for the findings, and there, too, cooperative relationships have been underestimated.

Thus Darwin's theory of natural selection as the creative force molding diversity seems certain to continue as the major element of evolutionary theory, even as discussion continues about the source of variation. By contrast, the third component of Darwin's theory, sexual selection, should not, in my opinion, be resuscitated.

DARWIN'S SEXUAL SELECTION

I appreciate the gravity of discrediting a discipline's master text. However, I doubt that the factual difficulties in Darwin's theory of sexual selection can be easily smoothed over. I also believe that this theory has promoted social injustice and that overall we'd be better off both scientifically and ethically if we jettisoned it. I am far from the first to call for a thorough overhaul of sexual selection theory. I join a tradition initiated in the courageous studies by Sarah Hrdy of female choice in Indian monkeys and continued today in the writings and the experimental and field studies of Patricia Gowaty.[5] I am, I confess, more extreme than they in calling for the outright abandonment of sexual selection theory.

Darwin's sexual selection is evolutionary biology's first universal theory of gender.[6] Darwin claimed, based on his empirical studies, that males and females obey nearly universal templates. He wrote, "Males of almost all animals have stronger passions than females," and "The female . . . with the rarest of exceptions is less eager than the male . . . she is coy."

Darwin offered sexual selection as an explanation for why males and females should obey these universal templates. Whereas artificial breeding for yield was the model for natural selection, artificial breeding for show was the model for sexual selection. Darwin proposed that females, like the farmer, choose showy and virile males. Females choose males who are, he wrote, "vigorous and well-armed. . . . Just as man can improve the breed of his game-cocks by the selection of those birds which are victorious in the cock-pit, so . . . the strongest and most vigorous males, or those provided with the best weapons . . . have led to the improvement of the . . . species." Beauty, too, could be a factor. In particular, "Many female progenitors of the peacock must . . . by the continued preference of the most beautiful males, [have] rendered the peacock the most splendid of living birds." Thus Darwin imagined that males come to be the way they universally are because these males are what females universally want, and the species is better off as a result.

Darwin further proposed a universal template for social life in animals: "It is certain that amongst almost all animals there is a struggle between males for the possession of the female. . . . The strongest, and . . . best armed of the males . . . unite with the more vigorous and better-nourished females . . . [and] surely rear a larger number of offspring than the retarded females, which would be compelled to unite with the conquered and less powerful males." In these writings, Darwin pejoratively viewed diversity within a species as a hierarchy beginning with superior individuals and winding down to the "retarded," a view that is diversity-repressing and elitist, stressing a weeding out of the weak and sickly and naturalizing male domination of females. In his earlier writings, however, Darwin viewed diversity favorably across species within an ecological community, imagining that each species fills a special niche in nature. The contradiction evident in Darwin's attitude to diversity within species, as opposed to diversity between species, plagues our society today, from biology and medicine to politics and law.

However, Darwin didn't ignore diversity altogether. Juxtaposed with universalist claims are acknowledgments of "exceptions" to the general pattern. In some species, males "acquire" females by defeating their rivals. In other species, males cannot unilaterally capture females but must allow for female choice instead. "In very many cases the males which conquer their rivals do not obtain possession of the females, independent of the choice of the latter." In such cases, "the females . . . prefer pairing with the more ornamented males, or those which are the best songsters, or play the best antics . . . [and] at the same time prefer the more vigorous and lively males." In still other species, males and females are equals, and male choice of females is as important as female choice of males. Darwin wrote that in the "much rarer case of the males selecting particular females . . . those which . . . had conquered others . . . would select vigorous as well as attractive females." Darwin was an experienced naturalist who knew of diversity in mating behavior but dealt with this diversity by privileging the narrative of the handsome warrior, relegating everything else to exceptions. Darwin made no attempt to explain why "exceptions" occur or why species vary in the balance of power between the sexes. His labeling of this diversity as exceptional sidestepped the need to explain.

Darwin also acknowledged that many animals do not align with a simple sexual binary. Although Darwin worked at length on barnacles, which are simultaneously hermaphroditic, he never tried to fit them into his theory. Instead, he simply set barnacle-like species aside and asserted that all the remaining species do obey the universal male and female templates: "On the whole there can be no doubt that with almost all animals, in which the sexes are separate, there is a constantly recurrent struggle between the males for the possession of the females."

Similarly, Darwin knew of sex-role reversal but offered no explanation other than to say that such reversals are rare: "With birds there has sometimes been a complete transposition of the ordinary charters proper to each sex; the females having become the more eager in courtship, the males remaining comparatively passive, but apparently selecting the more attractive females. . . . Certain hen birds have thus been rendered more highly colored or otherwise ornamented, as well as more powerful and pugnacious than the cocks." After reviewing the sex-role reversed cassowary, emu, tree-creeper, and nightjar, Darwin concluded, "Taking as our guide the habits of most male birds . . . [females] endeavor to drive away rival females, in order to gain possession of the male. . . . [Here] the males would probably be most charmed or excited by the females which were the most attractive by their bright colors, other ornaments, or vocal powers. Sexual selection would then do its work, steadily adding to the attractions of the females; the males and the young being left not at all, or but little modified." Even today, sex-role reversals are "explained" as resulting from a higher parental investment from males than females in raising the young. Yet even today no theory has been proposed that explains when this transposition of the sex-role binary occurs.

Darwin does not appear to have been aware of natural same-sex sexuality, or of gender multiplicity in the sense of coexisting alternative reproductive and/or life history strategies within each sex. Nor does Darwin consider any functions for mating that are not directly linked to reproduction. Yet Darwin did anticipate the theory of parental investment based on the relative cost of egg and sperm: "The female has to expend much organic matter in the formation of her ova, whereas the male expends much force in fierce contests with his rivals, in wandering about in search of the female, in exerting his voice. . . . on the whole the expenditure of matter and force by the two sexes is probably nearly equal, though effected in very different ways and at different rates."

Darwin should be credited for distinguishing between traits contributing mostly to survival in the physical environment and those contributing mostly to reproduction in the social environment, for acknowledging many exceptions, and for anticipating many of the concepts still employed today. Darwin should also be credited with attributing evolutionary status to females. The possibility that females were even capable of choice was controversial at the time. Yet Darwin wrote, "Females have the opportunity of selecting one out of several males, on the supposition that their mental capacity suffices for the exertion of a choice. . . . No doubt this implies powers of discrimination and taste on the part of the female which will at first appear extremely improbable; but by the facts . . . I hope . . . to show that the females actually have these powers."

What then are we to make of Darwin's theory of sexual selection? The matter comes down to whether the underlying metaphor is correct. Is selection in a social context the natural counterpart of artificial selection for show? Does social life in animals consist of discreetly discerning damsels seeking horny, handsome, healthy warriors? Is the social dynamic between males limited to fighting over the possession of females? Does diversity within a species reflect a hierarchy of genetic quality?

Is today's sexual selection theory any better than Darwin's? No. Today's theory makes matters worse by adding new mistakes, morphing what Darwin actually wrote into a caricature of male hubris. According to today's version, males are supposed to be more promiscuous than females because sperm are cheap, and hence males are continually roaming around looking for females to fertilize. Conversely, females are supposed to be choosy because their eggs are expensive, and hence they must guard their investment from being diluted with bad genes from an infe-rior male. A male is naturally entitled to overpower a female's reluctance lest reproduction cease, extinguishing the species. In fact, Darwin's writ-ings do not endorse the expensive-egg-cheap-sperm principle. Today's sexual selection lore is based on an accounting mistake that Darwin did not make. Darwin referred to the total energy expended by each sex in reproductive effort over a lifetime as being equal.[7]

The second contemporary mistake is elevating deceit into an evolu-tionary principle. Darwin claimed that warfare to secure control over fe-males is the universal social dynamic among males. Therefore, cooperat-ive relations, especially those between members of the same sex, appear to falsify the social template that Darwin claims is universal. The con-temporary work-around is to postulate deceit. Today's sexual-selection-ists have produced a proliferation of "mimicries": sexual mimicry, fe-male mimicry, egg mimicry, and so forth. By postulating these types of mimicry, the spirit of warfare and conflict is preserved but driven un-derground, turned into guerrilla combat. Yet in no case have any of the mimics been shown to be fooling any other animal, and the circum-stances suggest that the animals are in fact perfectly aware of what is happening. The sexual-selectionist picture of nature is not pretty. Not correct either.

Darwin conceived his theory in a society that glamorized a colonial mil-itary and assigned dutiful, sexually passive roles to proper wives. In mod-ern times, a desire to advertise sexual prowess, justify a roving eye, and disregard the female perspective has propelled some scientists to continue championing sexual selection theory despite criticism of its accuracy.

SEXUAL SELECTION FALSIFIED

Contemporary sexual selection theory predicts that the baseline outcome of social evolution is horny, handsome, healthy warriors paired with dis-creetly discerning damsels. Deviations from this norm must then be ex-plained away using some special argument. But is the theory that makes this prediction correct to begin with? How many exceptions are needed before sexual selection theory is itself seen as suspect?

The time has come to set the glass on the table: to declare that sexual theory is indeed false and to stop shoe-horning one exception after another into a sexual selection framework. We need to face the fact that sexual selection theory is both inaccurate and inadequate. To do otherwise suggests that sexual selection theory is unfalsifiable, not subject to refutation.

The universal claims of sexual selection theory are inaccurate. Males are not universally passionate, nor females universally coy. The social dynamic between males is not universally combat to control females. Diversity among males and among females does not universally fit a hierarchy of genetic quality. Females do not universally select males for their genetic quality. Moreover, sexual selection theory is inadequate to address the diversity in bodies, behaviors, and life histories that actually exists. Darwin didn't bother to explain the exceptions he recognized, and as data on diversity in gender and sex continue to accumulate, sexual selection theory, which addressed only a subset of the facts to begin with, becomes increasingly inadequate.

Let's record, then, the many ways we've seen in which real species depart from the sexual selection norm:

1. *Bodies do not conform to a binary model.* Gametic dimorphism doesn't imply a binary of body types. The individuals in many species don't make only eggs or sperm for the duration of their lives. In most species, distinct "male" and "female" bodies are undefined or unstable. Sexual selection theory doesn't apply to many species because distinct male and female individuals as envisioned in the theory simply don't exist in those species, a point Darwin recognized.

2. *Genders do not conform to a binary model.* Gametic dimorphism does not imply a binary of gender roles either. The two sexes, even if located in separate bodies, may each entail more than two genders, defined as distinct morphologies, behavioral roles, and life histories in sexed bodies. Societies with one, two, and three male genders, together with one or two female genders, have been extensively described. However, sexual selection theory is a two-gender theory.

3. *Sex roles are reversible.* Even when distinct male and female bodies exist, with one gender per sex, the behavioral roles these genders carry out may be the reverse of what sexual selection theory envisions. Pipefish and jacana sperm are tiny and their eggs

large, just as in other metazoan species, yet the overall parental investment by the male exceeds that of the female in these species, resulting in a reversed operational sex ratio leading to female-female competition for males and male choice of females. Neither today's extensions to sexual selection theory nor Darwin's original treatment offer any prediction for when this occurs.

4. *Sperm are not cheap.* According to well-known primatologist Meredith Small, "Non-human primates show us what many single women in America today already know—sometimes it's very hard to get a date. Female rhesus monkeys and baboons often present to males, a clear sign of preference and choice, but males regularly refuse. Lion-tail macaque females, especially subadults, share this rejection. Females of this species initiate almost 70 percent of the copulations but only 59 percent end up in mounts. No one is sure why these males refuse, inasmuch as sperm is supposed to be so cheap, but males often ignore estrous females."[8] Why should males refuse the invitation to sex when sperm are supposedly so cheap, as sexual selection theory requires? Because sleeping together is meaningful in itself. Animal sex is not anonymous. Mating is a public symbol. Animal "gossip" ensures everyone knows who's sleeping with whom. Therefore, mate choice, including male mate choice, manages and publicizes relationships. A male may not want the commitment that accepting a new girlfriend entails.

5. *Females do not choose "great genes."* Females choose mates for many reasons, but rarely or never to acquire the great genes that a male is supposed to have according to sexual selection theory. Low-ranking males have offspring just as capable as those of high-ranking males. Females select for males who deliver on their promises of parental care and spread the probability of paternity among males to ensure offspring safety. Physical characteristics in a male serve to endow offspring with the bodily markers of a powerful lineage, not to acquire attractiveness; females are buying their offspring membership in the old genes club.

6. *Family size is negotiated.* Egg and sperm production are not necessarily independent, as sexual selection theory envisions. Males don't have to run around trying to fertilize a fixed number of eggs. Males and females can negotiate to increase the number

of eggs a female produces beyond those she would make if she were to raise them by herself. In addition, males need to make sure the eggs they do fertilize are successfully raised—it doesn't matter how much sperm they produce if the quality of parental care is compromised.

7. *Social deceit is not demonstrated.* The deceit required by sexual selection theory has never been demonstrated. Despite scientists' invention of many categories of social deceit, such as sexual mimicry and egg mimicry, it has never been proved that the mimetic traits are not simply social symbols. Perhaps animals do lie to each other now and then, but biologists have yet to catch them in a lie, so a presumption of honesty is appropriate.

8. *Same-sex sexuality is common.* Same-sex sexuality is contrary to sexual selection theory, so the existence of homosexuality must be explained away as either an aberration or a deception. Instead, the extensive documentation of same-sex sexuality among vertebrates rules out any further denial of homosexuality and contradicts sexual selection theory.

9. *Mating is not primarily for sperm transfer.* The purpose of mating, both heterosexual and homosexual, is more often to create and to maintain relationships than to transfer sperm. Sexual selection theory requires that mating be primarily about sperm transfer, whereas the amount of mating that actually takes place is a hundred to a thousand times more frequent than that needed for conception alone.

10. *Secondary sex characteristics are not just for heterosexual mating.* Sexual selection theory limits the meaningfulness of secondary sex characteristics to heterosexual mating. In species with common homosexual matings, secondary sex characteristics, including genital geometry, are shaped to facilitate all types of mating, including homosexual matings.

The sheer number of difficulties with sexual selection theory precludes plugging all the leaks. An occasional leak might be fixable, but this many leaks make repair impossible. The theory of sexual selection was taking on water long before evidence was found of widespread homosexuality, but homosexuality is the final torpedo.

The uncritical acceptance of sexual selection theory has led to under-estimation of the extent of cooperation among animals, forcing scientists to construe all interactions between organisms as somehow competitive. From a scientific standpoint, sexual selection theory is inaccurate in its claims and unable to account, even by extension, for the diversity of bodies, genders, sexualities, and life histories.

Most important, sexual selection theory is diversity-repressing. Sexual selection theory envisions male-male competition as weeding out the frail and sickly, and female choice as welcoming to bed the winners of male-male competition so that their children may inherit great genes. This elitist, regressive stance incorrectly views gene pool diversity as consisting of mostly bad genes that males must eliminate and females avoid.

SEXUAL SELECTION CORRUPTED

Sexual selection theory has long been used to perpetuate ethically dubious gender stereotypes that demean women and anyone else who doesn't identify as a gender-normative heterosexual male. By hesitating to declare sexual selection theory scientifically false, scientists prolong the injustice that emanates from this theory, as the writings of contemporary evolutionary psychologists illustrate.

Evolutionary psychology extrapolates the cheap-sperm-expensive-egg principle of today's sexual selection theory to "explain" human desire. One psychologist writes, "Because women in our evolutionary past risked enormous investment as a consequence of having sex, evolution favored women who were highly selective about their mates. . . . A man in human evolution history could walk away from a casual coupling having lost only a few hours of time. . . . A woman in evolutionary history could also walk away from a casual encounter, but if she got pregnant as a result, she bore the costs of that decision for months, years, and even decades afterward."[9] This view implies that motherhood is a punishment for sex rather than a desirable end in itself. If women do wind up having to abide with more severe consequences from a casual encounter than men, this reflects a social inequity in the division of childcare, not some universal difference between the sizes of egg and sperm. We thus see how psychologists attempt to naturalize gender inequality.

Another psychologist writes, "Differences in mating strategies can be traced to the minimum 'parental investment' required to produce an off-spring. In our species, parental investment required to produce offspring is much greater for females (i.e., nine months for females vs. minutes for males). Given that females can only produce a maximum of 20 offspring in a lifetime, having sex with a relatively large number of males is unlikely to have adaptive advantages. It is generally far better to invest more in each offspring by carefully selecting a mate with good genes who will participate in the raising of the offspring. For males, having intercourse with a larger number of fertile females was likely correlated with reproductive success since in ancestral environments contraceptive devices were not available."[10] Apart from asserting a natural right to promiscuity, this quotation also manages to suggest that nonprocreative sex awaited the invention of condoms. Drawings on Greek pottery, not to mention the behavior of our primate relatives, demonstrate many nonprocreative heterosexual positions.

These quotations illustrate how Darwin's theory, which might otherwise be written off as merely incorrect, is open to corruption by psychologists, yielding a stimulating fantasy. The assertions by psychologists claiming to speak biological truth have finally come to the attention of professional evolutionary biologists and are being refuted with uncharacteristic vehemence. One of the most accomplished experimental population geneticists today, Jerry Coyne, writes, "Evolutionary psychologists routinely confuse theory with idle speculation. . . . Evolutionary psychology . . . is utterly lacking in sound scientific grounding." Its "stories do not qualify as science, and they do not deserve the assent, or even the respect, of the public."[11]

What provoked such an unusual declaration? The recent publication of yet another theory of the naturalness of rape supposedly based on evolutionary biology.[12] The idea is that men unable to find mates in the "usual way" can reproduce through rape. Genes for rape then increase, leading to the brain's acquisition of a "rape chip." All men are therefore potential rapists, although they do not necessarily act on this potential, depending on external circumstances. Coyne points out that this I-can't-fight-evolution theory is falsified by the facts that one-third of all rapes are of women too young or too old to reproduce; 20 percent do not involve vaginal penetration; 50 percent do not include ejaculation in the vagina; 22 percent involve violence in excess of that needed to force copulation; 10 percent of peacetime rapes are in gangs, thus diluting each man's chance of reproducing; wartime rapes usually culminate in the murder and sexual mutilation of the victim; some rapists are wealthy, giving them access to women without coercion; and many rapes are homosexual. So many rapes are nonreproductive that rape can't plausibly be viewed as a means of sperm transfer for disadvantaged men to achieve reproduction. Like other mating acts, rape is about relationships—in this case, domination.

The assertion that all men are potential rapists is offensive enough to make men angry about the misuse of sexual selection theory—as women and others outside the sexual selection templates have been for years. Coyne has been prompted to say publicly what many have already observed: that evolutionary psychology "is not science, but advocacy," that evolutionary psychologists "are guilty of indifference to scientific standards. They buttress strong claims with weak reasoning, weak data, and finagled statistics . . . [and] choose ideology over knowledge." Coyne points out, "Freud's views lost credibility when people realized that they were not based on science, but were actually an ideological edifice, a myth about human life, that was utterly resistant to scientific refutation. . . . Evolutionary psychologists are now building a similar edifice. They, too, deal in dogmas rather than propositions of science." Worse even than being theorized as a latent rapist, the misuse of science offends Coyne: "To a scientist, the scientific errors . . . are far more inflammatory than . . . its ideological implications."

Thus Darwin's sexual selection theory uses an incorrect model of social life in animals—that when not busy looking for food or escaping from predators, discreetly discerning females are busy selecting for horny, handsome warriors. This theory that social life boils down to a selection for showy traits is both inaccurate in its universalist claims and inadequate to address the diversity of bodies, gender expression, and sexuality that actually occurs in nature. Furthermore, the theory has been corrupted by evolutionary psychologists and others to naturalize injustice and deny freedom of expression.

Still, some may feel that denying sexual selection theory is too drastic. I get responses like "She throws out a very healthy baby with some slightly soiled bathwater"[13] to my proposal that sexual selection theory should be discarded. Couldn't we just substitute new wording for Darwin's—invest the theory of sexual selection with new meaning—much as we have done with the theory of natural selection? Well, from my perspective, the crux is that the underlying model of sexual selection—selecting for show—is incorrect. To me, all that's floating in the dishwater of sexual selection theory is dirt—no baby there, never was.

I invite you to make your own judgment on retaining sexual selection theory as a scientific principle. I've been clear about where I'm coming from. I'm a transgendered woman; I have standing, as lawyers say, to sue for damage against this theory: it denies me my place in nature, squeezes me into a stereotype I can't possibly live with—I've tried. For me, discrediting sexual selection is not an academic exercise. By now, nearly everyone can claim to be misrepresented by sexual selection theory. Today we have a call-to-action from society to scientifically audit sexual selection theory. I have done this audit, and found the books cooked. If we're serious that scientific principles are open to falsification by facts, then I believe we're compelled to rule that sexual selection theory has now been discredited. I propose a different theory.

SOCIAL SELECTION

My underlying assumption is that animal species with distinct males and females interact socially to acquire opportunities for reproduction—that is, through trade or other exchanges, they obtain access to resources that enable the production and survival of young. Animals are not seeking each other's genes; they are seeking access to the resources that each controls. Each animal has a time budget to allocate among between-sex and same-sex relationships. Together, these relationships further the expected number of offspring successfully placed in the next generation.

Females may be thought of as starting with total control of reproductive opportunity, and males none, because an egg can potentially develop without any male contribution (as in the case of parthenogenesis). What benefit, then, do males offer to make sexual reproduction advantageous to females? They allow a continual rebalancing of the species' genetic portfolio. This benefit must be substantial, because—instead of producing 100 percent daughters, each of whom can lay eggs—females dilute their future reproductive rates by one-half, producing 50 percent sons, who don't lay eggs, along with 50 percent daughters, who do lay eggs. However, by negotiating male parental care in return for male input into the offspring, a female can increase, even double, the number of offspring that she could produce by herself, thus partly compensating for the 50 percent loss that the invitation to sex originally cost. Courtship therefore consists of exchanging information about ability to pay, likelihood of payment, and transfer of control. Meanwhile, the ability to pay, for both males and females, depends on the same-sex relationships each is engaged in. Males interact with one another to acquire and defend the resources they pay out as parental care, and females interact with one another to acquire the circumstances in which they can safely rear the young under their control.

The packaging of male and female functions in one body type—as seen in plants, many invertebrates, and coral reef fish—may be thought of as the initial and more general condition. Confining one sex to one body emerges as a specialization for the "home delivery" of sperm. Wind-pollinated plants and broadcast spawners like sea urchins suffer substantial sperm loss, opening a niche for specialized delivery systems. Barnacles glued to rocks in the intertidal zone, for example, remain simultaneously hermaphroditic but have evolved a very long penis, typically three or more times the body diameter, to deliver sperm to adjacent barnacles without losing any to the pounding surf. Plants, which are sessile and can't carry out home delivery by themselves, contract with insects and birds to deliver their sperm to other plants. Mobile animals have the option of locating sperm in a separate body type for delivery to females. But once males exist as separate bodies, they assume an agenda of their own. Males may find their interests furthered by offering parental care to females to increase successful paternity. Because males must negotiate with females and with one another, the delivery of sperm itself can assume a secondary and almost incidental function to the act of mating. Mating is then more about maintaining the between-sex and same-sex relationships needed to provide food and safety for the young than about sperm transfer as such.

If social life in animals is primarily about acquiring and trading the opportunity to reproduce, then the dynamics of animal societies are complex, nonlinear, and unpredictable. I'm struck by the unpredictability of how social evolution has played out in closely related species. Take our two closest relatives, the bonobo and the common chimp: they differ slightly in chin hair and habitat. Yet one is peaceful, the other violent. Female spotted hyenas have a penis, but their closest relative doesn't. The Idaho ground squirrel performs mate guarding, Belding's ground squirrel doesn't. These pairs of very closely related species have developed societies with diametrically opposed power relations. Why? Traditionally, it is thought that a society's organization reflects properties of the environment, that a society is somehow put together for overall efficiency, a great machine organized for a collective function.[14] Instead, I suggest that social evolution is turbulent, that an animal society is throbbing, vibrating, and energetic, and that the unpredictability of the power relations emerging in closely related species is the evolutionary signature of turbulent social dynamics. The outcome of social evolution seems as uncertain as where a white-water stream deposits a floating leaf.

If social evolution results from complex nonlinear dynamics, then phenomena like sex-role reversal, which Darwin noted in passing, are not so anomalous. A common feature of nonlinear systems is the presence of alternative multiply stable attracting states. The axes of gendered morphology and gendered behavior may each have two simultaneously stable evolutionary states and give rise to many combinations of morphology and behavior that are evolutionarily stable. Some would be sexually monomorphic, some dimorphic, some with typical gender roles and others with reversed roles. Similarly, various family arrangements, either monogamy, polyandry, or polygyny, may emerge as the diverse outcomes of social negotiations about how to control access to various kinds of resources needed for reproduction and safety. This suggestion is pure conjecture on my part, but I believe this is the direction in which we should start thinking.

When we focus on social life as a continual exchange of control over resources to reproduce, then complex multigendered societies are not anomalous. The genders emerge as occupational categories, with gendered symbolism to signal occupational roles in bringing about matings, raising young, or tending resources, much as a worker's uniform does in human society. The payment for services rendered is in terms of increased opportunity to reproduce. While some genders reach a market-based accommodation of their needs, others linger on the outside of their political economy, taking the opportunity to reproduce by force and aggression. Social violence is not nature's baseline state, but a special case of failing to strike a successful bargain in an animal society's marketplace for access to reproductive opportunity.

As ever-increasing similarities between animals and humans are revealed, do animal societies become more relevant to human societies than previously believed? Should political science and sociology, basic subjects in the human social sciences, be widened to include investigations of how animal societies function? I think so. People are not demeaned by the comparison with animals, but animals are elevated by the comparison to people.

SOCIAL-INCLUSIONARY TRAITS

Finally, we are left with the one issue on which many feel that Darwin's sexual selection theory was correct—the peacock's tail, an example of a so-called secondary sex characteristic. Other supposed examples would include the long nose on an elephant seal, the antlers on a deer, and countless other male ornaments. As Darwin wrote, the female peacock's "preference for beautiful males, [has] rendered the peacock the most splendid of living birds." Is female preference for beautiful tails why male peacocks have them? Even if one grants that Darwin's sexual selection theory is inaccurate in its claims of universality and inadequate to address the diversity of bodies, gender expression, and sexuality that actually exists, perhaps Darwin is still correct about peacocks. Perhaps sexual selection applies solely to those few species like the peacock, in which the males, and the males alone, are highly ornamented, and where the males actually do display these ornaments to a female during courtship.

If I were settling out of court with Darwin's lawyer, I'd happily concede peacocks to obtain a compromise. Someday, though, someone will challenge Darwin on peacocks too, and I'll bet they'll win. Here's the problem. Let's turn our gaze for the moment from ornamented males to species where females are the sex with unusual structures. Some species with female ornaments are sex-role reversed, like the pipefish and the jacana. However, other species with female-limited ornaments are not sex-role reversed. Take the spotted hyena, in which females all have penises.

No one suggests that females have these structures because male hyenas prefer females with a large penis. The female penis in hyenas is used for social interaction among females and has nothing to do with what males want. This case raises the possibility that some structures are used as a condition for inclusion in the same-sex social groups that control the resources needed to reproduce. If a female hyena lacks a penis, she has no chance of effectively interacting with other females. She would therefore be excluded from the all-female groups that control resources in hyena society: she would not be able to reproduce, the evolutionary equivalent of death.

Candidates for social-inclusionary traits include the masculine genitals on female spotted hyenas, female same-sex sexuality in bonobos and Japanese macaques, and the human brain (as we will see in chapter 12). Social-inclusionary traits evolve fast because, once a trait takes hold, anyone without it is excluded from the group—a lethal situation. Unique to the group in which they occur, they are a bodily manifestation of animal prejudice. Social-inclusionary traits are to social selection what secondary sex traits are to Darwin's sexual selection, but social-inclusionary traits pertain to both within- and between-sex social dynamics, and to relationships distributed across many individuals, not just dyadic relationships. Selection for social-inclusionary traits would seem to account for traits found solely in females of species that are not sex-role reversed, traits that presently lack any explanation. The idea of social-inclusionary selection thus fills an explanatory vacuum.

Social-inclusionary traits also provide an alternative explanation for many, if not all, of the traits conventionally interpreted as secondary sex characteristics in males, which, like the peacock's tail, females are supposed to prefer. The problem is that the traits of the males with whom females wind up mating may be intended more for the attention of other males than for display to the females. Antlers, for example, serve as weapons by which males can physically beat up other males, but they may also be symbolic to other males of what they seek in companions and allies. In short, these traits may be "medals" valued by other males rather than ornaments valued by females. A female might not necessarily care if a male held another male in high regard, unless that regard correlated with the amount and reliability of parental care he would provide her. But a male not held in high regard by other males might never have the opportunity to court a female. Thus an illusion emerges that the female prefers the male who is victorious, or otherwise held in high regard among males, when she is in fact indifferent to those characteristics except insofar as her own direct reproductive success is affected. Here it is male-male social dynamics that determine who qualifies as an eligible suitor. Thus the test of whether a male's showy trait is an ornament resulting from sexual selection or a medal resulting from social-inclusionary selection is whether the trait is valued by the females or by other males, and not whether males lacking the trait don't mate. For this reason, I wouldn't bet money that Darwin is correct about peacocks, because we don't know how male peacocks regard each other's tails—whether male peacocks require beautiful tails on each other as a condition for participating in whatever male-male social dynamic establishes eligibility to become a suitor.

Social-inclusionary medals are within-species counterparts of what evolutionary biologists call premating isolating mechanisms. Animals use color spots and vocalizations to tell what species they belong to and avoid hybridizing with other species. These traits reinforce the distinction between species. Biologists have long wondered how species become distinct from one another. The selection pressure to reduce hybridization gradually disappears as species become more distinct from each other, stalling the evolution before completion and leaving a residual hybridization rate. If the traits that separate species also function as social-inclusionary medals, then selection for social inclusion augments selection to lower hybridization and propels the evolution of species distinctness to completion. Species are more distinct in animals than in plants, where extensive hybridization takes place across the species in many genera. If premating isolating mechanisms in animals are also social inclusionary medals, then animal species should evolve sharper between-species distinctions than plant species for this reason.

This review of diverse gender expression and sexuality among the vertebrates demonstrates that biology need not tell one single, simple, and boring story. Biology need not be a purveyor of essentialism, of rigid universals. Biology need not limit our potential. Nature offers a smorgasbord of possibilities for how to live, and an endless list of solutions for every context, some of which we'll wish to reject, and others to adopt or modify.

The true story of nature is profoundly empowering for peoples of minority gender expressions and sexualities. Yet this truth has been suppressed by biologists, and the few accounts that do surface are embedded in pejorative language. To remove the conceptual rot, we've had to excavate deep into the foundation of evolutionary theory, identify the collapsing member—Darwin's theory of sexual selection—and replace it with new ideas that may be better able to carry the load as the future unfolds.

PART TWO

HUMAN RAINBOWS

10

An Embryonic Narrative

Many developmental mechanisms must exist to produce the diversity of bodies, gender expressions, and sexualities so evident among the animals we've just visited. What are these mechanisms? How do two fertilized eggs that start out looking just about the same wind up producing two adults as different as a lion and a lioness, or a man and a woman? How does one fertilized egg grow up to become a corporate CEO, while another grows up to be a drag queen? This part of the book is about the developmental mechanisms that bring about diversity.

The story of development is told by molecular genetics, cell biology, embryology, physiology, as well as developmental psychology—areas I refer to collectively as developmental biology. Developmental biology has fallen into the same trap that sexual selection theory has: it assumes that one master template is the norm, and that variety reflects a defective deviation from that ideal norm. Although early scientists could equally well have approached developmental biology with an open heart, ready to embrace the diversity of molecular mechanisms that produce bodily and behavioral diversity, the party line has instead been to sound the alarm at any hint of diversity, then to label diversity as disease and "cure" it. Of course, disease does sometimes occur, and cures for true diseases are needed, but the disease model of diversity fundamentally misrepresents human nature, inflicting needless procedures or actual harm on people in the name of "curing" them.

The fundamental mistake made in developmental biology is privileging a master controlling narrative for genes. According to developmental biology, genes recline on chromosomal thrones in a nuclear palace, from whence they direct subcellular minions to accomplish their selfish ends. That story is still taught today in biology class, forming the intellectual basis for medicine. Is that story true? How much, if anything, do genes control? In this and subsequent chapters, I will challenge the master controlling narrative for genes and suggest instead a story of human development that emphasizes relationships among the organic components that make up our bodies. I believe that my story more accurately reflects nature and that its adoption will lead to more economically efficient medicine, more profitable biotechnology, and a more just society.

The model for developmental biology imagines that a master gene triggers a subordinate gene, which cascades to downstream genes in a descending hierarchy of control. In this picture, bodies develop as though a bowling ball were accurately rolled to hit the genetic kingpin at just the right spot and cause all the genetic bowling pins behind to fall down in perfect order. Producing a normal baby is bowling a genetic strike.

Instead, imagine that genes are like mice released at the top of the bowling lane, who scurry down the lane, bumping into genetic pins as they go and eventually knocking down all the genetic pins in a variable, but directional, clamor. In my picture of how development works, diversity figures from the very beginning.

The narrative I tell of development emphasizes the interrelatedness of gene function and avoids exaggerating the role of genetic control. My model of how a gene works is the "genial gene"—a gene that cooperates with other genes, in contrast to the well-popularized concept of the "selfish gene."[1] In other words, my narrative of development deemphasizes individualism. The fertilized egg begins as a genetic partnership between egg and sperm, and the sperm part of the genome isn't expressed until after a few zygotic cell divisions. In the beginning, the egg alone carries the ball, both genetically and in the cytoplasm, and the zygotic cell is a team of egg and sperm parts. Eventually the embryo transforms into an individual from its original egg-sperm partnership. Furthermore, the mother plays an active role in how the egg-sperm partnership is formed:

she screens the chemical eligibility of a sperm before allowing it to fertilize one of her eggs, and she transports sperm to the back of her oviduct, where her eggs are waiting, rather than relying on the sperm to swim there. As the embryo develops, tissues talk to each other and cells shuffle around to build the body of a child. Thus a newborn child is already socialized and chemically experienced, the proud graduate of a biological education from many organic teachers. We should think of our biological education as being continuous with our social education— merely a school we attend before going to kindergarten.

Much has been made of gay genes, gay brains, and transsexual brains, as though such organic differences among people, whenever they exist, were somehow anomalous. Instead, we all naturally differ from one another materially. Someday we may even know how our brains change after reading a novel, for instance.

I will now tell the story of human development as though my genes, my cells, and my tissues could talk. I trust my genes to provide my body with the parts I need. I cooperate with my body to live a good life. I believe agency extends throughout my entire being, and I see no grounds for splitting off my biochemical functioning from my deliberative actions. This revised narrative of human development offers a new foundation for understanding organic variation as a healthy, joyous alternative to the medical monotony of disease.

MY EGG PART

My egg part began when my mother was four weeks old and tiny, only 0.2 millimeters long.[2] She was still an embryo herself, sprawled out across the ball-like yolk sack that nourished her. I was one of the earliest cells my mother produced—a primordial germ cell.

I started life at the edge of the yolk sac, plump and round, not yet within my mother's developing body. To join my mother in the core of her body, I moved by sticking a piece of me out in front, then I flowed into the piece and reabsorbed my trailing parts. Step by step. I entered her where her tummy would be, near the bottom, where her colon would form. I moved up toward her head and inward toward where her backbone would be. Yes I did.

By the time my embryonic mother was six weeks old, I had come to rest in ridges along either side of her spine, called the genital ridges. This would be my home while I remained within my mother's body. By the time my mother was twelve weeks old, she had prepared her genital ridge as an ovary. I then began transforming into a form called an oogonium and lived within my mother's ovary, as though her ovary were my nursery.

I divided many times, making sister oogonia. At eight weeks, we numbered over five hundred thousand, and between the second month and the seventh month of my mother's life as an embryo, my sister oogonia and I divided so often that our peak numbers reached over seven million. This was far too many, and we would decline to two million by the time my mother was born.

Some of my genes were outside of my nucleus, in my cytoplasm, within my mitochondia, and the rest were in my nucleus. I now had to eliminate half of my nuclear genes to make room for the genes that would come from my sperm part. I started the process of setting aside half of my genes, becoming an oocyte, then put the process on hold, suspended. I sat on the bench, waiting for her nod to continue. Meanwhile my mother was born, and she grew up in the world to become a young woman.

When my mother was twelve, she gave us the first call to action. At staggered times, she would ask one of us to warm up. We oocytes numbered five hundred thousand at the start, a strong bench. Only about four hundred of us would ever get the chance to play—one per month for about forty years, between my mother's puberty and her menopause.[3]

After years of waiting, when my mother was over twenty years old, the action suddenly picked up. I quickly grew over twelve days to become five hundred times bigger. I made a sheath around me, called the zona pellucida. My mother offered an extra coating of cells outside my sheath, called granulosa cells. I was making myself ready to become an embryo. In my cytoplasm, I stored materials, enzymes, messages from my genes, and ribosomes to synthesize proteins. I prepared enough to last for a week on my own as an embryo. During that week my genes alone would be expressed, until the additional genes I obtained from my sperm part could be used.[4] Biologists call the combination of me an oocyte, with my sheath and surrounding coat of granulosa cells, a follicle.

My mother had prepared the playing field. About two to three weeks earlier, she had cleaned her uterus by shedding the old lining of tissue and blood vessels.[5] While I was growing within my follicle, she added a fresh uterine lining. The mucus near the back of my mother's vagina then thinned, making it possible for sperm to move into her reproductive track, where I was waiting. Indeed, sperm did pass through my mother's oviducts at this time and waited for me to appear.

I resumed the process of setting aside half of my genes, the process I had suspended for many years. I then transformed from an oocyte into an egg. My diameter was now almost visible to the naked eye, a little less than a tenth of a millimeter (80 microns).

I was ready. I burst through the granulosa coat and out of the follicle into the oviduct that leads from the ovary to the uterus. I was not left alone. I was accompanied by cells from my mother, called cumulus, named for the cottony clouds in the sky. As I emerged into the oviduct, I met many suitors. I fused with my sperm part right then and there! I was still in my mother's oviduct, near her ovary. I was now a zygote, residing in the oviduct, prepped, warmed up, and ready for life!

MY SPERM PART

My sperm part began when my father was only a few weeks old, an embryo just 0.2 millimeters long, sprawled out across his yolk sack. I too was a primordial germ cell. I too migrated to the genital ridges along either side of his spine.

My father, working fast, outfitted his genital ridges as testes by the time he was seven weeks old in utero. The testes would be my nursery as I matured. I became a spermatogonium, destined to become a sperm, slightly smaller than I looked when I was a primordial germ cell.

I then sat dormant in my father's testes while much happened around me. By four months the major pipe that would someday carry me to the outside, called the vas deferens, was completed. Sometime between the sixth and eighth month of my father's embryonic age, my whole nursery began to move. My father's testosterone caused my nursery to descend into his empty scrotum on the outside of his body. My father pushed the testes in which I was living through his inguinal canal and through his abdominal wall with pressure from his breathing and an occasional hiccup.

My father was then born and grew to puberty as a gangling teenager, when testosterone cruised through his veins. The cells of my nursery, my father's testicles, now turned into an extensive set of tiny pipes, called seminiferous tubules, that flowed like streams down a mountainside into tributaries, winding up in the vas deferens. It was my time to spring back to life.

I multiplied several times. My brothers and I stayed close to one another. Even though we each had a separate nucleus, our cell bodies remained connected to one another with bridges through which we interchanged molecules—our version of holding hands in a football huddle.

I divided my genes in half, becoming a spermatid. I stripped down and split off from my brothers. At one end I developed a cap with chemicals to serve as a name tag identifying me to my intended when we met so that I might dissolve her veil, the zona pellucida sheath surrounding her. At the other end I formed a flagellum to propel me through the birth canal and up to the location where we'd meet. And I flattened my nucleus and jettisoned any remaining cytoplasm. I was lean and ready for action!

I went from a spermatogonium to a well-dressed sperm in two months. Each day one hundred million of my brother sperm matured, and my father released us in batches of two hundred million at a time. During my father's life, 10^{12} to 10^{13} (over a million million) brother sperm would mature. It's hard not to feel insignificant, but I do have half of my father's nuclear genes. In contrast to my mother, who was born with somewhat over a million oocytes, my father would host over a million million sperm, a ratio of sperm to eggs of about one million to one.

My father could have used me in two ways: to join the nucleus of an egg or, while mating, to form relationships with other adults, relationships to promote his survival or his fertility at some later time. As it turned out, I was destined to join the nucleus of my egg part.

OUR CELLULAR COURTSHIP

When my sperm part was released into my mother's vagina, I thought I was lean and fast, proud of being in good shape and ready for the race to meet my egg part. I was in for some big surprises. I didn't call all the shots. Thirty minutes after my release into my mother's vagina, I found myself in her oviduct—much faster than I could have swum under my own power.[6] In fact, my mother transported me to her oviducts with the muscles of her uterus. Once there, I was able to cover short distances by myself. Of the 280 million brother sperm released with me, only two hundred of us, about one in a million, reached the end of my mother's oviduct, where my egg part was waiting.

I needed lots of chemical help. I found I was incapable of joining with my egg part the way I was. My mother modified my cell membrane, shaved off some unsightly molecules coating my surface, changed my attitude by siphoning off some salt ions and opening me to the prospect of accepting some calcium ions in their place, and dressed up some of my proteins with phosphorus. Without this chemical endorsement, called capacitation, other sperm were held up by the cumulus cells that my mother provided as guards surrounding the egg.

I acquired these chemical stamps of approval in different places along my mother's reproductive tract on my way from her vagina to the end of her oviduct. As a leading developmental biologist has clarified, "The female reproductive tract, then, is not a passive conduit through which the sperm race, but a highly specialized set of tissues that regulate the timing of sperm capacitation and access to the egg."[7]

As I approached my egg part, I was humbled. She was like the earth, huge and colorful. I was tiny, like her moon. I swam up to her, approaching at a tangent. I nuzzled against her velvety surface, called microvilli, which reached out to me.

QUITTING MY SINGLES CLUB

As I (the egg part) felt a sperm touching my microvilli, I wondered if he was the one for me. I felt the surface proteins on his head and checked for signs of my mother's endorsement. This was the guy. I allowed his enzymes to dissolve a tiny hole in my sheath, the zona pellucida. In about twenty minutes, I accepted him into my body, restoring the genetic quorum in my nucleus.

Now I had to work fast. Some 199 attractive sperm were nearby. Alas, no space remained in my nucleus for more genes, and I needed to broadcast the message that I had made my decision and was no longer eligible. Right beneath my cell membrane, I had stored lots of tiny granules containing enzymes that would make my microvilli unattractive to sperm and prevent any sperm from attaching to the outside of my sheath. I allowed my tiny granules to fuse with my cell membrane, turning themselves inside out to release the enzymes contained within. My granules released their enzymes one by one in a wave across my cell, starting where my chosen sperm had entered. In one minute, the wave spread across my entire cell. Now my surface was no longer attractive to sperm, and any sperm still nuzzled up in my microvillae floated away.

So there I was at last. A fertilized egg, a zygote, soon to become a baby. I felt like I'd lived a long life already. My egg part had already lived ten to twenty years, and the clan from which my sperm part came had also lived many years, yet a new phase was about to begin. The clock had started for my life as an embryo.

As I and my sperm part began our life together, there was lots to do. We had to find one another within my huge cell wall. He unwrapped his DNA and expanded his nucleus, while I completed the last stage of setting aside half of my genome and got ready to include his genes in their place. Then we looked for each other across the crowded cell. Already, while we were still traveling toward each other within our common cell, we began duplicating our genes. When we met up, all the preparation was done—we fused our nuclei, combined our chromosomes, and immediately divided. We never existed as a single cell with a single diploid nucleus. Instead, we lived for a day as one cell with two nuclei and then quickly became a two-celled embryo, each cell carrying a copy of our combined genes.

For our honeymoon, we rafted down my mother's oviduct into her uterus. As we went along, we became more multicellular. Because we were still living within my sheath, the zona pellucida, our total volume didn't expand; we were simply dividing ourselves into smaller and smaller cells. When we reached sixteen cells, we were numerous enough that we didn't all have to do the same thing, and we began to specialize among ourselves. By the time we had divided into sixty-four cells, we had segregated ourselves into two groups: an "inner cell mass" (ICM) of a dozen cells that would continue our embryonic development and a surrounding layer of helper cells that would interface with my mother, obtaining nutrients and unloading waste products. As a sixty-four-cell embryo, we were what is called a "blastocyst," about one week old.

We were getting too big for our raft, and as we rounded the turn out of my mother's oviduct, heading into a large bay (her uterus), we decided it was time to jump off and swim for shore. We dissolved a small hole in our sheath, the zona pellucida, and squeezed ourselves out. As we approached the wall of my mother's uterus, called her endometrium, we found a likely spot to beach, and our outer ring of helper cells pitched a tent within her uterine lining. As we settled into my mother's endometrium, our helper cells became our contribution to the placenta, called the chorion. My mother's contribution formed from her uterine lining, called the decidua. My mother and I shared in producing the placenta.

At this point I was getting more comfortable thinking of myself as a single individual. The genes from my sperm part and my egg part were working together.

BECOMING A BABY

Our two-week honeymoon as egg-sperm newlyweds was over. Now began the excitement of growing into a baby.[8] My first task was getting oriented. I spread myself out and located top and bottom, front and back, and right and left. I consulted my committee of genes for directions, and placed genes named Hox in charge of sections of my body, starting from where my head would be and working down. Other genes took charge of telling left from right, so that my heart would wind up on my left side and my liver and large intestine up on my right. By three months I had transitioned from a sphere of cells, the ICM, into an embryo with three distinct axes.

As I was determining my major body axes, my cells were taking up stations for the future. Have you ever seen the Stanford marching band? After the band plays a number, the trombone players run off in one direction, the drums in another, and the flutes in a third, with everyone scurrying to new locations. A few moments later a new formation materializes, ready to play another number. The USC marching band, in contrast, changes from formation to formation by moving in stately columns. National television rarely shows the Stanford band, afraid its chaotic appearance will frighten off viewers who long for an orderly world. But I got to enjoy both shows. As I developed into a baby, my cells did both the Stanford scurry and the USC procession. Cells piled up in a ragged line where my body was to form. This line, called the primitive streak, had a groove running down the middle called the primitive groove. The cells at my primitive streak were continually changing as they dived into the interior of my body along the primitive groove. Later, other cells moved as a sheet.

About two weeks after pitching tent in my mother's uterus, I had come to consist of three cell layers: the ectoderm on the outside, the mesoderm in the middle, and the endoderm on the inside. (The cell movement leading to these layers is called gastrulation.) The ectoderm would become my skin, nails, hair, eye lens, ear linings, mouth, anus, tooth enamel, pituitary gland, mammary glands, and all of my nervous system. The mesoderm would become muscles, bones, lymphatic tissue, spleen, blood cells, heart, lungs, and reproductive and excretory systems. The endoderm would become my lung lining, tongue, tonsils, urethra, bladder, and digestive tract.

These tissues talked to each other as they developed. The mesoderm induced the ectoderm above it to develop the central nervous system, as though the mesoderm said, "I think I could use a brain," and the ectoderm replied, "OK, coming right up." Three weeks after my egg and sperm parts had fused, I had already started on my nervous system and had developed some blood—and I was only 1 to 1.5 millimeters long.

By four weeks, barely after my mother knew she was pregnant, I had grown to 5 millimeters long, my spinal cord was developing, my brain had enlarged and begun differentiating into three parts, my heart was beating, my eyes and ears were started, and my arms and legs were starting to stick out from my body. The highlight at four weeks was when I formed my own primordial germ cells. They were initially outside my body. Two weeks later, at six weeks after fertilization, they migrated into my body to live within my gonads. From this time until my birth, my mother embodied three generations at once—her own generation, my generation, and my child's generation. Although I had lived for so long as a germ cell within my mother's body, I could now see the other side: my own germ cells were living within me, trusting in me for their future.

Over the next two weeks—eight weeks after fertilization—the first signs of electrical activity in my brain and muscles would start, and I began to outfit my gonads to host the recently arrived germ cells. I realized I had a Y chromosome with a noisy gene on it called SRY that directed me to outfit my gonads as testes. In two more weeks, my external genitals started to become recognizable. By the end of the first trimester of my life in my mother's body, all my organs were in place, although they were still rudimentary.

As the second trimester began, week twelve, I ceased being called an embryo and was now called a fetus. I looked human. I started to move around on my own within my mother's uterus. My task was to grow and mature. As I grew, the placenta surrounding me grew too. At five months, my testes started to descend into my scrotum, and my legs and arms approached their final proportions relative to the rest of my body. During this trimester, my body assumed the shape that would determine whether I would be eligible to become a sumo wrestler or a racehorse jockey. Also in the fifth month, I began to recognize sounds, such as my mother's breathing, heartbeat, voice, and digestion. At six months, my eyes responded to light. I was thin, with zero baby fat.

As the third trimester began, my brain woke up, and by seven months my brain waves, or EEG, had attained the form they would have at birth. I started fattening up, getting ready for birth. By eight months, I was up to 2 to 3 percent body fat; my body growth slowed, but my brain kept expanding. My eyes were open when I was awake and closed when I slept. I started developing my immune system. During this trimester, my brain developed and my temperament started forming.

At nine months, I was nearly ready to be born. My body fat was 15 percent of my weight, insulating me and raising my body temperature higher than my mother's. My skull was not fused, but rather consisted of five bony plates that would allow my head to elongate while squeezing through my mother's birth canal.

I am ready! I feel it coming!! I start heading down my mother's birth canal. I see light at the end of the tunnel. I feel my mother pushing, pushing. I hope I'm not too big. I'm sorry I'm hurting her. Pushing, pushing. I squirt through, into the hands of a doctor. I look him in the eye. He slaps me on the back, congratulates my mother, and says to me, "Be a man."

This narrative of life as an embryo is only one of many possible narratives. The narrative differs for a baby boy and a baby girl—indeed, it differs for each and every person, because our individuality begins at conception, if not before. The life narrative, including the embryonic phase, varies for people who become basketball players, football linemen, long-distance runners, corporate CEOs, musicians, high-fashion models, elementary school teachers, parents with six children, drag queens, and all other individual expressions of body type, temperament, and inclination.

11

Sex Determination

The single biggest difference among people is sex—traits related somehow to the size of the gametes they make. Yet the overall difference between human males and females is moderate compared to other vertebrates. Mice males and females are nearly identical, except for gamete size and related genital plumbing, whereas lions have conspicuous male/female differences. Other than gamete size, our statistically valid sex differences are few and small, and our distributions overlap extensively. How do such differences between males and females develop?

WHEN SEX IS DETERMINED

Accounts of how male and female differences develop in mammals usually begin with gonadal differentiation—the genes that determine whether gonads mature as testes or as ovaries. Yet the gonad is merely a nursery for the germ cells. The germ cells are a different tissue from the gonad in which they reside and are descended from the primordial germ cells, which differentiated in the very early embryo before the tissue that gives rise to the gonads differentiated. The gonad is the site where hormones like testosterone and estrogen are synthesized, and once the gonads form, many other aspects of the body develop a gendered morphology. The hierarchical bowling-a-strike view of sexual differentiation is that a master gene starts the gonad, and then the gonad propels the rest of the body into a male or female template. However, this controlling narrative is simply not accurate, even though widely believed and often taught.

The sex of an embryo—whether its primordial germ cells mature as eggs or as sperm—is evident before the gonads start transforming into testes or ovary.[1] In mice, male embryos grow faster than female embryos even before the gonads differentiate into ovary or testis.[2] In marsupials, both male and female external genitals start to develop before the gonads form.[3]

In humans, the Y chromosome has a version of the gene for ribosomes, the cell's protein-making module, that is not found on the X chromosome. The cells of XX and XY embryos thus differ in their ribosomes, because females have one type of ribosome and males two types, long before any gonads develop.[4] Therefore, protein synthesis takes place slightly differently in male and female embryos from the time the sperm contribution to the embryo's genome is first expressed, a few divisions after fertilization. This sex difference is manifest even before the first primordial germ cells differentiate and long before the gonads differentiate.

Thus gonadal differentiation, while important in gendered development, is not the stage at which sex differentiation occurs, because sex differences are already manifest before gonadal development starts. The genes that determine gonadal differentiation do influence an individual's ultimate bodily and behavioral presentation. I will call the key genes determining gonadal differentiation the gender-determining genes, or "gender genes" for short.

WHEN GENDER IS DETERMINED

Even though a primordial germ cell may already have a good idea, so to speak, of whether to mature as a sperm or an egg, a germ cell that reaches the gonad is influenced by whether the gonad there becomes a testis or an ovary. Sex (whether the primordial germ cells mature as a sperm or egg) and bodily gender (starting with whether the gonad differentiates as a testis or an ovary) are subject to biochemical negotiation.

In mammals, a key player in this negotiation is a gene called SRY on the Y chromosome. SRY redirects and accelerates gonadal differentiation in the direction of a testis; in its absence, the gonad differentiates more slowly into an ovary.[5] Biologists often describe SRY as the master gene controlling sexual differentiation, the essence of maleness. When present, SRY is said to take over an embryo, commanding it to develop into a male; without it, an embryo develops "by default" into a female.

Well, not so fast. As we've seen, the gonads develop after some sex differences are already determined, so SRY doesn't fully control sex differentiation; it can only influence gendered presentation to some degree. Moreover, SRY doesn't act alone. That's not to say SRY isn't important. SRY produces a protein that binds to DNA, causing sharp bends, which in turn affect whether the genes in the bent area of DNA can be expressed. SRY censors the DNA, determining which genes get their messages published throughout the cell.

In one experiment, an SRY gene was introduced into XX mice.[6] About 30 percent of these mice went on to develop testes, as well as male external genitalia and some male mating behavior. The germ cells in these male XX mice, which would have otherwise become eggs, started to develop as sperm but couldn't finish without the right accessories. Key wardrobe instructions are needed from the full Y chromosome—SRY is not sufficient. However, SRY can direct at least some female embryos to develop a masculine presentation, and can coax the testis into convincing some primordial germ cells to mature as sperm rather than eggs.

In another experiment, an SRY gene was deleted from the Y chromosome of XY mice. These mice went on to develop ovaries, as well as other feminine traits. The germ cells in many of these female XY mice, which would otherwise have become sperm, developed as eggs, even resulting in some litters.[7] Thus, the absence of SRY leads male embryos to develop a feminine presentation and can coax the ovary into convincing some primordial germ cells to mature as eggs rather than sperm.

In mammals, then, gonadal gender is determined in large part by the presence or absence of the gene SRY. For this reason, SRY has assumed its legendary status as *the* sex-determining gene—if the Y chromosome is the marker of maleness, its power depends on SRY's presence on it. But is SRY really in a one-way controlling position at the top of the bowling-a-strike hierarchy? In fact, the gonads can develop at least partially into

testes on their own, even without coaxing by SRY. In XX female wallabies, genital ridges develop partially as testes in the absence of germ
cells.[8] Thus, in the absence of germ cells that would turn into eggs within
the ovary, the gonad moves on its own toward becoming a testis, even
without SRY present.

The narrative of hierarchical control by a master gene, SRY, is thus an
oversimplification. SRY is only one player in the germ-cell-to-gonad negotiation, with the role, as a genetic lobbyist, of coaxing the gonad in a
male direction. SRY doesn't unilaterally control sex determination, because sex is already determined before SRY is expressed.

BEHIND THE POWER STRUGGLE

Okay, so SRY has a loud voice, but does it really bring to the table important information about how to be male? No. SRY turns out to be
heavy on influence, light on substance. Here's how SRY has wormed its
way onto the genetic committee that determines bodily gender.

Everyone has the genes to make both ovaries and testes, but which we
make depends on some network of intergene negotiation. One key gene
at the conference table is SOX9, which is located on a nonsex chromosome: it holds the basic testis recipe for all vertebrates. SOX9 is expressed in the developing gonads of male mammals, male birds, and
male alligators.[9] Other genes at the conference table, in addition to
SOX9 and SRY, are WT1, SF-1, and DAX-1 (or DSS) on the X chromosome.[10] Here's how the interaction goes:

1. WT1 prepares the genital ridge and adjacent kidney area. Then
   SF-1 and WT1 together urge SOX9 to make a testis.

2. But DAX-1 intervenes, preventing SF-1 and WT1 from activating
   SOX9, so an ovary forms instead.

3. In males, SRY inhibits DAX-1, permitting SF-1 plus WT1 to
   activate SOX9, which in turn produces a testis.

Thus SRY stops a gene, DAX-1, which itself was stopping testis development according to SOX9's recipe. Wow, not simple. Notice that SRY

and DAX-1 don't contribute materially to the recipe for making a testis. They are at the conference table just to argue, like genetic lawyers.

Nothing is universal about SRY or DAX-1; these genes don't appear in other vertebrates, including some other mammals. Species without SRY or DAX-1 have testes and ovaries, implying that different types of genetic negotiation can also produce gonadal differentiation. SOX9 is the only one of these genes with any claim to universality, at least among vertebrates. Even in species whose gonadal differentiation does emerge from the SRY-DAX-1-SOX9 committee, multiple alternative forms, or alleles, of SRY, DAX-1, and SOX9 exist, so the genetic narrative leading to bodily gendering in each individual differs depending on the precise alleles an individual has at these three genetic loci. The genes at the gonad-determination conference turn up at other conferences as well, and their expression is noted in many other tissues. These genes are a basic source of diversity in bodily aspects of gender.

SRY has pulled a coup d'état in our genetic palace, acquiring the power to preempt the differentiation of gonad into testis. But SRY is a loud-mouthed bandit living on a puny chromosome, Y, which itself recently degenerated during evolution from the X chromosome.[11] The how-to-do-it capability for testis construction resides in chromosomes other than Y.

Thus genes, including even the noisy SRY, work together during the body's development. The selfish gene is a sound byte, not science. Genes occupy a common body, their lifeboat. A selfish gene had better know how to swim. Survival for a gene means being genial—the genial gene. Not only do genes work together to jointly construct a pathway of consecutive steps in biochemical pathways, but they also collaborate in the synthesis of single enzymes. Some enzymes have multiple subunits that come from distinct genes. Furthermore, an enzyme called cytochrome c oxidase even has some subunits coded by genes in the nucleus and others by genes in mitochondria, so that making this enzyme involves the cooperation of nuclear and mitochondrial genes.[12]

A subversive philosophical shift is occurring in how biologists think about genes. As a student, I was taught that genes come first and the phenotype second. We live our lives with whatever traits our genes stick us with. A new view, from evolutionary developmental biology (fondly known as evo-devo), states that the traits come first. The need for a trait appears in the world, like the ability to make a testis to contain germ cells maturing into sperm. Then genes, like SRY and DAX-1, compete to deliver that trait during development. A takeover artist like SRY with no information of its own can evolve by promising to deliver the trait faster. Ecology writes the specifications and places an order for a trait. Genes compete to deliver the trait—ecology the consumer, genetics the producer, in a client-server relationship. This new view empowers the context in which phenotype is meaningful.

WHEN Y DOES NOT EQUAL MALE

SRY's power, as we've seen, is far from absolute—it must negotiate with other members of the gender-gene committee to effect the differentiation of a gonad into a testis. Sometimes SRY is dispensable altogether, as in the case of *Ellobius lutescens,* a mole-vole. This burrowing mammal is 10 to 15 centimeters long with a velvety cinnamon coat and lives in semi-desert areas of the Caucasus, eastern Turkey, Iraq, and Iran, where it feeds on underground plant parts. Males of this species have no Y chromosome, nor any SRY gene anywhere. Yet *E. lutescens* males are still real males: they make sperm in testes.[13] Males of another mole-vole, *Ellobius tancrei* from Uzbekistan to Sinkiang, China, also don't have Y chromosomes.[14]

In other cases, SRY and the Y chromosome may be present but be completely overridden by other genes, outvoted on the gender-gene committee. In four species of South American vole mice of the genus *Akodon,* 15 to 40 percent of the females have both SRY and a Y chromosome, yet they are still female and make eggs. These females evidently have genes that silence the noisy SRY.[15]

That SRY can be completely outvoted foreshadows discussion later on that genes on the X chromosome in humans control how much effect testosterone has on tissue development. Thus, even if SRY succeeds in obtaining legislation to produce a testis, resulting in the synthesis of testosterone, the genetic bureaucracy has a say in whether the legislation is implemented. The genetic bureaucracy may partially implement the legislation by ensuring that testosterone has only little effect, or it may fail to implement the legislation at all, as in the case of complete androgen insensitivity.

Therefore, among mammals, a Y chromosome and an SRY gene are neither necessary nor sufficient to determine male sexual identity. Admittedly, in some species, including humans, SRY is a major player on the gender-gene committee. Yet even in species where SRY is empowered to cause testis development, SRY does not control how much effect the hormones secreted by the testis have on the body's adult morphology.

Thus, the development of even as basic a difference as that between males and females does not follow a standard template either across or within species. The bowling-a-strike view of development as the unfolding of a hierarchy of successive genetically mandated decisions simply doesn't occur. Instead, every individual has his or her own unique and equally valuable narrative of how the gender-gene committee fashioned the compromise that became that individual's embodiment of gender and sexuality.

WHEN OVARIES AND TESTES COMBINE

Every aspect of the body is on the table for the gender-gene's committee to negotiate, even the structure of the gonads themselves. Most gender-gene committees, with or without the presence of SRY, pass a resolution creating only a testis in males and only an ovary in females. In some species, though, even this most elemental aspect of bodily gender has been given a different configuration.

Among *Talpa occidentalis*—another burrowing mammal, an old world mole from the Iberian peninsula—all females have ovotestes, gonads containing both ovarian and testicular tissue.[16] The ovotestes occur at the site in the body where simple ovaries are found in other species. *Talpa* XX individuals have ovotestes and make eggs in the ovarian part of their ovotestes. They don't make sperm, but they do have both sperm-related and egg-related ducts. The testicular part of these ovotestes secretes testosterone. XY individuals have testes only and make sperm.

Four species of old world moles are now known whose females have ovotestes instead of ovaries.[17] Yet when a human is born with ovotestes, bells and whistles sound in the hospital as though a law of nature had just been threatened. Old world moles would view modern medicine as primitively mistaken. Thus, even gonadal structure doesn't follow a standard template across mammals.

WHAT HAPPENS TO THE "EXTRA" X

In mammalian species where females are XX and males are XY, females have an embarrassment of riches—two X chromosomes where one suffices. Expressing both X chromosomes would presumably provide an overdose of the enzymes tweaked to work at the lower concentrations produced by a single X chromosome (as occurs in males). The workaround for females is to make one of the X chromosomes inactive. One of the Xs scrunches up, becomes unavailable for transcription into protein, and appears under the microscope as a speck in the nucleus called the Barr chromatin body.[18]

Which of the two X chromosomes is inactivated in a cell is pure chance, a flip of the coin. Hence one cell might use the X chromosome inherited from Dad, while the cell next door uses the X chromosome inherited from Mom.[19] If, however, one X chromosome contains a poorly functioning gene, females have an alternative. Cells with that chromosome can be weeded out and replaced by cells expressing the other X chromosome. The advantage of diploidy is maintained across whole cells, rather than between genes within a cell.

WHEN TEMPERATURE DETERMINES SEX

Although testes and ovaries are much the same across all the vertebrates,[20] the negotiations that lead to whether a gonad is furnished as testes or ovaries differ among the classes of vertebrates. Among reptiles, specifically turtles, crocodiles, and some lizards, gonadal identity is determined by the temperature at which eggs develop, not by chromosomes.[21] The eggs are usually laid in the ground and covered with sand or moist dirt from which they absorb water, swelling in size as they age. Reptile embryos start developing within their egg, and after a while primordial germ cells form. When reptile primordial germ cells move to the genital ridges of their parents, both the germ cells and the parental embryo presumably experience the same environmental temperature. Both germ cells and parent therefore receive the same message about which sex to develop as, and their agendas automatically agree.[22] But could there be a difference of biochemical opinion? Might an intersexed phenotype result if the temperature were changed between the time the primordial germ cells first differentiated and the time the gonad differentiated into a testis or an ovary? Such experiments don't appear to have been tried yet.

Temperature-dependent sex determination is unavailable to warm-blooded animals, such as birds and mammals, who live at only one temperature. Instead, birds and mammals concoct genetic schemes to determine sexual identity. We've already seen the mammalian schemes, which usually involve the X and Y chromosomes. The story is reversed in birds.

ZZ MALES AND ZW FEMALES

In birds and snakes, the sex chromosomes are called Z and W instead of X and Y. Males have ZZ chromosomes, and females are ZW, the opposite of mammals, where females have two identical sex chromosomes and males two different ones.

In birds, something on the W chromosome convinces the gonad to turn into a single ovary on the bird's left side. Something else on W tells the gonadal ridge to start synthesizing estrogen, which makes the gonadal ridge continue along its path of turning into an ovary. In the absence of estrogen, the genital ridge turns "by default" into two testes, one on the right and one on the left.[23]

The avian negotiation is thus the mirror image of that in mammals. In mammals with XY males, the genital ridge heads off to become an ovary unless the noisy SRY speaks up to argue for a testis. In birds, the genital ridge heads off to become testes unless a militant counterpart of SRY on the W chromosome insists on an ovary.

Did chance create birds and mammals with mirror-image schemes for determining gonadal identity? Or are the opposite schemes somehow adaptive? Birds and mammals have fundamentally different social lives. In mammals, females carry and control the young, whereas in birds, both males and females control the young, who reside in a common nest.

Could these differences have worked their way back to the genome, determining how natural selection molds the rules for gonadal sexual identity? I don't know. My conjecture is that birds and mammals differ in whether their social systems require the male or female to develop a gendered presentation first. For mammals, intrasex competition for access to reproductive opportunity may be higher in males than females, and in birds the reverse. If so, the sex experiencing the higher intrasex competition may evolve the accelerated development of a gendered presentation.

THE COSTS OF GENETIC MYTHOLOGY

Development begins and ends with egg and sperm, one big gamete and one little gamete. Although this overall beginning and end point may be the same in many species, we see no standard templates for how female and male development are accomplished. How an animal's sex is decided, whether it will make eggs or sperm, varies among species. The decision is genetic in some species, physiological in others; even when it is genetic, various genetic criteria apply, depending on the species. And the individual's development is no more ordered or predictable than the outcome of a day's parliamentary debate. A diversity of people emerges from a cacophony of developmental histories. No one or two developmental narratives can be privileged as a standard against which to judge the rest.

Why is the bowling-a-strike metaphor, the cascade of successive downstream genetic decisions culminating in birth, so persistent in developmental biology, in spite of so much contrary evidence? I suggest that its persistence stems from a desire to own and control development. If a master gene produces some trait, then anyone who owns the patent for that gene controls the trait. But if traits emerge from a committee of genes, then what's to own? Buy out the whole committee? One can patent a gene, but not a relationship between genes. If the body emerges more from intergene relationships than from the individual genes themselves, then control of development moves beyond human ownership.

The material consequence of trying to own human development is evident in the money lost by biotechnology's efforts to patent the human genome. For the most part, the patent on a gene is worthless, because no gene works alone. My emphasis on genetic cooperation is not an ode to natural harmony. Rather, I argue that the present emphasis on individualism in science—from the selfish gene to the selfish organism—is empirically misleading, one result being that genetic engineering investors are wasting real money.

12
Sex Differences

Suppose now that a baby's sex has been determined—the baby's gene committee has met during early development and somehow managed to come to a decision. What next? How does development continue? Will all males graduate from uterine school wearing the same coat and tie, and all females the same dress? Obviously not. How much biological difference is there among males, and among females, and between males and females?

A motivation behind this chapter is that genetic and anatomical differences are increasingly being detected between gay and straight people, between transgendered and nontransgendered people, and between and among intersexed and nonintersexed people. These differences are publicized as anomalous deviations from the supposed norm set by straight males and females. But differences between gays and straights, or between any two groups, must be assessed relative to differences within the groups. If straight males, for example, show great biological variation among themselves, then the difference between a straight male and a gay male may be no more than that among straight males anyway, and therefore not biologically noteworthy.

Similarly, in education the unspoken assumption is that people are, for the most part, biologically the same, and that with instruction everyone can acquire the same skills and knowledge. Of course, everyone knows that people differ genetically from one another, but "normal" people are assumed to be more or less the same biologically, so that a one-size-fits-all approach to education can be applied, except for "special" cases. But what happens if we realize that people are as different biologically as they are culturally?

GENETIC DIFFERENCES

Research on the human genome is beginning to clarify how genetically different people are from one another. These differences can be divided into those that arise from the nonsex chromosomes and those that come from the sex chromosomes. Furthermore, males have their own ways of differing from other males and females from other females.

We each have about thirty thousand genes total. I estimate that each of us differs from the next person by about sixty genes, not including genes on the sex chromosomes, X and Y. Thus in this sense we're all very much alike: we differ by only sixty out of thirty thousand genes, or 0.2 percent. Yet this difference is enough to make for lots of biochemical variation among us because of the ripple effect of how these genes interact. Moreover, any two people differ from each other by a particular set of sixty genes that are different from the set by which two other people differ from each other. Thus people differ from one another in different ways.[1]

VARIETY FROM THE X CHROMOSOME

An X chromosome has about 1,500 genes, and two people differ from each other in about three of these. XX people express only one of their chromosomes, and XY people have only one to begin with. Thus people differ genetically from each other, on the average, by sixty genes on the nonsex chromosomes, plus three more from genes on whichever X chromosome is being expressed.[2]

Humans with XX chromosomes are typically women, and most differences among women are from the sixty genes that have nothing to do with sex. However, the additional three genes from X can lead to further differences that are unique to women. Males have an X chromosome too, so three genes of difference coming from the X chromosome could apply to men as well as to women. However, the phenomenon of X chromosome inactivation is unique to XX people and provides a way that women may differ genetically from one another that is unavailable to men.

Two women may differ in genetic expression because different Xs remain active in different cells. One woman may have X chromosomes from her dad active in her kidney cells, while her sister has X chromosomes from her mom active there instead. This variability in the way genes are expressed adds to the underlying variability in the genes themselves. If one of the three genes that differ between the two Xs happens to be harmful, then the body can prune some of the cells expressing the bad X and use cells with the good X instead. Males carrying the bad X suffer the most severe disease, whereas women carrying the bad X suffer only in those cells that haven't been pruned.[3] Furthermore, women can differ from each other in the severity of a disease, depending on how many of the cells with the bad X were successfully pruned.

The incidence of autoimmune disease is higher in women than in men.[4] I conjecture that susceptibility to autoimmune disease is a side effect of X inactivation and the ability to prune cells that express harmful genes. The immune system faces the challenge of detecting which cells are self and which are foreign, and removing the foreign ones. In XX bodies, two types of cells are self, depending on which X chromosome is active. Having two types of self cells may make the discrimination of foreign cells more difficult, leading to more autoimmune response in XX people than in XY people.

Women may also differ from one another because of which specific genes on the X remain active and which become inactive. The big-picture view of X inactivation is that one X chromosome is completely active while the other is all scrunched up in a ball. In fact, only 80 percent of the genes on the inactive chromosome are truly turned off, while 15 percent are still expressed. These 15 percent are said to escape inactivation. The remaining 5 percent are especially interesting: they are expressed from the inactive chromosome in some women and not others. And finally, one gene is known that is expressed only from the inactive chromosome and not from the active one, the reverse of the typical pattern.[5]

Thus the distinction between active and inactive does not apply to an entire X chromosome, but rather applies selectively to various parts of both X chromosomes. As a result, the cells within the body of an XX person can be quite heterogeneous, and women whose genes are similar can still differ a lot biologically because of genes being active or inactive.[6] All of these ways in which XX bodies differ from one another guarantee that women differ from each another in ways unavailable to men.

VARIETY FROM THE Y CHROMOSOME

Based on statistics for the other chromosomes, a Y chromosome potentially has about five hundred genes and is one-third the size of the X chromosome. Two people with Y chromosomes would be expected to differ from each other in about one of these genes.[7] Yet only about two dozen genes have been identified so far on Y, well below the estimated five hundred, leading some biologists to describe the Y chromosome as a "genetic wasteland," a "degraded relic."[8] These two dozen genes come in two functional clusters. One cluster contains genes for male versions of cellular biochemistry, such as the gene for the male form of ribosome mentioned in chapter 10. The other cluster consists of genes expressed only in the testes. They affect sperm development, and their absence leads to male sterility.

Humans with XY chromosomes are typically men. Most genetic differences among men come from the sixty genes that have nothing to do with sex, plus the three from the one X chromosome males have. XY people might conceivably also differ by an additional gene from their Y chromosomes. If such differences exist, they would provide ways in which men uniquely differ from one another that are unavailable to women.

However, the variation on the Y chromosome is low compared to the variation on other chromosomes. Most genes on the Y chromosome are bundled into a large unit called a linkage group. Except at tiny spots at both ends of the chromosome, the genes on Y don't pair to recombine with genes of any other chromosome—Y goes alone; it is haploid. Therefore, the genes on Y, except the few at the ends, all stand or fall together.

At any particular time, a single version of the Y chromosome is temporarily the best, and all others are quickly weeded out within any well-mixed local population. But when times change, what was once a weed comes rushing back, displacing the previously best Y. At any particular time, then, Y doesn't have much variation within a species, but can vary across species and over time. Some variation in Y is always waiting around, ready to mount a triumphant return, but there is not as much variation as on the other chromosomes. A possible implication is that XY bodies are more uniform than XX bodies.

Here SRY—a major gene affecting masculinity in mice and men, one of the gender genes—comes into play. If males differ in SRY, they differ in an influential gene for how male gender is embodied. And SRY is one of the fastest evolving of all known genes. This gold standard of masculinity differs greatly across species.[9] SRY is also variable across populations within a species, so the expression of masculinity is not constant from place to place within a species either.[10] Primates have undergone especially fast evolutionary change in SRY, implying that the embodiment of masculinity is not static but rather quickly changes over evolutionary time. This evolution is clearly caused by natural selection, not random genetic drift, because new DNA molecules replace old ones faster at sites where the difference affects SRY's protein than at sites where the substitution doesn't change the protein.[11]

The protein made by SRY consists of a central portion called the HMG-box. The portions to the left and right, the flanking regions, are called the N-terminal and the C-terminal regions. The HMG-box portion doesn't change much—this conserved part binds to the DNA and allows the SRY protein to affect how the DNA is translated. The evolutionary action is in the flanking regions, particularly the C-terminal region.

Variation in SRY causes variation along a masculine-feminine body continuum. Laboratory mice have three types of SRY genes, which cause different body types. In one type, XY bodies are male; in another, XY bodies are intersex as embryos and male as adults; and in a third, XY bodies are female or intersex as adults. The differences among these SRY genes turn out to be simply eleven, twelve, or thirteen repeats of a certain sequence, CAG, in the C-terminal region of the DNA.[12] Thus evolutionary changes in SRY outside of the HMG-box affect the gendered body.

Moreover, SRY can directly influence many parts of the body other than the gonads and the reproductive track. SRY indirectly influences the entire body because the testes whose differentiation SRY helps initiate secrete a hormone (AMH) that affects nearby cells, as well as testosterone, which affects distant cells. SRY probably influences many tissues directly, without bothering to use hormones as an intermediary. Depending on the species, SRY is expressed in tissue from bone to brain.[13] Quite possibly, SRY influences gendered bone growth and brain development.

SRY's impact on gendered embodiment is perhaps why it's evolving so quickly. If SRY is a gene for male gender, it would evolve in response to the turbulent winds of social change, endowing males with the latest body style and pickup lines for success in the "meet market."

GENETICS AND THE GENDER BINARY

As already mentioned, people differ from each other in sixty genes, on the average, from the nonsex chromosomes. They also differ in the genes provided by the sex chromosomes. Let's compare these sources of genetic variation among people. Because of the Y chromosome, XY people have two dozen genes not in XX people. Because of having two X chromosomes, XX people have about 225 genes (the 15 percent of the inactive X chromosome escaping inactivation) that are expressed in a double dosage relative to the single X chromosome of XY people. Considering both the presence or absence of Y and the presence or absence of the genes on X that escape inactivation, an XX person and an XY person differ by about 250 genes total from their sex chromosomes, or about four times their difference of 60 genes from the nonsex chromosomes. Thus a male and female differ from each other on the average by about four times as many genes as two males, or two females, do from each other.[14] These data suggest a clear-cut genetic binary distinction between males and females. In fact, though, this genetic system allows for a great deal of overlap at the whole-body level.

The XX/XY system of sex determination is widely believed to define a biological basis for a gender binary. Yet this system allows for both a sharp gender binary and great overlap between XX and XY bodies, as well as gender crossing. The details of what's actually on the X and Y chromosomes, and which tissues respond to the products of these genes, determine the degree of male/female difference at the whole-body level, as well as allowing for transgendered bodies.

The bodies of males and females in mammalian species may be very similar overall, even though they differ in gamete size and associated plumbing, or they may be very different. Compare guinea pigs, where females and males can't be distinguished without plopping their genitals under a magnifying glass, to lions and lionesses, who personify the distinction between masculinity and femininity in the popular imagination. Clearly, XX/XY sex determination doesn't dictate any fixed difference between male and female within a species.

How can XX bodies and XY bodies vary from being nearly identical in some species to strikingly dimorphic in others? First, look at Y. Natural selection can tune the duration and number of tissues in which gender genes like SRY are expressed. If SRY is expressed for a few hours only in the gonad, as in mice, then its impact is limited. If SRY is expressed for months and in many tissues, as in some marsupials, then dimorphism becomes widespread throughout the body. Natural selection can also tune how closely the genes governing cellular biochemistry on Y resemble their ancestral counterparts on X. The gene on Y that determines the male ribosome could be replaced by other alleles, either more or less similar to its ancestor on X, thereby affecting the gendered difference in how proteins are manufactured. Thus evolution on the Y chromosome can affect the average difference of males from females.

Next, look at X. The extent of X inactivation controls how many genes in XX bodies are expressed in different dosages than in XY bodies. If X inactivation is 100 percent, then both male and female bodies will see only one X chromosome, minimizing dimorphism. At the other extreme, if there is no X inactivation, XX bodies see all 1,500 genes in a double dose compared with XY bodies, potentially leading to huge body differences between the sexes. Natural selection can modify the percentage of genes that escape X inactivation, as evidenced by the 5 percent of the genes on X that are inactive in some people and active in others. This represents genetic variation in the extent of X inactivation, variation that can be acted on by natural selection. Natural selection can also tune how closely the dozen genes on X for cellular biochemistry resemble their counterparts on Y. The gene on X controlling the female ribosome could be replaced by other alleles either more or less similar to its descendant on Y, thereby affecting the gendered difference in how proteins are manufactured. Thus evolution on the X chromosome can affect the average difference of females from males.

Together, evolution on X and evolution on Y control the overall degree of sexual dimorphism—the sum of the X-determined differences of females from males and the Y-determined differences of males from females.[15] Further, variation on X can produce transgender expression. To obtain feminine males and masculine females, the X and Y chromosomes must first have genes that maintain enough average body difference between the sexes so that feminine and masculine are statistically well defined. Then, because X occurs differentially in males and females, allelic variation on X at loci that escape X inactivation can hypothetically cause some males to resemble females, and vice versa.

To obtain feminine males, imagine that five units of pigment produce pink, and ten units make red. Quantities of pigment above ten are also red because ten units are enough to saturate the color. Suppose X has a locus for pigment that escapes X inactivation with two alleles. One allele makes only five units of pigment and appears in 99 percent of the gene pool; another allele makes ten units of pigment, but is rare, at only 1 percent. All females, then, are red because, with any of the alleles from two Xs, they make either ten, fifteen, or twenty units of pigment, all of which appear simply as red. Among males, 99 percent are pink, because they have only the allele for five units, and 1 percent wind up red with the allele for ten units. These 1 percent of males appear as feminine males with respect to the trait of redness. To obtain masculine females, imagine now that twenty units of blue are automatically made in both males and females, and that X has a locus for degradation of the pigment that escapes X inactivation. The pigment saturates at navy blue with ten units. Suppose one allele degrades none of the pigment but is rare at 1 percent of the gene pool, whereas the other degrades 50 percent of the pigment and is common at 99 percent. All males are then navy blue because they have either twenty or ten units of pigment. Among the females, 99 percent are light blue, as they have only five units of pigment, because one of their alleles degrades half of the original twenty units and the other the remaining ten. The remaining 1 percent are navy blue because they have either ten or twenty units of pigment. This 1 percent of females appear as masculine females with respect to the trait of blueness.

Although the genetics of transgender expression are unknown and, in humans at least, may be superseded by late embryonic and early postnatal developmental experience, transgender bodies are fully consistent with an XY system of sex determination. Indeed, feminine males might easily be feminine enough, and masculine females masculine enough, to count socially and to identify as women and men, respectively, even though they possess XY and XX bodies.

Thus genetics does not dictate a gender binary. Although the mammalian system of sex chromosomes produces a binary based on gamete size, the gendered bodies that make those eggs and sperm are not constrained by the genetics of sex determination; they are free to adapt evolutionarily to local context. Indeed, research on the human genome is revealing that all people are genetically different. Individuality is not skin-deep, but extends deep into our DNA. When any two people are compared, genetic differences can be found. And if people sort themselves into social categories that reflect innate inclinations, then genetic differences will also be found between the people of those categories. "Normal" people are not a sea of homogeneous genotypes, bodies, and brains. "Normal" people are as genetically diverse as snowflakes.

HORMONAL SEX DIFFERENCES

A major source of diversity in the developmental narratives of people is their differing hormonal experiences. Yet the story of hormones begins not with diversity, but with failed attempts to define a binary "male" and "female" in terms of chemicals. An authoritative summary in 1939 asserted, "As there are two sets of sex characters, so there are two sex hormones, the male hormone . . . and the female."[16] But was this any more than a wish? In 1927 female hormone was extracted from pregnant women's urine, and in 1931 male hormone from men's urine. Already in 1928, however, reports of female hormones in men elicited scientific rejoinders of "disconcerting," "anomalous," and "somewhat disquieting." In 1935 a cherished symbol of male virility, the stallion, turned out to have large amounts of female hormone in his urine, eliciting remarks such as "surprising," "curious," "unexpected," and "paradoxical." Conversely, the male hormone was shown to operate on female tissue. Testes transplanted into female rabbits whose ovaries had been removed induced uterine growth. In the 1930s the male hormone was shown to increase mammary glands, enlarge the uterus and the clitoris, and prolong estrus in female rats.[17] So both sexes possessed and responded to "male" and "female" hormones.

In fact, everyone possesses testosterone, estrogen, and all the other "sex" hormones. Sex hormones are instruments in the body's chemical orchestra. The body's score calls for all these instruments at various times, and together they help make the body's music.

Once isolated, the sex hormones turned out to be chemically similar to one another—as close as, say, the sugars in honey and sugarcane. Humans, both males and females, can synthesize all the sex hormones starting from cholesterol. The recipe goes through two steps to progesterone, then three more to testosterone, the hormone that signals an embryo to make male internal plumbing plus male secondary sex characteristics at puberty. An additional step leads to dihydrotesterone, the hormone that shapes the male external genitals. A different step converts testosterone to estradiol, which interconverts with estrogen to produce female secondary sex characteristics at puberty, bone growth, and masculine brain anatomy.[18] These hormones, called steroid hormones, are widespread throughout the vertebrates. Each person has different amounts of each hormone, forming part of our biochemical individuality, but everyone has at least some of every type of sex hormone.

Making hormones is half the story; the other half is whether cells have receptors for them. All the hormones in the world will have no effect unless cells contain certain substances that chemically bind to the hormone. The overall impact of a hormone depends on both how much has been made and how much receptor is present to respond. Thus, the committee of genes that composes the body's sex hormone symphony includes gender genes like SRY, genes for sex hormone receptors, and genes for the many enzymes that catalyze sex hormone synthesis and interconversion. Quite a large committee.

UTERINE ENVIRONMENT

At birth, a baby is chemically experienced. While still in the uterus, an embryo makes hormones both in the gonads and in the adrenal glands. The placenta, a structure jointly made by baby and mother, is also a site of hormone synthesis, and the mother contributes some hormones of her own. All these hormones irreversibly influence behaviors in later life.

In species that give birth in litters—producing a group of fraternal "twins"—brothers and sisters influence each other's development because of the shared effect of their hormones.[19] Certain behavioral inclinations of rodents—such as mice and rats, for example—have been measured in the laboratory. A male mouse was offered a choice of two females, one who had lived in the uterus next to two sisters (a two-sister female) and one who had lived next to two brothers (a two-brother female). The male mated twice as often with the two-sister female as with the two-brother female. Male rats who developed next to two sisters in the uterus (a two-sister male) had a higher sexual appetite than male rats who developed next to two brothers (a two-brother male). When paired with a receptive female, the two-sister males mated and came to climax more often than the two-brother males. In mice, the two-sister males had a higher appetite for same-sex matings as well. When two-brother males and two-sister males were paired with a reference male, the reference male mated more with the two-sister males than with the two-brother males. Lots of differences that can be detected in mice and rats point to an effect of their embryonic hormonal environment on temperament later in life.

The data for humans are scantier, in part because humans typically give birth singly rather than in litters. One behavioral trait has been studied, however: a peculiar trait called inner-ear clicking. Believe it or not, clicking sounds are generated inside the ear. These sounds travel out of the ear (instead of into the ear like most sounds) and can be recorded with a microphone. People don't hear the sounds made in their own ears—they get used to them. These clicking sounds are made more often by women than by men. A female with a twin brother, however, doesn't produce these sounds. Apparently, the brother's hormones masculinize his sister's ear development, as indicated by the absence of click production.[20] Lesbian and bisexual women also produce fewer clicking sounds than straight women (see p. 246).

Another clue that hormones from twins influence each other's later development comes from asymmetries in the teeth. Teeth are generally more asymmetrical in men (the right jaw has larger teeth than the left jaw) than in women. A woman with a twin brother has a more asymmetrical jaw than other women.[21]

MATURATION

Hormones have long been known to have a large effect on morphology, during embryonic growth as well as in puberty. In males, external body changes begin while the person is still embryonic. At three months, a typical penis might be 0.3 centimeters in length, growing 0.7 centimeters per week until birth, when it reaches 3.5 centimeters. Penis growth is caused by dihydrotestosterone, converted from testosterone circulating in the blood. While the penis is growing, the vaginal pouch is reabsorbed, although some men still retain a small pouch called a prostatic utricle.[22] Puberty starts with growth of the testes at about eleven years, just a few months after the first signs of puberty in females.

In females, the average age at which breasts start to grow is 10.6 years for girls of European descent and 9.5 years for girls of African descent, with a range from 6 to 13 years. The first menstrual period begins about two or three years later, at 12.9 years in women of European descent and 12.2 years in women of African descent.[23] Estrogen secreted by the ovaries causes breast growth. At the same time, testosterone secreted by the adrenal glands and the ovaries causes the pubic hair to grow.

Before puberty, boys and girls grow at about the same rate. At puberty, boys wait two years later than girls for their spurt, winding up about 12.5 centimeters taller than girls on the average. Their greater height results from starting their spurt at a taller height and having a faster maximum growth speed during the spurt. The growth spurt in both girls and boys results from estradiol. One of the "female" sex hormones, estradiol has long been known to produce the growth spurt in girls, but it also causes the spurt in boys. Boys' testosterone from the testes is converted to estradiol in the bones, where the growth occurs. Indeed, testosterone often has its effect only after conversion in local tissues to estrogen and/or estradiol.[24]

HOW HORMONES MAKE US FEEL

In 1889 the physiologist Charles Edouard Brown-Séquard injected himself with extracts from crushed animal testicles and claimed renewed vigor and greater mental clarity. A decade later, he admitted that the effects were short-lived and he couldn't rule out that he had been mistaken all along.[25]

In what I first assumed was a spoof, around April Fool's Day in 2000, the *New York Times Magazine* printed an homage to testosterone.[26] The author, a man taking testosterone as part of HIV therapy, offers this description: "It has a slightly golden hue, suspended in an oily substance and injected in a needle about half as thick as a telephone wire. . . . I push the needle in . . . [and] as I pull it out . . . an odd mix of oil and blackish blood usually trickles down my hip." "Within hours," he declares, "my wit is quicker, my mind faster, but my judgment more impulsive." A transgendered man the author interviewed adds, "My sex-drive went through the roof. I felt like I had to have sex once a day or I would die,"[27] and a forty-year-old executive taking testosterone for body-building gushes, "I walk into a business meeting now and I just exude self-confidence." The author credits the big T for increasing his weight from 165 to 185 pounds, his collar size from 15 to 17.5 inches, and his chest from 40 to 44 inches.

The article continues, "Men and women differ biologically mainly because men produce 10 to 20 times as much testosterone as most women do and this chemical, no one seriously disputes, profoundly affects physique, behavior, mood and self-understanding. . . . It helps explain . . . why inequalities between men and women remain so frustratingly resilient in public and private life." This claim is misleading. Testosterone doesn't stand alone; by itself testosterone doesn't do anything and needs receptors to have any effect.

The author declares that affirmative action for women is impossible, and we "shouldn't be shocked if gender inequality endures" because of the hormone differences between men and women. Instead, the "medical option" is to give "women access to testosterone to improve their sex drives, aggression and risk affinity and to help redress their disadvantages." So, to cure women of their womanhood, testosterone should be administered, although "its use needs to be carefully monitored because it can have side effects . . . but that's what doctors are there for." Rectifying social injustice by giving women testosterone to convert them into men is, shall we say, inadvisable.

Why would someone write such an irresponsible article? An answer is suggested in the concluding sentence: "It seems to me no disrespect to womanhood to say that I am perfectly happy to be a man, to feel things no woman will ever feel . . . to experience the world in a way no woman ever has. And to do so without apology or shame." Male-male posturing.

The article details a stereotypical view of how testosterone affects behavior: "I feel a deep surge of energy. It is less edgy than a double espresso, but just as powerful. My attention span shortens, . . . I find it harder to concentrate on writing and feel the need to exercise more. . . . Lust is a chemical. It comes; it goes. It waxes; it wanes. You are not helpless in front of it, but you are certainly not fully in control. Then there's anger . . . mere hours after a T shot . . . I had nearly gotten into the first public brawl of my life." The article seems only dimly aware that it subverts the value of manhood, even as manhood is being championed. Men are portrayed as irrational creatures, ricocheting from one impetuous mistake to another, as testosterone propels their quest for sex. Women have long borne the brunt of criticism as irrational creatures, victims of a monthly hormone cycle, monsters on "bad hormone days." Men apparently have bad hormone days every day of their lives, suffering mental cramps, not menstrual cramps.

Transgendered people have much to contribute on how hormones "feel." Transgendered people tell of great variability in hormone sensation, probably reflecting differences in hormone receptors as much as hormone production. Perhaps most interesting is the seemingly unanimous report from transgendered men that testosterone calms them. The *New York Times Magazine* article provides only a partial picture of the transgendered man quoted as saying testosterone gave him an enhanced libido. In fact, that man, Drew Seidman, also stated that it is a "myth" that testosterone is a cause of undue aggression, that testosterone has a "calming effect" on him, and that he "feels much better" with it.

Similarly, Patrick Califia, a prominent transgendered writer who transitioned recently, said in an interview, "I've been much more comfortable on T. I feel like a calmer and more reasonable human being. Men are supposed to be more angry, but I just keep getting more mellow and loving and sweet, and I think it's because I'm happier. This chemical balance just feels right."[28] The distinguished transgendered leader Jamison Green writes, "The initial effect of testosterone was that it allowed me to feel 'normal' for the first time in my life. It allowed me to feel calm, balanced, centered, the absolute antithesis of the clichés about . . . testosterone poisoning. And once I got comfortable with that feeling . . . along came libido."[29] I have spoken with other transgendered men about testosterone. They all confirm the reports of Seidman, Califia, and Green that testosterone has been calming. All also confirm a large increase in libido and speak of how they've accommodated this new, happy sensation into their lives.

What transgendered women say about testosterone differs markedly from what transgendered men say. Concerning the male libido, transgendered women talk about freedom from the burning in their groin, from the never-ending need for relief; they talk about pacing in romance, not wham-bam. For me, testosterone was a triple-espresso buzz, razzing, annoying, clouding thought, taking me where I didn't want to go. When I replaced testosterone with estradiol, within a day I felt a deep calm and happiness.

Transgendered people speak of hormones as the most important step in gender change, tipping the balance of subtle signs connoting gender identity as a man or woman. Transgendered people tell too of losing their partner or spouse soon after starting hormones because intimate relations were fundamentally altered.

A major reason for individuality, for the emergence of diversity in body and temperament, is the effect of hormones and their receptors. Hormones early in life cause irreversible effects on temperament later in life, and hormones later in life can reversibly affect mood and activity.

MENTAL SEX DIFFERENCES

Our quest for the developmental sources of human diversity leads now to the brain, the most mysterious of all our organs. Here lies the circuitry that activates sex drive, hunger, temperament. Here, too, lies creative spirit, free will, love, humor. Somehow our personhood emerges from the substance of our brains. Clues about what kind of people we are reside in brain size and shape, in its pattern of electrical discharges.

Brain and behavior work together in a back and forth. Just as weight-lifting strengthens the biceps, and big biceps then allow heavier weights to be lifted, parts of the brain shrink and expand with use, especially early in life. People's biceps start at different sizes, before any weight-lifting. Likewise, brains differ at birth, reflecting an inherent disposition to different behaviors. How, then, are our brains involved in our sex lives, in the disposition we have to express gender and sexuality?

The brain listens to sights and sounds from outside, as well as to the music of the hormones within the body. The brain secretes hormones too, playing in the hormonal orchestra—it does its listening as a performer in the orchestra pit, not as a spectator in the audience. The brain "hears" the body's hormones using receptors located in the preoptic area of the hypothalamus, running from the back of the brain, near the spinal cord, along its bottom, to the front near the eyes. The brain also listens directly to genes, such as gender genes like SRY in the male, without going through hormones as intermediaries.[30]

BIRD BRAIN ANATOMY

Biologists who study brain anatomy are used to looking for fine details, differences between a few cells here and there. In 1976 brain anatomists were amazed by what they found. It was known that while male canaries and zebra finches sing, female canaries sing only a little, and female zebra finches don't sing at all. It was also known that in both species males learn their song from listening to other males.[31] The surprise was that the brains of the males and the females in these species are so different that they can be told apart with the naked eye.[32] Place the brains from a male and a female zebra finch next to one another, and with practice, you can tell their sex just by looking. In the upper part of his brain, a male bird has extra nerve cells, which occur in clusters containing extra hormone receptors as well—hormone receptors in addition to those along the base of the brain.[33] These upper-brain nerve cells enable the male bird's singing.

Although the neurobiology of avian brains isn't directly comparable to that of mammalian brains, avian brains set valuable biological precedents.[34] Here's a list:

1. The brains of males and females can differ, and differ substantially, as in canaries and zebra finches.

2. The degree of difference between male and female brains correlates with the degree of difference in their behavior. In species where males sing and females don't, the difference in the size of nerve-cell clusters responsible for learning and producing bird song is most marked. In dueting species, where males and females sing to each other at courtship with interlocking songs, the nerve-cell clusters that control song are the same size in both sexes.[35] A survey of twenty songbird species spanning six families shows that the degree of dimorphism in brain anatomy correlates with the degree of dimorphism in both variety and quantity of song.[36]

3. Testosterone organizes the brain of newly hatched male chicks to develop song-control clusters of nerve cells. These clusters don't develop in the absence of testosterone, and they can be caused to form in females if testosterone is administered.[37]

4. Hormones activate nerve-cell clusters in adults, expanding and shrinking the size of the clusters to match the breeding season.[38]

5. The brain is masculinized by estrogen that has been converted from testosterone by the enzyme aromatase. Because some aspects of the male body are masculinized directly by testosterone, other parts by estrogen that has been converted from testosterone, and still other parts by direct expression of genes without any involvement of hormones, reconstructing the developmental pathways of sexual differentiation is complicated.[39]

6. Personality differences can be traced to brain differences. Reproductively active male Japanese quail vary in aggressiveness by fourteenfold. This variation relates to the amount of aromatase in the hypothalamus of their brains—the more aromatase, the more aggressiveness.[40]

7. Female parents influence the temperament of their chicks by introducing estradiol or testosterone into the egg yolk. Birds given more testosterone in their yolk are more aggressive. A female bird deposits increasingly more testosterone in the eggs as the egg-laying season progresses, so the birds hatched last in the nest are the most aggressive, presumably to defend themselves against their older siblings. This effect is comparable to the role of maternal hormones in the developing mammal fetus.[41]

8. The direction of partner choice for adult female zebra finches shifts from between-sex to same-sex after estradiol is

administered to hatchlings. Estradiol masculinizes the brain and changes the direction of sexual preference.[42]

9. The white-throated sparrow has a territorial white-striped morph and a nest-tending tan-striped morph in both sexes (see chapter 6). This reciprocal transgender expression in body color and behavior extends to the brain too. Within a sex, the song-control clusters of neurons are larger in the territorial morph than in the nest-tending morph. Between sexes, a male nest-tending morph has larger song-control neural clusters than a female territorial morph, even though both sing about the same amount.[43]

Transgender gender identity apparently has not been investigated. In birds like canaries, males learn their song from male "tutors," often their fathers. How does a male chick know to listen to his father instead of his mother? Female canaries sing a different song from males. I wonder if an occasional male chick learns his mother's song, and an occasional female chick learns her father's song.

Birds have reverse chromosomal sex determination relative to mammals (see p. 204). In birds, estrogen from the ovaries feminizes a body that would otherwise be masculine, whereas in mammals, testosterone from the testes masculinizes a body that would otherwise be feminine. Although hormones work somewhat differently in birds than in mammals, they bring about the same overall result in gender and sexuality.

The connection between brain structure and behavior may be more direct in birds than in mammals, as though birds relied more on instinct and less on thought than mammals. However, I'm struck by how clear-cut the data are linking differences in bird brain structure to differences in gender presentation, personality, sexual orientation, and transgender expression. I wonder to what extent something similar happens in mammals.

RODENT BRAIN ANATOMY

Like birds, mammals have some sexual dimorphism in brain anatomy. Here's a list of brain dimorphisms in rodents:

1. Male rodents, including gerbils, guinea pigs, ferrets, and rats, have a larger cluster of cells in the preoptic area of the brain,

along the base of the brain toward the front. The clusters are about five times larger in males than females, and the difference can be seen with the naked eye.[44] Testosterone near the time of birth, converted to estrogen in the brain, organizes this difference. Testosterone given later in life doesn't cause these cell clusters to form.[45] Nobody knows what this cluster of cells does in a male. If they are removed surgically, little effect on behavior is noticed. However, if the whole preoptic area is removed, male copulation is affected, so the clusters may have something to do with male mating behavior.[46]

2. A nearby cluster of cells, called the bed nucleus of the stria terminalis, also shows a sexual dimorphism that is controlled by testosterone near the time of birth.[47] This cluster is of interest because recent work on transgender identity in humans has focused on the human equivalent of this structure.

3. A cluster of nerve cells in the spinal cord in the lower back also differs between males and females. In males, these nerve cells control muscles in the base of the penis. These cells and muscles are present in newborn animals of both sexes. In males testosterone prevents this muscle and the nerve cells that control it from shrinking, whereas in females testosterone causes the muscles and nerve cells to shrink. Testosterone produces this effect directly, without needing to be converted into estrogen.[48]

When pregnant female rats are stressed in the laboratory by shining bright lights on them all day long, the male embryos in the litter produce less testosterone during their fetal period. They wind up with smaller clusters of nerve cells in the preoptic area and fewer nerve cells in the spinal cord for control of the penile muscle.[49]

The scent of testosterone in a male offspring induces his mother to groom his genital region more often than the genital region of a female offspring. If the mother isn't able to smell, then she doesn't groom either sex very much. A young male offspring who isn't groomed winds up with fewer penile-muscle-controlling nerve cells in his spinal cord and takes more time to copulate than a male who is groomed.[50]

Thus, in mammals too, brain and spinal cord anatomy can differ between the sexes, and these differences partly reflect the hormonal and social environment in which the animals develop. Again, temperament and inclination originate near the time of birth.

HUMAN BRAIN ANATOMY

Human brains show few sex differences compared with other species. Lots of research has been directed to showing brain differences between males and females, and many small differences have been found. Overall, though, male and female human brain anatomy is very similar—the big story here is the overlap between the sexes, not their difference. Picture two bell-shaped curves placed on top of one another. Then *gently* nudge one to the right and the other to the left, so the curves are just slightly askew. That's how close together the two sexes' brain anatomy is.[51] Here is a summary of the small differences that have been found:

1. Males have a somewhat bigger total brain size, 120 to 160 grams in adults. The difference is almost absent at birth, becoming more pronounced during puberty, partly reflecting the overall size differential at that age. Even at birth, though, a group of newborns who weighed the same showed a 5 percent difference in brain weight by sex.[52] This 5 percent, although statistically valid, is tiny compared to the overall variation in brain size.[53]

2. Both male and female humans have the counterpart of the penis-controlling muscle found in mice. In human males, the muscle wraps around the base of the penis to aid in the ejaculation of semen. In human females, the muscle encircles the opening of the vagina and can constrict its entrance. The muscle is somewhat larger in males, although the male/female difference is not as great as in mice. Males have about 25 percent more nerve cells to control this muscle than females. As in mice, the cells are located in the spinal cord. As in mice, females and males have the same-size muscle and same amount of controlling neurons at birth, but testosterone directly causes the female muscle and neurons to shrink and die back.[54]

3. The preoptic area of the hypothalamus in humans features a cluster of cells called the SDN-POA (the sexually dimorphic nucleus of the preoptic area), which is the counterpart to that in mice. Both males and females are born with about 5,000 cells in

this cluster; both increase to about 50,000 by age four, and then a dimorphism develops as the number of cells in this cluster declines in females to about 25,000 by age twenty. This approximately twofold difference in cell number persists through adulthood. The function of SDN-POA remains unknown, although it is presumed somehow to influence mating behavior. SDN-POA is minute, a cluster of nerve cells the size of a grain of rice in a quart-sized brain.[55]

4. Another minute cluster of nerve cells that has been publicized in relation to gender and sexuality is found slightly above SDN-POA. It is called BSTc (the central subdivision of the bed nucleus of the stria terminalis). As in rodents, men have a larger BSTc than women, about 2.5 cubic millimeters with 35,0000 cells in men, and 1.75 cubic millimeters with 20,0000 cells in women.[56] (See also chapter 13, concerning transgender gender identity.) BSTc is part of a region called the septum, which is involved in sexual function. Electrical discharges in this part of the brain occur during orgasm, and electrically stimulating this part of the brain causes orgasm.[57] Altered sexual behavior, such as hypersexuality, as well as change of sexual orientation and fetishism, results from damage to the septum,[58] suggesting that natural variation in sexuality may be associated with corresponding variations in particular areas of the brain. Moreover, hypersexuality raises the possibility that typical human sexuality is not as intense as is biologically possible, that through evolution our sexuality has been set at some optimum intermediate level. The septum provides a natural veil of modesty covering a potential for increased sexuality.

5. Slightly below SDN-POA is a third, even smaller rice-grain of nerve cells, a 0.25 cubic millimeter cluster called VIP-SCN (vasoactive intestinal polypeptide containing subnucleus of the suprachiasmatic nucleus). After about ten years of age, a sexual dimorphism can be detected, wherein males have about 2,500 cells and females about 1,000 cells in this cluster. (See also chapter 14, concerning sexual orientation.)[59] But just how important could a difference in a teeny neuron cluster be? We don't know yet. They seem too tiny to account for much of the behavioral differences in gender and sexuality, but then even the small bite of a black widow spider or yellow-jacket wasp can pack quite a wallop. Let's keep an open mind (so to speak).

6. The human brain shows some right-to-left specialization, especially in right-handed males. Males process verbal information faster and more accurately on the left side, and spatial information on the right side.[60] Females don't show such a pronounced asymmetry. Strokes also reveal a male/female difference: females recover better overall, whereas the impact for males can be predicted by knowing which side of the brain the stroke is on.[61] The differences in brain symmetry have a slight anatomical basis.[62] The corpus callosum is a conduit of nerve cells that bridges the two sides of the brain. Males and females differ slightly in the shape, but not size or number, of neurons in their corpus callosum. The corpus callosum may be positioned slightly more toward the back in females than in males.[63]

7. Moving up from the base of the brain into the cerebral cortex at the top of the brain, dissections of brains from six males and five females show that males have more nerve cells with fewer connections among them, whereas females have fewer cells but more connections among them. Both males and females have the same overall amount of brain material. Males have about 115,000 ± 30,000 neurons per cubic millimeter, whereas females have about 100,000 ± 25,000 neurons per cubic millimeter. Moreover, males are more asymmetric between right and left sides of the brain. Males have an average of 1.18 more neurons per cubic millimeter on their right side compared with the left side, and females have an average of 1.13 more neurons per cubic millimeter on the right compared to left.[64] These differences between the brains of males and females follow the familiar script: small statistically valid differences in the averages, with a large overlap.

When the brain first forms, more nerve cells are produced than needed. The cells are pruned through a process of programmed cell death, called apoptosis. Testosterone slows the pruning. Females wind up with fewer but more selected neurons than males do at the end of this process, which occurs during the last ten weeks before birth.[65]

Although certain mental functions can be pinned down to specific locations in the brain, more general-purpose cognitive processes emerge from the collective activity of many neurons distributed throughout the brain.[66] Do human males and females differ in these general-purpose mental abilities too?

MALE AND FEMALE THINKING

Men and women differ in cognitive abilities and aptitudes, much as it might seem inflammatory to say so. Overall intelligence, whatever that is, doesn't differ between men and women, regardless of what you hear each gender saying about the other in moments of mutual incomprehension, but some specific mental skills do differ. As with brain anatomy, the differences are small, but statistically detectable nonetheless, and show great overlap.

Women test better than men, on the average, in verbal fluency, articulation, and memory. Fluency is measured in tests such as trying to think of all the words that begin with a specific letter (for example, every word that starts with "t," or words that rhyme with "mind," or all the words that pertain to some subject or category, like "ice cream"). Women can usually say a tongue twister, such as "Sweet Susie swept seashells," faster than men. Women can also more quickly scan an array of symbols or figures and remember which one matches a previous symbol or figure. This memory advantage applies to visual and spatial information as well as to letters and words.[67]

The enhanced aptitude of females for verbal fluency may result in part from estrogen. One clue comes from the female gorilla Koko, who was trained to communicate in American Sign Language. Both the number of different signs and the total number of signs she gave increased during the part of her monthly cycle when her estrogen level was highest. This effect of estrogen is temporary and does not affect brain structure (activational, not organizational).[68] In humans, too, performance on tests for articulatory skill improves, and performance for spatial ability declines, at the preovulatory phase, the time of highest estrogen, compared to the intervening time of lowest estrogen.[69] Postmenopausal women on estrogen hormone-replacement therapy show cognition benefits.[70] Young women may also receive cognitive benefits from estrogen. Dyslexia, a reading disability, was originally thought to reflect a visual difficulty. Instead, dyslexia results from not discerning the components of words correctly. About an equal number of boys and girls are born with dyslexia. Of those who develop the ability to compensate, 72 percent are girls, resulting in more adult males with the disability than adult females. Apparently the estrogen available to girls at puberty leads to their developing verbal abilities that permit them to compensate for the dyslexia.[71]

Men test better than women, on the average, at visualizing how to rotate a shape or object in two-dimensional or three-dimensional space.[72] This ability to visualize spatial relations shows up especially in tests of quantitative aptitude such as the SATs. The gender gap on this test progressively widens toward the high end. Boys outscore girls 2:1 at scores of 500 and above, 5:1 at scores of 600 and above, and 17:1 at scores of 700 and above. High scores on this test are particularly sensitive to performance in spatial relations.

Beyond these details, if you've sometimes felt that men and women just *think* differently, there is some hard evidence to back up your feeling. Functional magnetic resonance imaging (fMRI) now makes it possible to take a picture of the brain while it's thinking in real time. The picture lights up the places in the brain where thinking is going on. In one study a group of men and women were given the same verbal task: they were given a written list of nonsense words, like "lete" and "jete," and asked to say them out loud, making them decide if the words rhymed. As they thought about how to pronounce these words, their brains were photographed using fMRI. The results were astounding. For the same task, the men and the women used their brains differently. Men relied on only one part of their brain (the left inferior frontal gyrus), whereas women used two parts of their brain (both left and right inferior frontal gryi). This shows that female brains function more symmetrically than male brains, even though this claim has been difficult to demonstrate anatomically. The photographs of this result are dramatic: they show different parts of the brain lighting up.[73] These differences don't pertain to tiny clusters of nerve cells, but rather to large regions of the brain. An anatomical underpinning to these functional differences between how men and women think hasn't been discovered.

Men and women also think about spatial tasks in different ways. A group of men and women were asked to find their way out of a three-dimensional virtual-reality maze. To traverse the maze, men used one part of their brain (left hippocampus) and women two parts (right parietal cortex and right prefrontal cortex). This difference in where think-ing goes on may correlate with the observation that women rely on landmark cues rather than geometry for navigation.[74]

What are we to make of these mental differences between men and women? Such differences can be amplified by social convention. If the ratio of men to women who are excellent at an occupation requiring spa-tial rotation abilities is 60:40, then the occupation may acquire a mas-culine character, which discourages women from joining. If the ratio of men to women who are excellent at an occupation requiring verbal flu-ency is 40:60, then the occupation may acquire a feminine character, which dissuades men from joining. The social character acquired by an occupation may lead to the belief that an occupation is a "man's job" or "woman's work," far outweighing differences in native skill.

Each of us can compare ourselves to the averages for human males or females and find we don't completely match. Almost everyone seems to cross over the statistical norms for their sex in some way or another. The average values for the sexes don't have much meaning for us as individ-ual people: it's like saying the average American lives in Kansas, which doesn't apply to most of us.

The combination of average differences and a great overlap in men-tal characteristics sets the stage for many kinds of transgender expres-sion. The reality of differences between male and female averages means the statement that A is a male trait and B is a female trait is statistically valid. But the great overlap also means that many males will have B and many females will have A. Transgender presentations necessarily occur in all sorts of dimensions simply because everywhere the averages differ the overlap is also huge. The same can be said of gender identity.

Why should men have more ability at spatial relations than women, and women be more verbally fluent than men? Spatial rotation skills might help in throwing a spear or evading an attacker, as evidenced by the male high-risk life history (see pp. 236–37). Spatial skills are also needed to build structures to contain women, enabling mate guarding, and are used in constructing weapons too. Conversely, ver-bal ability is fundamental to teaching children, to nursing the mind. The higher degree of interconnections among neurons in women's brains may permit performing more simultaneous activities. Such con-jectures raise the major issue of human evolution—why we have the brains we do—because the human brain is perhaps the single trait that defines our species.

HUMAN BRAIN EVOLUTION

Our brain size has swelled by a factor of three in the last 2.5 million years. That's fast. Why? Evolutionary psychologists have developed a tortuous theory based on sexual selection to explain the brain's evolution. Their theoretical machinations illustrate how they have become intellectually addicted to sexual selection theory. We can't begin to account for something as basic as the human brain's evolution until we "just say no" to sexual selection.

The traditional explanation is that our brains evolved as we developed technology to solve ever more complex problems, allowing us to become tool-using animals. This view is seriously challenged by the observation that technological innovation was at a standstill during our brain's evolutionary expansion.[75] Only after all the evolutionary action was over did any cumulative tradition of technological progress emerge. Only then did any global migration take place from the tropics into colder climates, or any population spurt occur. Natural selection can't look ahead millions of years and produce a brain in the hope that it will be valuable in the future. How did natural selection propel the brain's evolution forward if it was only intended for future use?

Instead, the evolutionary psychologist Geoffrey Miller suggests that our brains serve to create "the more ornamental and enjoyable aspects of human culture: art, music, sports, drama, comedy, and political ideals."[76] His theory goes on to state that cultural products somehow promote finding mates, whereas tool use mostly promotes survival. Perhaps the brain evolved primarily to aid in reproduction rather than survival. As Miller explains, "The human mind and the peacock's tail may serve similar biological functions. . . . The peacock's tail evolved because peahens preferred larger, more colorful tails. . . . The peacock's tail evolved to attract peahens. . . . The mind's most impressive abilities are like the peacock's tail: they are courtship tools, evolved to attract and entertain sexual partners. . . . The mind evolved by moonlight . . . as an entertainment system to attract sexual partners." According to this theory, intelligence signifies the "great genes" that men supposedly have and women supposedly seek.

But wait a minute. Sexual selection theory only applies if the attracting trait is a male ornament that is not also possessed by females, and it must actually be preferred by females in heterosexual courtship. This brain-as-a-peacock's-tail theory is incorrect because men and women have nearly identical brains. In peacocks, only the male has large tail feathers; peahens don't bother with such decor. If the human brain were just a man's tail feather, then women wouldn't bother to develop a similar brain.

Miller has therefore suggested several modifications to the sexual selection theory to account for brain evolution. One modification postulates that both men and women use their brain to advertise the absence of bad genes rather than the presence of great genes: "Any deviation from the genetic norm is a deviation from optimality." According to this theory, if you're smart and witty, you're also healthy. "The human mind's most distinctive capacities evolved through sexual selection as fitness indicators. . . . The healthy brain theory suggests that our brains are different from those of other apes because . . . the more complicated the brain, the easier it is to mess up. The human brain's great complexity makes it vulnerable to impairment through mutations, and its great size makes it physiologically costly. . . . Our creative intelligence could have evolved . . . to reveal our mutations."

This modification is fatally flawed from the onset by its assumption that variation from the norm is suboptimal. Sexual reproduction exists to maintain genetic variation. This theory of brain evolution contradicts itself by postulating that the purpose of mate choice is to eliminate the very variation that sexual reproduction is there to promote. This flawed theory is diversity-repressing.

Yet another modification notes, "It takes a sense of humor to recognize a sense of humor. Without intelligence, it is hard to appreciate another person's intelligence." According to this view, women have brains in order to admire the brains of men. But as Miller finally acknowledges, "I do not think that female creative intelligence . . . arose simply as a way of assessing male courtship displays."

So we're back to where we started. Why our brains have evolved remains as mysterious today as in the past. To move forward, let's analyze where evolutionary psychology has gone wrong:

1. Evolutionary psychology overemphasizes the amount of cultural production that goes directly into mate choice. A peacock is believed to show his tail feathers to a peahen during courtship—this is why Darwin's theory seems plausible, even if not demonstrated, for the special case of peacock tails. Except for the occasional love poem, few cultural expressions seem intended for one-to-one heterosexual courtship in the way a peacock's display is claimed to be.

2. Evolutionary psychology accepts biological sexual selection theory too enthusiastically and uncritically. Sexual selection theory is an elite male heterosexist narrative projected onto animals. Basing a theory of human behavior on sexual selection theory naturalizes this narrative and transfers the narrative back to people, as though it were a theory of human nature.

3. Evolutionary psychology needs a deeper conceptualization of female perspective. Females are viewed as what males think females ought to be.

4. Evolutionary psychology is in denial about same-sex sexuality. Miller claims that "homosexual behavior is just not very important in evolution. . . . Its existence in 1 or 2 percent of modern humans is a genuine evolutionary enigma."[77] Homosexuality is a valid color in the human gender/sexuality rainbow. It needs explanation, not dismissal. Imagine if the theory of light had ignored some of the rainbow's colors—we wouldn't have both RGB color monitors and CMY color photographs.

I suggest the human brain is a social-inclusionary trait for membership in the community of humans. People require the modes of interaction that the human brain supports in order to be included in human society and to have access to the chance to reproduce and to survive as human beings. This function of the human brain may account for its rapid evolution in humans and for its uniqueness to people. Playing at being human involves finding mates, raising young, and surviving, all in a social context. Functioning as a human requires building relationships, both within and between the sexes, navigating social power networks, and teaching the young how to enter society. The complexity of our society reflects our complex brain, which in turn socially selects for an increasingly complex brain to be effective in an increasingly competitive society, leading to runaway evolution in brain size and complexity. The brains of men and women would seem to be mostly the same because we are both playing in the same society overall.

LIFE-HISTORY SEX DIFFERENCES

Is there any pattern to all the small sex differences we've just enumerated? We've seen in other animals that each gender has a characteristic approach not only to mating but to its entire life. The three male genders of bluegill sunfish, for example, differ not only in mating and social behavior, but also in body size and life span. The traits these fish exhibit come together as a suite of tactics that carry out a life-history strategy. Perhaps human males and females too have slightly different life-history strategies that tie together their differences.

One basic feature of a life-history strategy is life span. Before 1940, men and women had about the same life span. During the last sixty years, though, women have been living longer than they used to, and longer than men. By 1998 the expected life span of a baby girl was 79.5 years, while that of a baby boy was 73.8 years, about five years less. A woman of age 75 was expected to live an additional 12.2 years, and a man an additional 10.0 years. Overall, improved health care is revealing an inherent tendency for women to live longer than men.[78]

The immediate cause of higher mortality in older men is more heart disease and cancer, but males have a higher death rate from injury and illness across all ages. Rather than considering what men might learn from women about how to live longer, some have attempted to demean the quality of life that women experience: "The female longevity advantage, however, is not without cost. Although females live longer," they "experience more disabling problems than males." By adjusting for "quality of life . . . a 5.38-year advantage for women is reduced to 1.3 years."[79] What this statistic means is debatable. The surviving women are, after all, surviving, whereas their male counterparts have already died. If we average in the dead males as being seriously disabled, male quality of life may drop below that of women. One suspects too that men are likely to underreport health problems compared with women. Moreover, the health professions overall have emphasized male care more than female care, so disabilities more common among women, such as autoimmune diseases, are less well understood than those in men, and the treatment less effective.

Women live longer by about five to ten years in all ethnic and cultural groups. This difference in life span is substantial: 5 percent. Using the categories Native American, Native Hawaiian, Samoan, Guamanian, Hispanic, Puerto Rican, Black, U.S. Virgin Islands Black, Chinese, Japanese, Filipino, and White, researchers found that females lived longer than males within each category. The average life spans varied from a low of 65 and 74 years for male and female Blacks, to a high of 80 and 86 years for male and female Chinese.[80] That women live longer than men is undeniable.

Why is this? An ecological perspective offers a possible answer. Ecologists use the concept of a "life history" for the biologically programmed schedule of important events through life. Key events are when reproduction can begin, how many young are produced at the same time, how long reproduction can last, and when death is likely to occur. Ecologists observe that life-history traits usually come bundled in distinct suites. In dangerous environments, animals evolve an early maturation age, have large numbers of young at a time, and senesce early—the high-mortality suite. In safer environments, animals postpone the start of reproduction to a relatively late age, raise fewer young at any one time, and live longer—the low-mortality suite.[81]

If we reflect on the full life cycle of many mammalian males and females, including humans, sexual dimorphism in life history emerges. The dimorphism shows that males, on the average, may have more of a high-mortality suite of life-history traits, and females more of a low-mortality suite. Specifically, sperm are more numerous and senesce faster than eggs. Male embryos grow faster from conception than female embryos. SRY grabs early control of the gonadal ridge to accelerate male body differentiation. At puberty, males have their first intercourse about one year earlier than females,[82] and the male growth spurt is timed to yield a larger body size. The reproductive skew is more pronounced in males than females, with 14 percent of males not having any intercourse in a year, compared with 10 percent for females.[83] Adulthood ends with faster male senescence and a shorter life span. Thus, on the average, males exhibit a high-mortality suite of life-history traits compared to females. Not only do males on the average encounter more danger than females, as evidenced by their higher mortality rate in the population, but through evolution their life history has apparently become adapted to this higher danger.

13

Gender Identity

D o the brains of politicians and poets differ? Can we find a rice-grain of Beethoven's brain shared by all composers and a different rice-grain of Picasso's brain shared by all painters—anatomical markers of ability in performing and graphic arts? Perhaps. No one's looked. But it is known that the part of the brain controlling left-hand fingers is larger in string players than in anyone else.[1]

As we have seen, among our vertebrate relatives, the two male genders of plainfin midshipmen fish have different brains, and in tree lizards, the three male genders develop with different hormone profiles. Ample biological precedent supports the hypothesis that different behavioral temperaments in humans, including gender expression, could spring from differences in brain organization. Might we, for example, detect differences in the brains of transgendered and nontransgendered?

TRANSGENDER BRAINS

Three rice-grains of brain in and around the hypothalamus are sexually dimorphic in males and females—SDN-POA, BSTc, and VIP-SCN. Of these, only BSTc differs between trangendered and nontransgendered people—this rice-grain of brain is perhaps a gender-identity locus. The data supporting this claim may be thin but should be taken seriously.

Two studies have analyzed a total of thirty-four brains preserved in formaldehyde in a reference collection at the Netherlands Brain Bank.[2] The collection includes brains of people identified as heterosexual nontransgendered male, homosexual nontransgendered male, heterosexual nontransgendered female, and transgendered women with varied sexual orientations. Here's what was found:

1. Among nontransgendered heterosexuals, the males' BSTc was about 150 percent the size (2.5 cubic millimeters) and number of neurons (33,000) of females' (1.75 cubic millimeters and 19,000 neurons): straight males bigger than straight females.

2. For the homosexual nontransgendered males, the BSTc was the same as for the heterosexual nontransgendered males: gay males same as straight males.

3. Among the six transgendered women, the BSTc matched that of the nontransgendered women, not the nontransgendered men: transgendered women same as nontransgendered women.

4. For the one transgendered man examined, the BSTc size and neuron count fell squarely in the male range and outside the female range: transgendered man same as nontransgendered man.

The studies included photographs of the spots in the brain where the BSTc occurs, so comparisons are readily visible, supplementing the graphs and tables of data. Another finding was that the size and neuron count of the transgendered women didn't relate to the age at which they transitioned. The femalelike number of neurons in the BSTc of transgendered women is "related to gender identity *per se* rather than to the age at which it became apparent." The investigators suggest that the neuronal differences between transgendered and nontransgendered people are "likely to have been established . . . during early brain development,"[3] just as testosterone organizes BST dimorphism in rodents soon after birth.[4]

The results are announced as though the traditional gender binary had been upheld, as though transgendered people had at long last been revealed as a rare form of intersex: "The present findings . . . clearly support the paradigm that in transsexuals sexual differentiation of the brain and the genitals may go into opposite directions and point to a neurobiological basis of gender identity disorder."[5] Similarly, an earlier paper from the same laboratory states, "Transsexualism is biologically conceptualized as a form of pseudohermaphroditism limited to the central nervous system, as the biological substrate of gender identity."[6]

These studies actually subvert the gender binary rather than supporting it. The three sexually dimorphic neural clusters vary independently of one another, leading to eight brain types, not two. For example, let P stand for a large SDN-POA and p for a small one, B for a large BSTc and b for a small one, S for a large VIP-SCN and s for a small one. P might correlate with an XY chromosomal makeup, p with an XX chromosomal makeup, B with a masculine gender identity, b with a feminine gender identity, S with same-sex sexual orientation and s with between-sex sexual orientation (see next chapter). We can see that eight brain configurations occur: PBS, pBS, PbS, pbS, PBs, pBs, Pbs, and pbs. These might correspond to various arrangements of chromosomes, gender identities, and sexual orientations. Of course, many more brain varieties may be found if more size classes and more sections of the brain are counted. But even if brain variation is scored only in binary sizes, large and small, and only at three independent sites, then eight brain types result, not two. No scientific reason supports selecting two of these eight as normal while declaring the rest abnormal. Moreover, these eight types of brains can be plugged into bodies with at least two genital configurations—those with a penis, C, and those with a clitoris, c. Eight brain types and two body types then lead to sixteen people types: $PBSC$, $pBSC$, $PbSC$, $pbSC$, $PBsC$, $pBsC$, $PbsC$, $pbsC$, $PBSc$, $pBSc$, $PbSc$, $pbSc$, $PBsc$, $pBsc$, $Pbsc$, and $pbsc$. You get the idea: brain-body combinations are limitless.

The discovery of more of these variable brain features will fill out the rainbow of brain morphology, dissolving any belief in a binary brain, just as the research on hormones dissolved the belief in a binary biochemistry. Nonetheless, medical scientists presently envision unraveling a cornucopia of "neurobiological diseases and disorders," the "tip of the iceberg . . . for many sexually dimorphic brain areas . . . and related clinical disorders."[7] Not to worry—won't happen. Our task as informed readers of science is to extract as best we can the data from the layers of medical prejudice in which they're embedded.

The studies of transgender brains have revealed an organic counterpart to some of the variation in gender identity—a valuable finding. We transgendered people wish to say, "We told you so." Coming out as transgendered means realizing something deep about ourselves. Why do some people find us more believable once the organic connection is evident? Couldn't they have taken us at our word without dissecting six formaldehyde-soaked brains?

WHEN GENDER IDENTITY DEVELOPS

When does gender identity form during development? When and how do these BSTc regions form? Gender identity, like other aspects of temperament, presumably awaits the third trimester, when the brain as a whole is growing. In males, three periods during life show unusually high testosterone levels. One is in the middle trimester, when the genitals are developing; the second is around birth; and the third is at puberty. The time around birth may be when the brain's gender identity is being organized—cognitive lenses that instinctively distinguish who will be emulated from who will be merely observed.

To determine when gender identity develops, a good strategy is to find an early limit and a late limit, and work in from these two end points. One clue that gender identity can't occur much earlier than the third trimester of pregnancy is the absence of sex-hormone receptors from the brains of mid-trimester embryos. It has become clear that the external genitals differentiate before the brain does.[8] Penis development, for example, depends on the concentration of testosterone and its products during the middle trimester. In cases of hypospadia—a common intersex form in which the urinary opening is not at the tip of the penis, but at some location on the underside—boys have gender-typical male identity and male gender-typical forms of play and presentation.[9] Here penis morphology is not connected to male identification. The male hormones affecting gonadal morphology, which act in the middle trimester, apparently do not affect the brain's later acquisition of gender identity circuitry.

Similarly, boys whose genitals are ambiguous at birth—*guevedoche*—mature at puberty into men and affirm a male identity. Originally studied in the Dominican Republic, more examples have now been investigated in Papua New Guinea, Mexico, Brazil, and the Middle East.[10] Guevedoche also show that the period when genitals form precedes that when gender identity forms. A low testosterone level when the genitals are forming affects genital morphology, but a presumably high testosterone level later on leads to male-typical gender identity.

Some XY people become intersexed because they have receptors that don't bind very tightly to testosterone. This trait, called androgen insensitivity syndrome (AIS), is X-linked. The genital morphology in AIS is variable, but unlike hypospadics and guevedoche, many AIS people identify as female. Presumably, the body's partial unresponsiveness to testosterone is not restricted to the time of genital formation, but lasts throughout embryonic growth, allowing some AIS people to develop female and others male gender identity.[11]

In congenital adrenal hyperplasia (CAH), the adrenal gland produces more than the usual amount of androgen. In girls, these hormones masculinize the genitals, leading to a large clitoris and sometimes labia fused into a partial scrotum. CAH girls almost always mature identifying as female. The androgen produced from the adrenal gland, while higher than typical in girls, is apparently still less than that produced by the testes of boys. Hence a female gender identity develops, even though some impact on genital morphology is evident.[12]

Similarly, in other primates, genital masculinization and behavioral masculinization take place at different times.[13] Taken together, these results imply that gender identity develops sometime after the middle trimester of pregnancy, when genital morphology is taking shape. At the earliest, then, gender identity forms about three months before birth, when the middle trimester ends and the third trimester begins.

Turning to the late end of the range, anecdotes and case studies reveal that gender identity must already be determined by several months after birth. Attempts to change a person's gender identity by rearing the child in one direction or another have simply failed, often tragically. Textbooks in medicine once asserted that a child's gender could be "assigned" by the child's upbringing. In 1997, however, a bombshell exploded in the field of child psychiatry: the textbooks were shown to be based on fraudulent information.

Textbooks claimed that "(1) individuals are psychosexually neutral at birth, and (2) healthy psychosexual development is dependent on the appearance of the genitals."[14] The capstone evidence for this belief came from a case of a baby boy whose penis was irreparably damaged during a circumcision procedure at seven months and who was reassigned at the age of seventeen months to live as a female. The boy's name was changed to a girl's name (Bruce to Brenda, originally reported as John to Joan), and he was raised as a girl. His testicles were removed, and a semblance of female genitals was surgically constructed in their place at twenty-two months. Female hormones were prescribed to begin at puberty. Because the boy had an identical twin whose penis was not damaged during circumcision and who was raised as a boy, a comparison was possible. The supervising physician, John Money, a famous and powerful professor at Johns Hopkins University, reported in 1972 that the boy reassigned to live as a girl was developing into a perfectly typical girl, interested in "dolls, a doll house, and a doll carriage," in contrast to his brother's interest in "cars, gas pumps, and tools."[15] On this basis, the textbooks taught that gender identity is determined by the way the baby is raised.

Given what's known about sex hormones and development, the received psychological wisdom that babies are psychosexually neutral seems highly unlikely, and the theory that we figure out as babies what our gender is by looking at our genitals in the mirror seems farfetched. Nonetheless, this dogma went unchallenged in medicine for twenty-five years, until courageous researchers, together with the courageous boy himself (now a man), spoke up.[16] In fact, "Brenda" never accepted the identity imposed upon him and transitioned at age fourteen to living as a teenage boy, changing his name to David. The medical reports about his supposedly successful development as a female were not true. He is now married and the father of an adopted family.

One of the authors of the exposé admitted to being "shit-scared" in coming forward with the truth, and the man whose sex had been reassigned has been acclaimed as a "true hero" for agreeing to detail to investigators a long and painful period during his life. The human as well as scientific dimensions of the case, together with the evidence of cover-up and apparent fraud, have been beautifully documented in the investigative reporting of John Colapinto.[17]

In rebuttal, child psychiatrists produced a very brief report on one counterexample. An XY baby whose penis was irreparably damaged during a circumcision at age two months was reassigned to live as a female at age seven months. The girl was interviewed at age sixteen and at twenty-six and was living socially as a woman. She sought and obtained surgery to construct female genitals and completely identifies as female.[18]

Taken at face value, these gender reassignment cases imply that the late limit for gender identity development is sometime between seven months (when the assignment did work) and seventeen months (when the assignment didn't work). I'll take twelve months as a working figure. Combining the data for the early and late limits, gender identity appears to form sometime between three months before birth and twelve months after birth.

I envision gender identity as a cognitive lens. When a baby opens his or her eyes after birth and looks around, whom will the baby emulate and whom will he or she merely notice? Perhaps a male baby will emulate his father or other men, perhaps not, and a female baby her mother or other women, perhaps not. I imagine that a lens in the brain controls who to focus on as a "tutor." Transgender identity is then the acceptance of a tutor from the opposite sex. Degrees of transgender identity, and of gender variance generally, reflect different degrees of single-mindedness in the selection of the tutor's gender. The development of gender identity thus depends on both brain state and early postnatal experience, because brain state indicates what the lens is, and environmental experience supplies the image to be photographed through that lens and ultimately developed immutably into brain circuitry. Once gender identity is set, like other basic aspects of temperament, life proceeds from there.

14

Sexual Orientation

I f outside behavior matches inside morphology, then gay and lesbian people may have unique bodies. If string players have special brain parts for left-handed fingering, and race jockeys special genes for a short physique, then perhaps people of same-sex sexuality have special brain parts and/or genes for sexuality too. The search for biological aspects of sexual orientation often confuses sexuality with transgender expression.

GAY BRAINS

Remember the three rice-grains of nerve cells in the preoptic/hypothalamus area at the base of the brain? These grains, called SDN-POA, BSTc, and VIP-SCN, are sexually dimorphic in humans. VIP-SCN size seems to align with sexual orientation in males. I bet you're guessing that gay males have a female-sized VIP-SCN. Nope. Gay males have an even bigger VIP-SCN than straight males, which is in turn bigger than the VIP-SCN of females. So much for the belief that gay men have female brains![1] Specifically, straight males have about 2,500 cells, and females about 1,000 cells in this approximately 0.25 cubic millimeter cluster.[2] Gay males have a volume of VIP-SCN 1.7

times as large, and with 2.1 times as many cells, as that of straight males.[3]

Another possible difference between gay and straight men comes from an unconfirmed study of a fourth and rarely mentioned rice-grain, the tiniest of all—a cluster of cells in the hypothalamus called INAH3. In heterosexual men, this grain averages 0.1 cubic millimeters; in heterosexual women, 0.05 cubic millimeters; and in gay men, also 0.05 cubic millimeters. This tiny feature in gay male brains has been singled out as matching that in women.[4] Thus gay males are closer to females in this rice-grain (INAH3), but farther from females in the other (VIP-SCN).

The brains of lesbian women appear to differ from those of straight women. Recall that men produce fewer clicking sounds in their internal ears than women do (see chapter 12). Lesbian and bisexual women produce fewer clicking sounds in their ears than straight women do, but more than men do.[5] Thus lesbian and bisexual women are intermediate between straight women and men in this regard. Indeed, ear clicking can change in an adult as a result of taking hormones. A transgendered woman who began taking estrogen prior to her sex reassignment surgery developed the ear clicking. So ear clicking does not necessarily say anything about how brain structure is organized.[6]

All in all, variation in the many rice-grains of nerve cells shows that brains vary with sex, gender identity, and sexual orientation. Further analysis of brain states may reveal as many differences among people's brains as among people's faces.

GAY FAMILIES

Did Dad go fishing? Do you go fishing? Did Mom bake cookies? Do you bake cookies? Lots of traits run in families, like hobbies and styles of food preparation. Like hair color and eye color. Hobbies and cooking styles reflect shared environment; hair and eye color, shared genes. Being gay and lesbian runs in families too. Does shared sexual orientation in families reflect shared environment or shared genes, or both? The answer isn't clear. Here are some clues.

If a man is straight, there is a 4 percent his brother will be gay, the same percentage as in the general population. If a man is gay, the likelihood increases fivefold, to 22 percent. Whether a man is straight or gay has no statistical effect on whether his sister will be straight or lesbian.[7] These figures show that gay men cluster in families but do not say whether this stems from shared genes or a shared environment. Similarly, if a woman is lesbian, her sister is about twice as likely to be lesbian, but whether a woman is lesbian has a very small or undetectable statistical effect on whether her brother is gay or straight.[8] Gay men and lesbian women cluster independently.

Comparing identical and fraternal twins suggests some genetic component. In a 1991 study, 52 percent of identical male twins were both gay, while only 22 percent of fraternal twins were both gay.[9] In a 1993 study, 65 percent of identical male twins were both gay, and 29 percent of the fraternal twins were both gay.[10] Similarly, a 1993 study reported that 48 percent of identical female twins were both lesbian, and only 6 percent of fraternal twins were both lesbian.[11]

The studies just cited come from the United States. A 1992 British study, which looked at males and females together, found that 25 percent of identical twins were both homosexual, but only 2.5 percent of fraternal twins were homosexual.[12] A 1995 Australian study used a different method.[13] Instead of inviting twins to participate by placing advertisements in magazines and other sources likely to be seen by gay readers, the study used a preexisting list of twins. Based on a strict definition of whether twins could be scored as both being gay, the investigators reported that 20 percent of the identical male twins were both gay, 0 percent of the fraternal male twins were both gay, 24 percent of the identical female twins were both lesbian, and 11 percent of the fraternal female twins were both lesbian.

The studies repeatedly show that identical twins are at least twice as likely both to be homosexual as fraternal twins. The chance that identical twins will both be gay ranges from about 25 percent to 50 percent, depending on the study, and is decidedly less than 100 percent. Thus, even though a genetic component may be present, other, presumably environmental factors account for 50 to 75 percent of the story.

Although comparisons between identical and fraternal twins suggest a genetic component in homosexuality, the possibility remains that identical twins are raised more similarly to one another than fraternal twins are, and that identical twins associate more closely and encounter more similar experiences while they are growing up than fraternal twins do. Further investigation of a genetic component should look at data from identical twins raised apart, because these data will show the effects of shared genes in the absence of a common environment.

A 1986 study located six pairs of identical twins who were raised apart and had at least one member who was gay or lesbian. In all four female instances, one member was lesbian and the other straight. In one of the male instances, both members were gay—in fact, they didn't know of each other's existence until they happened to meet in a gay bar where people had been mistaking them for each other. In the other instance of identical male twins raised apart, one member was bisexual until age nineteen and then became exclusively gay, whereas the other was homosexual between ages fifteen and eighteen, then later married and regarded himself as straight. In this instance, both members exhibited at least partial same-sex sexual orientation.[14] Thus the data on sexual orientation in twins reared apart are perhaps suggestive of a possible genetic component for gay male sexuality, but much less so for lesbian sexuality.

Nonetheless, an important contrary fact remains. The 1991 study mentioned above also showed that an adopted brother of a gay man is twice as likely to be gay (11 percent) as an adopted brother of a straight man (5 percent). So unless the adoptive parents are somehow selecting babies likely to become gay, something about the environment into which the adopted child is placed is contributing to sexual orientation as much as any genes are.[15]

Substantial evidence points to both genetic and environmental components in the development of same-sex sexuality. No one who pushes one factor to the exclusion of the other can be correct.

THE HAMER-PATTATUCCI STUDY AND THE QUESTION OF A GAY GENE

A milestone in the genetic analysis of gayness in males came in 1993 with the publication of a paper in the respected magazine *Science* by a team including a senior out gay scientist, Dean Hamer, and a young out lesbian scientist, Angela Pattatucci.[16] This work (hereafter referred to as HP) has since become quite controversial and must be considered carefully.

The paper confirmed yet again the tendency for gay men to cluster in families. The brother of a gay man had a 13.5 percent of being gay, whereas the brother of a straight man had only the baseline chance of being gay, which in this study was estimated at about 2 percent.

The researchers' distribution of men into the categories of straight and gay was claimed to be nearly absolute—bisexuals were almost completely absent. HP concluded that "it was appropriate to treat sexual orientation as a dimorphic rather than as a continuously variable trait." Although some other studies also report such a bimodal distribution (e.g., the 1986 and 1991 studies mentioned above), this claim has been seriously disputed.[17] In most cultures, same-sex sexuality is intermixed with between-sex sexuality. One anthropological study tabulates same-sex sexual practices from twenty-one cultures, and in fifteen of these homosexual practice was concurrent with heterosexual practice.[18]

The sorting into distinct categories is partly an artifact of present-day social pressures within the gay and straight communities. HP's subjects were self-acknowledged homosexual men recruited through outpatient HIV clinics in the Washington, D.C., area, and through local homophile organizations. The participants were 92 percent white non-Hispanic, 4 percent African American, and 1 percent Asian, with an average educational level of 3.5 years beyond high school and an average age of thirty-nine years. Among the gays, 90 percent said they were nearly exclusively homosexual, and 90 percent of the straight men said they were nearly exclusively straight, giving the impression of a clear-cut bimodality. On reflection, though, bimodality emerges in such a sample as a result of social pressure and isn't necessarily representative of the human population. A contemporary gay man can't admit to being sexually interested in a woman any more than a present-day straight man can admit to being sexually interested in a man. The organizations and magazines that offer safe space for those who insist on a bisexual identity were not solicited.

The response of the investigators to this criticism has been evasive. Hamer states, "I didn't tell these men to answer 0 or 6 [on a scale between heterosexual to homosexual], it's just that almost all of them did. Am I supposed to pretend the trait is continuous?" He continues, "Well, how many truly bisexual men have you ever met? I have no theoretic argument with bisexuality. It's just that before I started doing research, I'd never met any. Of the men we've interviewed, most identify themselves as either gay or straight. A handful identified themselves as bisexual, and we did not include them . . . for simplicity. But of the few who said—even insisted—they were bisexual and made their case with the fact that they were also sleeping with women, it would become clear with most of them after just a couple of casual questions that they were really only attracted to men but were in the process of coming out. . . . Besides, as a geneticist, to be blunt about it, I don't really give a damn what label anyone uses, or even what they do, or with whom. I care about what they feel inside."[19]

Pattatucci too is skeptical that bisexuals exist: "The most illuminating experience for me has been discovering that the way we ask the questions reveals what sexual orientation truly is. . . . I would often preface the central question by saying 'I want to know what's in your interior. . . do you feel that who you are now, your homosexual orientation, has always been part of you, part of who you are . . . even though your sexual behavior might have been with members of the opposite sex?'. . . It actually is fairly rare, even when I talk to people who identify as bisexuals, that they say their interior, true sexual orientations have changed. Their behavior may have changed, but their homosexual core has always been there. That's the important thing. The behavior is irrelevant compared to the core."[20]

These quotes show that homosexuality was defined in the study as a form of self-identity, irrespective of sexual practice. One wonders, though, whether any hypothetical gene for homosexuality would pertain more to perceived identity than to practice. Homosexual practice has existed throughout the evolutionary history of our species, whereas the assertion of homosexuality as an identity is located in our particular culture. Mischaracterizing the phenotype can render subsequent genetic analysis meaningless.

Pattatucci is selective about who counts as lesbian: "A relatively small number of women will say in the interview, 'I'm not a lesbian, I just fell in love with this one woman' and it's apparent that their feelings are . . . basically heterosexual. They perceive themselves as having had a serendipitous experience. They fell in love and committed to this particular woman, and sex became part of the relationship." Pattatucci excludes these women from the study because "you'd better narrow your field as much as you can. The best place to look is at people who show the greatest amount of expression." Another group Pattatucci excludes are "women who identify themselves as lesbian for political and ideological reasons with little or no evidence of a romantic or sexual attraction to women. . . . Am I sympathetic to those goals on a purely political level? I absolutely am. I'm a woman, and I'm a lesbian. But . . . I'm researching a scientific question, not a political one."[21]

This selection of subjects forces a bimodality between lesbian and straight by throwing away the data for people who would score in between. Scientifically, picking and choosing data in this way is a red flag. Politically, this winnowing of the social category of lesbianism in search of the truest lesbians of them all leads down a slippery slope, culminating in an biologically based hierarchy.[22] For these reasons, the assertion of HP that gays and straights sort cleanly into a dichotomy is dubious.

HP provides valuable demographic data on the life history of gay men, showing the ages when they first experienced same-sex sexual attraction, when they self-acknowledged their sexual orientation, and when they publicly acknowledged their orientation (came out). The average age of first same-sex attraction was ten years, two years before their average age of puberty at twelve. The average age of self-acknowledgment was fifteen years, and the average age of coming out was twenty-one years.[23]

HP claims that gayness in males is maternally inherited and linked to the X chromosome. The paper reports that maternal uncles and sons of maternal aunts (first cousins) of a gay man had a 7.5 percent chance of being gay, higher than the baseline chance of 2 percent. That is, out of one hundred maternal uncles and cousins of maternal aunts, only about seven are likely to be gay. Although seven is higher than the two who are likely to be gay on the paternal side, the number is still quite small. The strongest scenario from proponents of a genetic basis to homosexuality shows only a rather mild effect.

Building on the premise that a genetic component of gayness is maternally inherited, HP looked for spots on the X chromosome that might be statistically correlated with gayness. Such a spot could be called a "gay gene." HP reported that a section on the X chromosome at the tip of its long arm, a section called Xq28, was statistically related to gayness. This finding attracted enormous attention.

HP located forty families in which two brothers were gay, the father was straight, and not more than one of the sisters was lesbian. By looking at families in which the homosexuality was expressed mostly in males, they felt they had achieved a sample "enriched" for a hypothetical gene for gayness that was maternally inherited. Again, the sample had been picked and chosen.

HP devised a genetic test to detect a gay gene on the X chromosome. The question is: do gay brothers share the same X chromosome 100 percent of the time, or do gay brothers share X chromosomes at random (50 percent of the time)? A finding of 100 percent would mean the X chromosome was needed for brothers to be gay—it contains a gay gene—whereas 50 percent would mean the X chromosome was irrelevant to whether brothers were gay and there is no gay gene on the X chromosome.

Well, out of 40 pairs of brothers, 33 shared the Xq28 section of chromosome X, and 7 did not. This result is intermediate. If something in Xq28 were absolutely needed to be gay, then all 40 brothers would share this chunk of DNA, whereas if only 20 brothers shared Xq28, then it would be irrelevant to gayness. The figure of 33 out of 40 is statistically significant, and it was therefore concluded that some gene in Xq28 tends to produce gayness in males. Notice how modest this claim actually is. A gene in the Xq28 region of the X chromosome is neither necessary nor sufficient for gayness in males. Because identical twins aren't always both gay, genes alone don't guarantee gayness. Conversely, a male can be gay without the gene, because 7 of the 40 pairs were gay but didn't have this gene. Nonetheless, this gene would be part of some biochemical pathway occasionally involved with the development of gayness in males.

HP then followed up their own study and claimed to confirm their earlier work. In this case, 22 out of 32 pairs of gay males shared the Xq28 section of the X chromosome. Again, the result is intermediate. If Xq28 were irrelevant to gayness, then 16 of the gay-brother pairs would share this section of X, and if Xq28 were necessary for gayness, then all 32 pairs would share the section. The figure of 22 is not too far from 16, overall a weaker result than the original study.[24] Is even this limited claim for a genetic basis to gayness in males correct?

The Hamer-Pattatucci study has not received any further confirmation, and it has even been directly refuted by subsequent work from other investigators. One follow-up study did not detect any evidence of maternal inheritance for gayness, stating that "none of the samples showed a significantly greater proportion of maternal than paternal homosexual uncles or homosexual male maternal first cousins."[25] This follow-up study did confirm (yet again) that homosexuality runs in families: a brother of a gay man has about a 10 percent chance of also being gay, about two to three times the baseline chance of being gay in that study. A sister of a gay man has a 4 percent chance of being lesbian, also about two to three times the baseline chance of being lesbian in that study. Thus, the family clustering was confirmed, but the claim of maternal inheritance was not. A Canadian team repeated the attempt to detect a gay gene on the X chromosome, using the same overall design as HP.[26] Advertisements were placed in the Canadian gay news magazines *Xtra* and *Fugue* for families in which there were at least two gay brothers. Forty-six families with two gay brothers and two families with three gay brothers were studied. The sexual orientation was confirmed for each subject by direct questions from a "gay interviewer"; each subject read gay magazines and volunteered that he was gay; and his self-report was corroborated by interviewing the gay brother.

Of the 46 brother-pairs, 23 would be expected to share Xq28 if this section was irrelevant to gayness, whereas all 46 would share Xq28 if it was necessary. In fact, only 20 of the pairs shared Xq28, suggesting that Xq28 is irrelevant to whether a male becomes gay. The results demonstrate that there is no gay gene in Xq28. The Canadian investigators conclude, "It is unclear why our results are so discrepant from Hamer's original study. . . . Nonetheless, our data do not support the presence of a gene of large effect influencing sexual orientation at position Xq28 . . . [although] these results do not preclude the possibility of detectable gene effects elsewhere in the genome."

The failure of the Hamer-Pattatucci study demands a postmortem. What went wrong? Why was a widely publicized and believed report from a credible laboratory at the National Institutes of Health directly contradicted by later research? The most striking difference between these studies is the way in which people were identified as gay. The Canadian team apparently did not demand that a gay person affirm sexual orientation as their personal identity to the extent that the HP study did;

sexual practice may have been sufficient to qualify the person as gay. This sensitivity to how homosexuality is defined was anticipated by Pattatucci: "People hear you say 'How the question is asked will determine the answer' and they think this means they can discount the result. What it really means is that one needs to ask the right questions." When you do "and the data starts coming back, you're thrilled because you realize you've tapped the vein, you're on the right track . . . and you're simply thrilled."[27]

So who asked the wrong question? Was it HP, with their demand that people define their homosexuality as a core identity before being counted as gay? Did HP manufacture a trait that doesn't exist biologically, and select subjects in such a way as to fabricate the appearance of a result? Or did the Canadian team, with their apparently looser interview criteria for homosexuality, lump different types of homosexuality into one, obscuring a true underlying pattern? Did the Canadian team overlook a genuine finding by not being rigorous enough in their selection of subjects? The jury's still out.

I support the conventional wisdom, which is suspicious of any result highly dependent on how a question is posed. In my experience, a strong and robust result is not extremely sensitive to methodological details— different people, both friends and foes, get more or less the same answer, whether they like it or not, even with somewhat different approaches. I believe that if a gay gene were a major phenomenon, its detection wouldn't be so tricky.

WHY BOTHER WITH A GAY GENE?

One might wonder why so much effort has been invested in the scientific chase for a gay gene. Who cares whether a gay gene exists? Scientists and the general public have a big disconnect here, trapping gay scientists in the middle. From a scientific perspective, sexual orientation is a fundamental feature of mating behavior, and a task of basic research is to understand how this trait forms, what the relative contributions of direct gene products are, and how early hormonal and childhood experiences enter the picture. From a policy perspective, the issue is different: it is focused on whether gayness is a matter of choice, whether gayness is learned and can thus be unlearned.

Media interviews with Hamer illustrated this disconnect. On July 16, 1993, Hamer was interviewed by all the major television networks. The *Today* show announced, "There is new evidence that homosexuality may be inherited in some cases and not a matter of choice." Tom Brokaw opened with, "There's new medical evidence that homosexuality is genetic, not acquired behavior." On *Nightline,* Ted Koppel announced, "Tonight: the genetic link to male homosexuality. More authoritatively than ever before, a scientific study is suggesting that a man's homosexual tendencies may not be a matter of choice. . . . Think about the legal implication . . . it is not constitutional to make status such as race illegal. If the findings of this study are confirmed, it will not quite raise homosexuality to the same legal level as race, but it moves it a lot closer."[28]

Koppel asked Hamer directly, "If the findings of the study, Dr. Hamer, are confirmed, will it then be accurate to say that homosexuality is not optional behavior?" Hamer repeatedly answered a different question, that his work points to a gene in a particular region of the X chromosome, but stressed that other genes are involved too. After consultation during a commercial break, Hamer finally stated, "I think all scientists that have studied sexual orientation already agree that there's very little element of choice in being gay or homosexual. The question is whether there's a defined genetic component to homosexuality." For Hamer, the question is technical. According to alternative hypotheses, sexual orientation could become fixed for life because of hormones or other environmental factors impacting how brain circuitry develops, without necessarily being genetic. Hamer is taking the absence of choice in sexual orientation as a given and asking specifically how genes might be involved.

Agreement that choice is absent from sexual orientation is not as widespread as Hamer indicates. Many lesbian histories show transitions back and forth between straight and gay lives, whereas other lesbians stay in one sexual space for their entire life.[29] Gay male histories, which Hamer seems primarily to be discussing, are less fluid, at least in today's culture. Transgender narratives also show variability in sexual orientation. A recent study reports that about 30 percent of transgendered women changed their sexual orientation after they transitioned.[30]

The enthusiastic reception of early evidence for a gay gene has spawned an industry of genetic crystal-ball gazing by both scientists and science reporters. One geneticist states, "I expect sexual orientation will come down to just one or two genes. Sexual orientation is a simple trait. Everyone says it's complex, but it's not complex at all."[31] Another scientist speculates about the number of genes determining sexual orientation: "I imagine it will come down closer to one. I'd speculate that sexual orientation is linked to a very early event in embryogenesis and thus possibly could involve just a few fundamentally important genes that start the process unfolding. I would really be surprised, for example, if we were to learn that the gene turned on late in fetal development."[32] Still another scientist, "one of Hamer's colleagues, who would not allow himself to be identified, [is] much more frank. 'Look, you'll never get me to say it publicly, but I think it's clear that this is really a pretty simple trait . . . if you look at where the data are going, there's not much question.' "[33]

Alas, this molecular bravado must face up to contrary data. Science reporters seem especially taken in. A reporter compares the inheritance of gayness in males to the inheritance of short height in pygmies, a trait brought about by a single major gene for the growth-hormone receptor: "If a Pygmy has a child with an African of average height, all the children are either of average height or of Pygmy height. The trait is one or the other, clear cut, and the reason is that the trait is controlled by one gene."[34] This pattern is precisely *not* how sexual orientation is inherited, as the many studies of gayness in families have shown; gayness is clearly not inherited as a single major gene.

Similarly, a reporter compares gayness with two very rare genetic diseases, ocular albinism and menkes disease, which occur on the X chromosome near Xq28. These diseases are decidedly *not* comparable to sexual orientation. Gay sexual orientation is far more common than a genetic disease, and it is not associated with any physical disability (see p. 284).

I believe sexual orientation develops analogously to an accent in speech, which also develops early in childhood. Some people don't deviate even slightly from the accent they learned as a child, although a thick Russian accent isn't genetic. Other people easily acquire a new accent—

I can change mine in hours. Some people's sexual orientation is immutable, whereas other people's shifts. Not only is sexual orientation part of one's temperament, but so is the degree of flexibility, just like an ability to alter one's accent.

WHEN DOES SEXUAL ORIENTATION DEVELOP?

As we did for the timing of gender identity, let's look for an early limit and a late limit for the development of sexual orientation, and work in from there. An early limit would seem to be a year or so after birth, for two reasons. First, sexual orientation would seem to require the mental lenses that distinguish gender and thus would develop only after gender identity develops. Because gender identity seems set by about the first year after birth, sexual orientation could begin to form then. Second, adopted boys with a gay brother are more likely to be gay than adopted boys with a straight brother. This line of thought also suggests an early limit near a year or so after birth, when adoptions typically take place.

A late limit is indicated by the average age of first awareness of same-sex sexual arousal, which is about ten years of age, two years before puberty. So the window for developing one's sexual orientation would appear to be from infancy to very early childhood, say a nine-year window from one to ten years of age, as a first guess. The window is probably much narrower, and further study of when gayness develops in adopted children might be very revealing.

GAYNESS AND EVOLUTIONARY THEORY

If there is some genetic component to homosexuality, we may wonder how homosexuality fits into ideas about human evolution. Until recently, scientists have taken for granted that homosexuality is a deleterious trait within the framework of Darwinian fitness and have looked for theories, often far-fetched, to explain how a "bad gene" could somehow become common. But who says homosexuality is deleterious?

For lesbian women, a 1988 U.S. survey reported that the mean number of children born to women with homosexual experience was 1.2

compared to a mean of 2.2 for women without homosexual experience.[35] A 1994 survey reported that 67 percent of lesbian women were mothers, compared with 72 percent of straight women,[36] and a 1995 study of contemporary British women showed that bisexual women have a higher fecundity to age twenty-five and no significant difference in lifetime fecundity compared to heterosexual women.[37] From these studies, lesbian and bisexual woman apparently have somewhere between the same and one-half the reproduction rate of straight women.

For gay men, the 1994 study showed that 27 percent were fathers, compared with 60 percent for straight men.[38] On the other hand, of 655 homosexual and bisexual men in contemporary Japan, 83 percent had offspring.[39] Thus, gay and bisexual men also apparently have somewhere between the same and one-half the reproduction of straight men.

These references scrape the bottom of the barrel. One would have thought that if homosexuality was deleterious, the evidence would be plentiful and easy to find. Furthermore, fertility is only one component of Darwinian fitness, and fertility must be multiplied by the probability of living long enough to reproduce when computing the overall fitness relevant to natural selection. A disadvantage in fertility could be offset by a higher survival rate. No data are available on a trade-off between survival and reproduction for homosexuality in humans. Today's society is certainly not amenable to the survival and health of gay and lesbian people, but the matter may have been entirely different at other points in human evolutionary history.[40] All in all, the data do not support uncritical acceptance of homosexuality as deleterious.

One early theory postulated that gays and lesbians were like avian helpers at the nest, people who remain with their nuclear family to help raise brothers, sisters, and cousins, who would then go on to do the reproducing. This theory, which positively values the contributions to family and society that gay and lesbian people can make, was a step forward in depathologizing same-sex sexuality.[41] As one gay scientist notes, though, "Homosexuality is not simply the abandonment of sex in favor of altruistic behavior toward one's relatives; rather it involves the adoption of a different sexuality, one that can be quite costly in terms of time and resources."[42] Nor does helping at the nest account for why such helpers would specifically be gay or lesbian. Important though this suggestion has been, helping at the nest doesn't appear to hold the answer to why homosexuality has evolved in humans.[43]

A more devious theory making the rounds is based on the premise that a gay gene on the X chromosome, say in Xq28, causes homosexuality in males and not in females. If the gay gene escapes X inactivation (see p. 209), then females express two copies of the gene and males only one. If females benefit from this gene in some presently unknown way, then men might carry the gay gene as a side effect of the gene's double-dose benefit for women.[44] This idea, called sex-antagonistic pleiotropic homosexuality, is theoretically far-fetched, has no supporting evidence, and relies on the false assumption that a gay gene lies within Xq28. One science reporter even found a molecular biologist who speculated, "Homosexuality may be a type of bacterial infection . . . we may eventually be able to eradicate with an antibiotic."[45]

The various theories advanced, some of them absurd, all suffer from an uncritical acceptance of homosexuality as deleterious and therefore must conjure up evolutionary pathways whereby a deleterious gene can become as common as homosexuality is. Instead, if homosexuality is an adaptation, then the commonness of homosexuality is no problem. Indeed, the question becomes why everyone isn't homosexual, as in bonobos. Overall, an evolutionary theory of human homosexuality needs to explain the polymorphism in sexual orientation among humans. Why are, say, 90 percent of men straight and 10 percent gay, and why are 95 percent of women straight and 5 percent lesbian? And today at least, why are lesbian women more likely to be bisexual than gay men? Finally, why does homophobia exist? No evolutionary theory has been proposed for humans that addresses this complete suite of issues, although I believe some promising first steps have recently been made.

One study offers a long-overdue extension of evolutionary psychology to include homosexuality in humans.[46] This investigation contends that the "long history of institutionalized homosexuality between higher status and lower status males," usually of different ages by five years or more, produces "relationships [that] tend to socialize the youths into the adult male role, nurture and protect the youths and provide the basis for life-long friendships, social alliances and social status. . . . Social status, a reflection of political strength and alliances, appears to have played a large role in the evolutionary history of human male reproductive success." The study goes on to suggest that homosexuality for women provides bonds of friendship that lead to mutual assistance in raising children, assuming paternal assistance is absent in primitive societies. Here too homosexuality is hypothesized to provide higher reproductive success. These conjectures about how homosexuality evolved logically feed back to determining the type of environment in which homosexuality develops during infancy: "Homoerotic behavior may be evoked as a normal response to placement in an environment which closely resembles the environment in which it evolved and was adaptive in the evolutionary past." In particular, homoerotic behavior in single-sex groups would reflect not an absence of partners, but the adaptive development of same-sex bonds and alliances in the conditions when they would be most useful, which may resemble the social structure of early hominids.

Another study presents an anthropological perspective focusing on how homosexuality leads to various types of alliances among males.[47] As already noted, heterosexual and homosexual practice occurred together in fifteen out of twenty-one cultures. Homosexual behavior has also occurred more often in agricultural than in hunter-gatherer societies, and more often in larger social groups.[48] Homosexual behavior may be more frequent when it empowers political networks rather than independent individuals, and it may be expressed more in industrial nations after their demographic transition from high reproduction to high survival.[49] A difficulty faced by a theory of homosexuality as a form of alliance-building, however, is that male-male alliances can be built without using sexuality. Data are needed that alliances bonded through homosexual behavior are in some sense stronger, better, or longer-lasting than bonds lacking this ingredient.

These new theories for the evolution of human homosexuality seem to be on the right track, but they may be too specific. Homoerotic attraction can have multiple functions, depending on context. Homosexuality need not be dyadic—such as an alliance between two people. Instead, I suspect homosexuality may also be a social-inclusionary trait, a ticket for admission to a collective.

What explains the polymorphism in sexual orientation—the *ratio* of gays to straights? I conjecture that a polymorphism in sexual orientation may indicate alternative strategies of same-sex relationships that are equally effective in achieving access to net reproductive opportunity. These alternative same-sex relational strategies are the counterpart of alternative between-sex mating strategies, such as the controller and cooperator morphs. Abstractly, members of the straight morph may bond through the exchange of power, whereas members of the gay morph may build alliances through the exchange of pleasure. Conflict is likely to occur between alternative same-sex relational morphs because they are playing by different rules. Homophobia may emerge from this conflict. Transactions based on the exchange of pleasure may be seen as subverting the power hierarchy, and be crushed by those in control. A balance may then result. At one extreme, if everyone is in continual conflict, a cooperator can benefit by avoiding the hazards of conflict. In this view, homosexuality emerges as a complex social adaptation, a product of positive evolution.

15

Psychological Perspectives

My approach to variation in gender expression and sexuality is biological and behavioral, not psychological. Since Freud, however, gender and sexuality have often been discussed in psychological terms. I'm skeptical of psychology and, as a transgendered woman, have found psychologists to be dangerous, like gays and lesbians before me did. Psychologists operate with a medical model that pathologizes diversity. These medical wannabes have long persecuted and abused gender- and sexuality-variant people from a position of authority.[1] Nonetheless, some reviewers felt a purely biological account of gender and sexuality was incomplete and needed to be rounded out with psychological perspectives. Reviewers felt that transsexualism in particular needed more discussion.

Well, okay. In my opinion, though, the source material for this chapter is academically sketchy. Instead of engaging in scientific inquiry, we're dealing with scanty anecdotes that must be spliced together into some composite picture. Furthermore, I'm not sure how many people really want to hear in explicit detail about sexual fantasies and practices that go on behind closed doors. This chapter invites the invasion of privacy. Also, the chapter risks imbalance by providing coverage of the fantasies of transgendered people without matching coverage of nontransgendered people. Although the written accounts of some transgendered people may seem unusual, how many nontransgendered people match standard templates? It's difficult to evaluate the diversity of transgender practices relative to the diversity of sexual practices in the general population.

Still, I wouldn't offer this chapter at all if I didn't feel there was some value in the material. The transgendered writer Patrick Califia has charted the literary progression of autobiographies, noting how the first narratives from transgendered women denied any sexual ingredient, focusing only on realizing gender identity.[2] Taking the sex out of transsexual sanitized the narratives to gain acceptance in sections of society where sex is a dirty word. Only recently have transgendered women started to put the sex back into transsexuality. Yet assuming that transgender expression is entirely sexual commits the reverse mistake: it takes the gender out of transgender. In contrast to the autobiographical narratives of transgendered women, those of transgendered men have never been squeamish about sexuality. This chapter visits some narratives of transgender experience, striking a balance between motivations based on gender identification and on sexuality, respecting both the sex in transsexuality and the gender in transgender.

INFORMATION ON TRANSGENDER EXPERIENCE

Transgender narratives come mostly from two sources: reports from therapists and autobiographies published as books or testimonials over the web. Direct quotations from transgendered people are the most valuable, because they contain only the speaker's own bias. Digested summaries from therapists are the most suspect, because a frequently misguided theoretical perspective is layered on top of the bias already present in the primary report. In addition, therapists labor with a conflict of interest, because they benefit financially from their role as gatekeepers who authorize access to medical technologies such as surgeries and hormones. At best, therapists collect information about a diverse people while offering comfort and guidance. At worst, therapists psychologically torture clients with POW-type behavioral modification tactics, physically maiming them with drugs and electroshock treatment and diminishing their self-confidence.

The sampling by therapists is exceedingly uneven. Therapists meet transgendered people who come to them as clients, but many transgendered people never seek a therapist. Those who do are more likely to be anguished, to identify as transsexual, and to be sufficiently affluent to afford therapy. Most important, therapists encounter transgendered people in only one window of life. They see firsthand the difficulties attending transition and hear recollections of life before transition, but they rarely conduct follow-up interviews after transition.

Autobiographies offer a perspective over an entire life, but they too must be read with care. Transgendered people continually need to defend and explain themselves, and they write while looking over their shoulder. All in all, though, some fragments are emerging of what transgendered people go through as they discover their "true selves" at various ages. The phrase "true selves" comes from the title of a book by Mildred Brown and Chloe Rounsley, which offers the most reliable account of transgender narratives collected by therapists that I have found.[3]

CHILDHOOD NARRATIVES: IDENTITY

Transgender experience begins with the earliest moments of consciousness. Brown reports that 85 percent of her clients recognized a serious discrepancy between body form and gender identity before grade school, many years before puberty. A transgendered woman recalls, "My mother knew that I was transgendered from age four on, and she was bound and determined to crush that and to make a male out of me." Similarly, a transgendered man remembers being reprimanded by Mom: "For heaven's sake, Lisa, you walk like John Wayne."[4]

In response to pressure for gender conformity, transgendered children put enormous energy into trying to conform to expectations, into being the good son or the perfect daughter, often believing they'll eventually get it right. A transgendered woman recalls, "I learned to become a chameleon, to fabricate little masculine selves that had nothing to do with me but that I could send out into the world." Yet not everyone is able to change colors. Another transgendered woman relates, "I wasn't liked for who I was. I would walk out in the street and express myself in the only way I knew how, with a very childlike innocence. Just being alive seemed to be enough to draw taunts."[5]

All children try on their parents' clothes—a boy sees how his feet fit in Dad's big shoes and a girl ties on Mom's scarf. Transgendered children typically try on the clothes of the parent whose gender they identify with. A transgendered woman recalls, "I used to try on my mother's clothes when I was about six years old by standing under the hanger of her dresses in the closet. The sensation was one of peacefulness and integration." Conversely, transgendered boys often flatly refuse to wear a dress or throw a tantrum when required to do so. One transgendered man reports always coming home with a dress "accidentally" ripped or stained.[6]

Transgendered children report violence from other children. A transgendered woman remembers, "All through my elementary school years, I was picked on by the other boys and called names like 'wimp,' 'fairy,' or 'Little Lord Fauntleroy.' Even though I was bright and had a high IQ, I didn't know what those words meant, much less why they were calling me that."[7]

Therapists have accumulated thousands of narratives like these, all tucked away in their client records. Transgender expression emerges early in childhood, along with other indicators of personality, temperament, and inclination. These narratives show that transsexualism begins with gender identity, not sex drive. Transgender expression appears before puberty and well before any conscious sex urges. The narratives all reveal a very strong sense of not fitting in, but they also show variation in awareness of gender identity, suggesting that not everyone has figured out what's going on. Some children are completely convinced their gender differs from their body, whereas others keep trying to conform, unaware of where their difficulties come from. The lovely movie *Ma Vie en Rose,* which portrays a boy playing with dolls and wishing to be a bride, seems an unusual instance of a young child who has already come to complete awareness of her gender identity.

Narratives quoted by therapists emphasize effeminate boys who are physically bullied. Therapists acknowledge that although not all "look or behave effeminately, some do."[8] Even with this qualifier, the quotations leave the impression of transgendered children as defenseless sissies. Nonetheless, I know many transgendered women who fought back effectively against bullies while they were living as males.

TEENAGE NARRATIVES: BODY

The narratives shift gear with the arrival of puberty. Testosterone and estrogen cause secondary sex characteristics to develop. At this time, many transgender children become seriously unhappy with their bodies, bodies that up to then had not been noticeably gendered in either direction. A transgendered man recalls, "I simply could not reconcile the physical image with my mental image of myself. . . . I quickly learned to disengage my mind from my body in order to get through." Conversely, a transgendered woman notes, "Puberty came, and I was scared and anxiety-ridden. Was I going to get large and hairy and ugly like all the other boys?"[9]

Some young transgender people react especially strongly to their bodies. Brown reports that teenaged transgendered men sometimes "pound or hit themselves in the chest area until they are covered with black-and-blue marks or in some cases cut their breasts." And a transgendered woman recalls that, as a teenager, "I took the scissors to my genitals, fully intending to cut them. But at the last minute, I couldn't go through with it. I guess I hoped my action would make my parents see the severity of my anguish."[10]

Therapists acknowledge that "while some experience feelings of disgust about their penis, others are merely indifferent to it."[11] In spite of this disclaimer, there is a tendency to emphasize extreme quotations. Of the many transgender people I know, no one has ever volunteered in conversation, or in response to a question, that they mutilated their genitals or attempted to do so. I'm sure some have, but the account of transgendered people as self-destructively hating their bodies has been greatly exaggerated by therapists.

The theoretical goal of therapists is to construct a picture of the "true transsexual" as the reference standard for a sick individual who needs medical attention in every aspect of his or her life. Other transgendered people can then be located on a continuum from normal, through the "mild" maladjustment of occasional cross-dressers, to the deep pathology of the true transsexual. In fact, though, the transgender people I know view their bodies in many ways. Many view their genitals not as a source of pride, but rather as a neutral piece of anatomy, like an earlobe. Their diverse attitudes toward the body are reflected in the differing attitudes of transgendered people toward sex reassignment surgery. About one-third, say, of transgendered women have had sex reassignment surgery (postoperatives), one-third plan to do so (preoperatives), and another third decline it (nonoperatives). Preoperative transgendered women may postpone surgery for many years after transitioning, often saying they don't have the funds. I found it interesting that after San Francisco passed legislation authorizing insurance coverage for sex reassignment procedures, some preoperative women who were now covered decided they were nonoperative after all. Even though the financial hurdle had been removed, they weren't interested in the operation.

Still others actively and overtly enjoy a dual-body form. Many of these people work in the sex trade as "she-males" or the transgendered male counterpart, "he-shes." In pornography, the she-male is portrayed with a proud, erect penis, large breasts, and flowing hair—a very sexually charged image. I don't believe enough dual-bodied people have been interviewed to determine whether they shaped their bodies after entering sex work to satisfy commercial demand, or whether they shaped their bodies first and then entered sex work as the best available employment. The narratives assembled by therapists rarely mention these people because their conspicuously sexual presentation lies outside the mold of "true" transsexuals, who are supposed to be concerned only with gender identity and not sexuality. Transgender people themselves may not feel safe being affiliated with overtly dual-bodied people, or they may be put off by their links to the sex industry.

Overall, the value of sex reassignment surgery depends on how one feels about one's body, how one plans to use it, and the social significance of the surgery. Transgendered men choose among various procedures depending on cost, appearance, and functionality—male genitals with or without urinary function, large or small scrotal regions, and so on, as provided by current technology. A recently transitioned transgender man featured in the *New York Times Magazine* put the matter succinctly, saying of the $50,000 surgery: "We've got college tuition to pay. . . . I'm not *that* interested in a penis."[12] This deliberate cost-benefit thinking behind body sculpting rules out any obsessive motivation.

ADULT NARRATIVES: MALE-TO-FEMALE SEXUALITY

At puberty, when voices drop and breasts swell, the sex drive awakens. For postpuberty transgendered people, the experience involves a complex mix of gender identity and sexuality. Some therapists' writings emphasize gender identity while suppressing sexuality, and others emphasize sexuality while suppressing gender identity. Actual narratives tell a more varied story. After puberty, the lives of transgendered men and women cease being more or less mirror images. As adults, transgendered men and women experience increasingly different hormonal as well as social environments.

At puberty and on into adulthood, cross-dressing by transgendered women becomes more frequent and deliberate. One states, "I was sixteen when I started cross-dressing. I'd just reached Mom's height, and she had a couple of wigs back then. So I'd wait until everyone was gone—I'd make excuses to be home alone—and then dress up and put on a wig. . . . There was an overwhelming sense of everything being right. I remember being dizzy with exultation. In fact, I was always happy when I cross-dressed."[13] This narrative fits the standard mold of transsexual cross-dressing as an expression of female gender identity. And I suspect this narrative is in fact the most common motivation for cross-dressing, leading sometimes to transition. Still, the coincidence of increased cross-dressing with puberty seems more than accidental. In fact, many complex sexual aspects enter the picture as transgendered people become adults.

The simplest explanation for the sexuality of transgendered people is that they are merely gay. The narratives of many transgendered women refute this belief. For example, one transgendered woman recalls, "I came to the common but false belief that feeling as if I should be a woman meant that I was gay. So I set out in earnest in my first semester of college to see if this was the case. Though I would place myself in situations where I could have easily had sex with a man I found attractive,

I never found myself ultimately able to be comfortable with the idea of being in the situation of a gay male and always backed out."[14] Nor do gay males regard a transgendered woman as a gay male. Gay men rarely proposition transgendered women. By contrast, drag queens, a constituent within the gay community, project enough male signals to retain credibility as males. Their male voice and caricature of female dress leave no confusion about their gender identity, and they are often romantically involved with other gay men. Alternatively, transgendered women prior to their transition often try to date other women for a while, attempting to live out the script of a conventional heterosexual male. But, as one person recalls, "I had to eventually stop dating [women] altogether because it felt so unnatural. . . . I didn't want to date women, I wanted to *be* like them."[15]

Some transgendered women meet wonderful men, fall in love, and raise a family together. Others maintain a career and share life with a steady boyfriend. I know this. I have met these people. My estimate is that the majority of transgendered women, maybe 60 percent or so, are sexually oriented to men, and have aspirations in life not much different than those of many other straight women. These women may not regularly consult with therapists, and so don't figure in the narratives compiled by therapists. These women also do not write autobiographies. They often live in "stealth," avoiding all traces of their earlier male life. And even if not living in stealth, the attention from an autobiography might bring discomfort to them and their loved ones.

Instead, most of the transgendered women who have written about their sexuality are those whose existence is denied by therapists' allegiance to the gender-only template. These women keenly feel the need to speak up. For them, sexuality has been of as much or more importance as gender identity, and they employ their gendered presentation to further the realization of sexual completeness.

FETISHISTIC BODY MORPHING

In sadomasochism (SM), by consent, one person inflicts pain on another to provide the recipient with erotic pleasure. In bondage-discipline (BD), by consent, one person humiliates another with spanking, verbal abuse, or scolding to provide the recipient with erotic pleasure. Thus, SM eroticizes pain, and BD humiliation. The submissive person, the bottom, is typically male, and the dominant person, the top, is female.[16]

Sometimes SM/BD is practiced with a submissive male who is cross-dressed as a maid, or nurse. This eroticizing of a subordinate female role would appear to devalue women. However, the reality is more complex. "The Mistress helps the transvestite build up a good positive image of himself as a woman . . . and makes a strong effort to treat the transsexual and transvestite in all respects as a woman—to impress upon her that to be female is good."[17] Why would a mistress, or dominatrix, who is herself female, cooperate in devaluing females? Apparently she doesn't. Instead, the dominatrix trains the submissive male to be a woman. The submissive male can then act out becoming a woman because he is coerced to do so. In this way, given an excuse for becoming female, the submissive person creates his own path for a change of gender.

The transgender activist and historian Susan Stryker, director of the LGBT Historical Society in San Francisco, describes assuming a transsexual identity: "In 1990 . . . I was neither a lesbian nor a gay man nor a transsexual. . . . My desire to be with women sexually was anchored by . . . identification with female morphology through lesbian fantasy," even though Stryker functioned during the day as a heterosexual man in straight society. Rather than identifying as a transsexual, "I found another set of technologies . . . to enact my sense of self—gaff and gauntlet rather than scalpel and syringe. In dungeons and drag bars I discovered both. . . . space . . . and an audience." Later, though, Stryker decided that "transsexual technology would be my vehicle for . . . an impulsive leap into the real. . . . Naming myself transsexual was only. . . . a useful move." Stryker then elected to tell the personal history of her "bodily inscription in a politically productive way."[18] Her narrative doesn't sound like the now-classic story of a straight woman trapped in the body of a male. Stryker's reason for identifying as transsexual and completing sex reassignment surgery was to acquire the body for both her sexual identity and her gender identity.

A different narrative of transgender sexuality pertains to cross-dressing: "For many of us, sexual desire is the origin and the kernel of our transsexual impulse," writes Anne Lawrence, a trangendered woman and medical doctor formerly with a practice in Seattle.[19] In 1996, at a conference of postoperative trans women on Ocracoke Island in North

Carolina, Lawrence asked if people were sexually aroused before undergoing sex reassignment surgery.[20] Many said they were. Apparently, one-quarter of the dozen people at this gathering obtained their sexual reassignment surgery in part to fulfill an autoerotic sexual desire. In a 1998 follow-up, Lawrence found that about half of nearly a dozen participants confirmed that before surgery their "favorite erotic fantasy was that they had, or were acquiring, some features of a woman's body." Another investigator also found that over a quarter of postoperative trans women described being sexually aroused by the prospect of surgery.[21] I've asked transgendered women friends about this. Some agree that the prospect of having female genitals is sexually stimulating in and of itself, independent of the potential for subsequent sex with a male partner. For some transsexuals, genital surgery is partly fetishistic.

Undergoing genital surgery to fulfill autoerotic sexual desire raises difficulties. After the sexual excitement has died down, then what? The person now resides in a female body, yet life goes on. Lawrence writes, "The qualities we need to cultivate to live successfully in female roles can be very rewarding in their own right. Learning to embody such feminine traits as gentleness, empathy, nurturance, and grace improves the quality of our lives, and simply makes us better human beings." For Lawrence, transition first fulfills an autoerotic sexual desire and is followed by efforts to become womanlike.

Fetishistic transgender expression was once viewed as being unique to cross-dressers. I've met men who have purchased one hundred pairs of women's shoes, one hundred bras, or if they can afford it, one hundred complete outfits. Fetishistic transvestites are heterosexual men erotically stimulated by wearing women's clothing. They might ask a girlfriend or wife if they can make love while cross-dressed. In a perfect "only in San Francisco" moment, I was once introduced to a couple where the woman loved to dress her man in female clothes and the man loved to be dressed up by her. She had trouble keeping boyfriends, and he trouble keeping girlfriends, until they met and voila!

Fetishistic desire must be managed, like alcohol. The classic movie *Days of Wine and Roses* describes a romance that begins with sharing wine and other drinks, but where eventually the alcohol becomes more important than the relationship. With fetishes, too, attention can focus on the fetish object to the neglect of human relationships. Lawrence acknowledges that the fantasy of acquiring female genitals "often does seem to compete with arousal toward other people." The "partner is almost superfluous, or merely acts as a kind of prop." Just as one prone to alcoholism needs to take special care in life, so does one drawn to fetishism.

Cross-dressers aren't necessarily fetishistic, although some are. Some cross-dressers enjoy a feminine identity part of the time, as a chance to get away from a super-male work environment, to dress with a splash of color, or for other reasons. I believe Lawrence's narrative brings out a valid distinction between transgender expression motivated by gender identity and that motivated by fetishism. Transsexuals can be motivated by either, and cross-dressers can too. On the gender-identity-motivated to fetish-motivated spectrum, transsexuals cluster more toward the former and cross-dressers toward the latter. Each transgendered person probably has his or her own personal mix of these motivations.

Although Lawrence's narrative demonstrates that an autoerotic component can exist in male-to-female transsexualism, I'm not persuaded that many people match this profile. Lawrence herself is certainly unusual. She has long been interested in body morphing, and has posted photos on the web in which she used PhotoShop to place an image of her own face in famous works of art, like the Mona Lisa.

Lawrence claims, though, that she can generalize from her own experience and that she wishes to take the gender out of transgender. She solicited narratives to reveal others identifying with transgender autoeroticism, posting twenty-eight responses on the web.[22] As I understand them, none of the responses states that an autoerotic sexual drive was the primary reason for transitioning and pursuing sex reassignment surgery, although many of them acknowledge some autoerotic sensation as part of their overall experience. Indeed, some of the narratives directly contradict the primacy of autoeroticism. Yet the narratives that Lawrence posted are the ones most likely to be supportive. Lawrence discourages counternarratives: "Please note that I am not interested in statements from persons who have never had such feelings, or who object to the idea that other people might have them. I have plenty of such statements already."

INCLUSION

As far as I can tell, the vast majority of narratives freely told by trans-gendered people among themselves, and those recorded by ethnographers cross-culturally and through history, demonstrate that actualizing gender identity—not sex drive—is the primary motivation for transgender expression. Narratives also show that body morphing practices, such as sex reassignment and facial surgery, are done primarily to promote relationships, making people more attractive to their sex partners or enabling them to join the social groups or occupations of their choice. Many transsexuals are appalled by the idea of autoerotic motivation and can remember only terror before their sex reassignment surgery: that surgery could be sexually arousing seems preposterous and insulting to them.

However, even a single case of autoerotic transsexualism raises the issue of inclusion. Here is a test of whether we're really inclusive ourselves. Do we really believe in diversity, or are we riding its bandwagon? After all, it shouldn't matter why a sister becomes a sister. I will love her and support her anyway. At the same time, the sensational publicizing of autoerotic transsexualism poses a threat to the future of transgendered people. Today we transgendered people may enjoy the best prospect we've had since the time of Jesus to enter mainstream Western society and live productive, normal lives. We don't want this prospect to be undermined by bizarre sexualities. We don't want to give ammunition to those who wish to pathologize us, and endorsing autoerotic transsexualism would seem to do just that.

Autoerotic transsexualism is at most a minority within a minority. Its inclusion within the transgender community reminds us of the dilemma faced by gay and lesbian organizations when deciding whether to include transgendered people. Those who struggled over thirty years for gay and lesbian rights didn't wish to see that work jeopardized by admitting a relatively small number of transgendered people. Similarly, transgendered people now don't want to see their work to secure rights and recognition jeopardized by a sexual minority within their midst. Yet gay and lesbian people *have* included transgendered people. Similarly, I believe we must include transsexuals motivated more by autoerotic impulse than by gender identity, however few they may be. To do otherwise cedes the moral high ground.

Biologically, an autoerotic component to adult transgender feminine expression comes as no surprise. A female persona in a male body must survive testosterone. This chemical induces male libido, complete with autoeroticism. Some 41 percent of American men, but only 16 percent of American women, each month purchase autoerotic materials, such as X-rated movies, books, magazines, or sex toys, or visit a nude club or call for phone sex.[23] Thus, about 40 percent of all male-bodied people act financially on autoerotic desires each month, and many more act in ways not involving a financial transaction. Autoerotic cross-dressing is one of many autoerotic activities that male-bodied people do on a regular basis. Inevitably, some adult transgendered women combine autoerotic activity with feminine identification.

ADULT NARRATIVES: FEMALE-TO-MALE SEXUALITY

Many transgendered men awaken to a romantic interest in women and try for some time to live as lesbians. Therapists report statements like "I had occasional lesbian encounters in high school and college, but I never felt lesbian. I felt male, and when I would go to mix with lesbian groups, their issues seemed vastly different from mine. It was one more place I didn't belong."[24] In 1977, in the first autobiography of a trans man, Mario Martino stated, "I was a boy. I felt like one, I dressed like one, I fought like one. Later I was to love like one."[25] Even before transition, Martino enjoyed arousal from "girlie magazines" in his cousin's bedroom. Pat Califia comments, "Sex is central to this narrative. . . . Having sex with and gratifying a heterosexual woman is even more important to Martino's gender identity than possession of a virile physique."[26] Drew Seidman, whose reaction to testosterone was noted earlier, said in an interview after transitioning and starting testosterone, "I was into porn as a girl, but now I'm really into porn. It really gives me insight on males."[27]

A steamy testimonial on sexuality comes from the well-known transgender activist writer Jamison Green. After transition, he started to move beyond the "pages of pornographic magazines that I kept beside my bed, in the car. . . . Gradually I began to see that real women were starting to notice me, . . . seeing me as a man at last, not as a boy, not as a lesbian, not as an androgyne, but as a bearded, hirsute, solid physical man with something to offer them. The first time a woman did suck my cock I was amazed by the feeling of it, by the sheer joy she gave me as she swirled her tongue around my cock's head, as she slid her lips along its shaft, as she looked up at me, pleased with herself. . . . She enjoyed every inch of her body under my touch."[28]

Increasingly, the adult narratives of transgendered men and women are acknowledging sexuality and putting the sex back in transsexuality. Still, sexuality is not a big part of the transgender experience for everyone. Although ignoring sexuality is inaccurate, exaggerating its role is a mistake too. I know transgender people who are simply not very interested in sex. Perhaps one-fourth are not involved in relationships and not sexually active, but their narratives are short and easily overshadowed by those with complicated fantasies and practices. Consider that among nontransgendered people, 10 percent of women and 14 percent of men aren't sexually active with a partner for an entire year.[29] These percentages seem appropriate for transgender people as well.

CARRYING ON ANYWAY

In spite of these swirling internal currents of gender and sexuality, transgender people usually try to live in the sex they were assigned at birth. As Brown and Rounsley report, "They dress the part, develop their body, join groups, immerse themselves in careers, date the opposite sex, get married, have children."[30] Many trans women prior to transition seek the most rugged, stereotypically male profession or job they can find—law enforcement, auto or airplane mechanics, driving big rigs, or working in steel mills, auto manufacturing plants, or heavy construction. Military service is also a popular route. Brown and Rounsley report that over half of their clients served in the military, often taking the most rigorous or dangerous missions they could find. I've personally met two former fighter pilots, as well as former marines, who are now transgender women. In addition, transgender women prior to transition may work out, grow a beard or mustache, and affect a hypermale image through haircut, clothes, and demeanor. This may all need to be undone, because bulking up through weight-lifting interferes with a feminine presentation later on.

Therapists report that many transgender people cope using stress reduction techniques such as visualization, guided imagery, and meditation. Excessive immersion in work and career is another survival tactic. As one trans woman recalls, "I was a super-computer-kind-of-fast-tracking character who went in five years from making twenty-two grand to one hundred grand. I found that I could do that by behaving in this stereotypical macho kind of way they seemed to expect from me. I threw myself totally into my work. I would fly around the world and speak to people about our products and develop relationships but not really be there. But it really affected me. . . . I felt like I was dying little by little."[31]

Most adult transgendered clients suffer in silence and isolation for a long time before letting *anyone,* even a therapist, know about their gender-related conflicts.[32] One reason for silence is not knowing what to say. Therapists sort their adult clients into the "knowing" and the "confused." The knowing are aware with certainty that they are transsexual and need advice on how to proceed and how to deal with accumulated personal issues. These people are also likely to have already participated somewhat in activities of the transgender community.

The confused are unsure of what a transsexual is: "I had no real way to define myself. I knew who I *wasn't,* but had no idea of who I *was.* I called myself a transsexual with no real understanding of what it meant to be transsexual, without ever having met people who thought of themselves as transsexual to see if we were talking about the same thing."[33] These transgendered people usually come directly to a therapist without having immersed themselves first in transgender culture. Their first point of contact with other transgendered people may be the support group meeting that the therapist holds so his or her clients can meet one another. A transgendered person's conception of what it means to be transgendered is influenced by whether the first encounter is with a therapist or through contact with the trans community. The therapy route is more stigmatizing, as a result of its framing of transgender expression as a disease needing a cure.

TRANSGENDER TRANSITION

For transgendered people, coming out is called "transition." This is the time when people switch from living as the gender they were assigned at birth to living in the gender of their identity. In the United States, the protocol for transition under medical supervision, called the Standards of Care, requires working with a behavioral therapist for at least three months. Then, on referral from the therapist, a physician prescribes hormones. One may continue to take hormones prior to transitioning until the physical effects become unmistakable.[34] "Passing"—that is, being recognized as belonging to the gender of identity—preoccupies everyone at this time. Passing is necessary for survival. Without passing, one is challenged in public restrooms and stared at on the street and in restaurants. One may not even be able to buy groceries. One may be attacked or ridiculed. The opposite of passing is being "read."

Nontransgendered people are often amazed at the lengths to which transgendered people go to fashion a gender-typical presentation. The steps include removing facial hair for trans women and getting rid of breasts, cosmetic facial procedures, and bodybuilding for trans men. Perfect passing isn't necessary for safety: one's presentation simply has "to work." But passing to some degree is necessary for living a productive life in today's society. People whose identity doesn't conform in large part to accepted gender norms shoulder an extra burden.

Living full-time in the gender of identity during a one-year trial period is called the "real-life test." After this year, one is eligible for sex reassignment surgery, carried out by a surgeon upon referral by two behavioral therapists. Transition brings insecurity and unknowns, plus gains and losses. A transgendered man comments, "Sometimes I wonder if my losses will be too many and too great to recover from them, especially when constantly facing an uneducated, unsympathetic world and ceaselessly wondering whether time and hormones will ever allow me to escape their scrutiny and judgments." Similarly, a trans woman says, "What really strikes home for me is that all transsexuals lose something once they come out. The question I ask myself is, 'How much will I lose?' I've laid everything I've achieved in life—job, relationship, family, health, future—on the table, and it seems fate will decide what I am allowed to keep, if anything. It's kind of like starting life all over again."[35] For most trans people, transition is a far bigger moment than sex reassignment surgery, which, if it occurs at all, is often viewed as merely "icing on the cake."[36] Transition is taking two steps backward for four steps forward. If one is near the edge to begin with, two steps are enough to fall off.

The major practical issues during transition are finding a social network to be part of, coming out to family, and coming out on the job. The therapy literature includes sample letters drafted for employers, letters to relatives, and so forth. Issues include whether one changes jobs or stays in the same position. In decades past, transition was modeled on the protected witness program. A transgendered person resurfaced in a new city with a new name, career, and fabricated past—ultimate stealth. Today, transgendered people increasingly transition in public, on the job, acknowledging a past life and carrying their abilities forward.

The tactic most commonly used today to inform one's boss plays the medical card. One transgendered man recalls, "I set up an appointment and met with my boss. I explained that I had been diagnosed as gender dysphoric [medicalese for being transsexual] and briefly outlined the Standards of Care my doctor was following in my transition from female to male."[37] The interview went well, and the transgendered man retained his job. Just about everyone I know has transitioned using a variant of this narrative. It works. This is no time to worry about whether gender dysphoria is a made-up disease. The feelings are certainly genuine; the issue is whether these feelings constitute a "disease."

Despite all these seemingly endless traumas and obstacles, transition is wonderful. Transgender people often become euphoric after transition, dancing around their apartment, in love with the world. I know I discovered great goodness in others when I transitioned. I felt I could open up to people and empathize better than ever before.

GAY AND TRANS NARRATIVES COMPARED

In view of how much variety there is in transgendered experience, one might wonder if there is any discernible difference between transgendered experience and gay and lesbian experience. Many people are somewhat gender-variant in their youth. Many girls enjoy playing sports and climbing trees as tomboys. Many boys enjoy pursuits that don't require roughhousing. Of the children who are gender-variant, most mature as heterosexual gender-typical adults, some as gay or lesbian, and a still smaller fraction as transgendered. In addition, many people who are gay and lesbian didn't show gender variance as children.

The narratives of gender-variant children who mature as gay or lesbian are perceptibly different from those of children who are transgendered, primarily in their emphasis on sexuality rather than gender identity. For gay and lesbian gender-variant children, gender is often instrumental to sexuality rather than an end in itself. A recent collection of essays and narratives about childhood gender nonconformity by gay and lesbian people suggests the differences.[38] For example, Michael Lassell, an adult gay man, recalls, "Did I actually or literally want to be a girl when I was a child? Yes. But only because if I had been a girl I could have done all the things I loved doing. . . . I certainly knew my feelings for Georgie Bowen in gym class had something to do with . . . his adult penis."[39] Similarly, Kim Chernin, a lesbian, writes, "I was a boy. I felt the desire to gaze, to pursue them [women], to possess them, to take them to me, as was my right, do you understand? To feel that you have a right to a woman's body? That is what I mean by being a boy."[40]

From these examples at least, it seems that when gay and lesbian children show gender variance, sexual arousal and sexuality figure more prominently in their narratives than they do in transgender narratives. Boys who grow up to identify as gay and those who come to identify as transgendered both typically dislike gym. For gays, the discomfort seems to be more about dealing with the sexual drive they feel in the locker room, and for transgendered people it's more about feeling they're in the wrong locker room to begin with. Still, edges of the drag-queen subculture among gay men and of the butch presentation among lesbians blend seamlessly into the transgender experience.

16

Disease versus Diversity

A major threat to the human rainbow is the misclassification of human diversity as disease. Conventional techniques, from surgery to brainwashing, are applied to diverse peoples, often maiming them. Even those who escape overt injury live stigmatized lives, believing something is wrong with them. How could these abuses happen in today's world?

Medicine's pathologizing of diversity springs from the absence of a scientific definition for disease. Medical dictionaries feature definitions like this: "Disease is an impairment of the normal state of the body that interrupts function, causes pain, and has identifiable characteristics."[1] The problem with this definition lies in the concepts of normal and function, both of which refer to data beyond the individual patient.

Medicine does not define normal. How common does some trait have to be to be considered normal? Medicine is silent on the cutoff point for normal. And what is normal anyway?[2] Was Einstein diseased? Nor does medicine define function. How people or certain aspects of them function becomes clear from observing them in their environment, not in a doctor's examination room. And which functions count? Is a hand diseased when it is too small to palm a basketball? The absence of scientific definitions for normal and function opens the door for societal norms to take over—allowing social values to masquerade as science.

Another part of the definition of disease refers to how an individual feels. If an individual feels pain and the symptoms fall into a familiar pattern, then a doctor is expected to prescribe some treatment. This criterion seems the most useful part of the definition. Normalcy alone, without consideration of pain, should not be used as the criterion for determining if someone is diseased, as though one should be feeling pain even when one does not.

THE CRITERIA FOR A GENETIC DEFECT

In contrast to medicine, biological science has clear criteria for a "genetic defect" and doesn't use the phrase "genetic disease." Any genetic trait is an investment that may pay a dividend in offspring at some time and place. But at any particular time and place, a gene can be down on its luck. Only an inherited trait deleterious under *all* conditions can be considered a genetic defect. Furthermore, a trait that is deleterious under all conditions is necessarily rare (because it's continually being opposed by natural selection). Thus, to be a genetic defect, two scientific criteria must be satisfied—the trait must be extremely rare *and* the trait cannot be advantageous under any condition. If one of these criteria is not met, then the trait cannot be considered a genetic defect.

NOT RARE ENOUGH

Genetic defects are automatically weeded out over time by natural selection. The only way defects resurface is by mutation from adaptive genes into deleterious forms. The degree of rarity for a genetic defect is set by a balance between two rates: the rate of formation by mutation and the rate of elimination by natural selection. This level of rarity is called a mutation-selection equilibrium.[3]

A simple table shows the relation between how defective a trait is and its rarity:

RELATION BETWEEN RARITY AND
SEVERITY OF A DISEASE

Births	Reduction in Darwinian Fitness
1 in 10	0.001%
1 in 100	0.01%
1 in 1,000	0.1%
1 in 10,000	1%
1 in 50,000	5%
1 in 100,000	10%
1 in 1,000,000	100%

If a trait is lethal, then it occurs only as often as the mutation rate, which is one in a million, as noted in the bottom row of the table. If the reduction in Darwinian fitness (the probability of surviving to breed times the number of offspring produced) is 10 percent, then the frequency of the trait rises to 1 in 100,000. If the reduction in fitness is only 5 percent, the trait occurs at a frequency of 1 in 50,000. I'll take this figure as the convention for the threshold rarity at which a trait can be considered a defect. Even if a trait isn't particularly harmful, and a 5 percent loss of fitness wouldn't be all that easy to detect, this degree of disadvantage, if sustained through all generations everywhere, would eventually lead to the trait becoming as rare as 1 in 50,000.

Meanwhile, relatively common traits, say those in the 1 in 10 to 1 in 1,000 range, can only be consistent with a tiny and undetectable loss of fitness. A trait causing a loss in fitness of, say, 0.1 percent to 0.001 percent cannot be considered a "disease" in any sense, because people with or without the trait are not detectably different, and differences this small are easily masked by the random differences faced in consecutive generations. This point is fundamental. The phrase "common genetic disease" is a contradiction in terms.

In summary, a relatively common trait cannot be classified as a genetic defect according to biological science, regardless of medical opinion. If the trait is, say, ten times more common than the cutoff value (1 in 50,000) and has been traditionally considered a "disease," then either the trait's overall disadvantage has been overestimated to begin with, or else the trait has some possibly unknown advantages in addition to the known disadvantages. And when genes have both advantages and disadvantages, the effects may be felt in the same or different individuals.

SOMETIMES ADAPTIVE

As indicated, a genetic trait also cannot be called a genetic defect if it's adaptive in some circumstances. To prove that a trait is a genetic defect requires showing that the trait cannot be adaptive under any conditions. Exhibiting a condition in which the trait is adaptive falsifies the claim that it is a genetic defect.

To test whether a genetic trait is a defect, one may hope to discover the precise circumstances in which a trait is beneficial. Unfortunately, the uses for traits pertaining to gender and sexuality are often unknown because the depths of human history are obscure. Still, exhibiting a natural function is the most informative way to falsify the claim that a trait is a genetic defect.

MISCLASSIFYING TRADITIONAL DISEASES

It isn't easy to tell whether a genetic trait is a defect. Indeed, many of the "traditional" genetic diseases may be partly misclassified. A table in *Time* summarized the frequency of "commonly inherited disorders,"[4] but how many of these are really genetic defects, which, as we've seen, must be exceedingly rare? Huntington's disease, with 4 to 7 occurrences per 100,000 births, is a likely genetic defect, as is perhaps the sex-linked hemophilia A, with 1 birth per 8,500 males. But others in the table are rather common, from 1 in 1,500 to 1 in 3,600 for cystic fibrosis, muscular dystrophy, fragile X syndrome, polycystic kidney disease, and Tay-Sachs disease in Ashkenazi Jews. Sickle-cell anemia is about 1 in 500 births in blacks. Calling many of these conditions genetic defects is premature, when we haven't accounted for the discrepancy between their apparent deleteriousness and their relative commonness.

It is important to ask why some of these conditions are so common because, if the genes are eliminated, whatever good they do will be lost along with their bad effects. Sickle-cell anemia is caused by a gene that is good in a single copy because it protects against malaria, but bad when paired with a copy of itself. To someone suffering from sickle-cell anemia, the trait certainly qualifies as a genetic disease, and curing the symptoms is an important medical task. Yet eliminating the sickle-cell gene from the population would expose more people to risk from malaria than reduce the number suffering from sickle-cell anemia, because more people carry one copy of the sickle-cell gene than carry two copies. Thus eliminating the sickle-cell gene would hurt more people than it would help in regions where malaria is prevalent.

Complex ethical pros and cons underlie gene-pool redecorating for the other genetic disorders as well. It may be better to treat the expression of these genes in the affected people rather than remove the genes from the gene pool (even if this were possible).[5]

HOW COMMON ARE HOMOSEXUALITY AND TRANSSEXUALITY?

Turning to LGBTI traits, could they be genetic defects? The fraction of people who are gay or lesbian is between 1 in 10 and 1 in 100, depending on how the category is defined. Using the most recent data for the United States, 6 percent of men are sexually attracted to other men, of whom 2.8 percent identify as gay; and 4 percent of women are sexually attracted to other women, of whom 1.4 percent identify as lesbian.[6] Let's take 5 percent as a working figure. Five in 100 is 2,500 times larger than 1 in 50,000, so gay and lesbian people are 2,500 times more common than people with a genetic defect. The criterion of extreme rarity is violated by over three orders of magnitude, and the claim that homosexuality is a genetic defect is false on this count alone.

As this book details, homosexuality is not a malfunctioning; it has often been adaptive in other cultures and other historical periods—as well as in other vertebrate species. Moreover, being homosexual is not disabling or painful in itself. Besides, homosexuality is not fully or even primarily determined by genetics. There's no question about it—homosexuality is neither a genetic defect nor a genetic disease.

What about transgendered people? Uncertainty surrounds the number of transgendered people. Until recently, the figures being bandied about were 1 in 10,000 for male-to-female transsexuals and 1 in 30,000 for female-to-male transsexuals, based on data from Holland.[7] These numbers are bigger than the 1 in 50,000 figure asserted by the earliest literature, but still small enough that transsexualism could be a border-line genetic defect. Furthermore, no one has ever suggested that being transgendered is somehow adaptive.

Although being transgendered is in itself neither painful nor disabling, the agony preceding transition is debilitating to many. Therefore, most transgendered people, myself included, have accepted that being trans-sexual means carrying a disadvantageous, and presumably biologically determined, trait—a genetic defect of some sort. Transgendered people then set out to live their lives as best they can with "their condition," much as anyone else with some genetic disease would.

However, new data increasingly undercut the interpretation of trans-sexualism as a genetic defect. Lynn Conway, a transgendered engineer, first raised a red flag simply by counting the sex reassignment surgeries (SRS) performed by leading surgeons, which turned out to be a large number.[8] According to Conway, one surgeon, Stanley Biber, has performed over 4,500 SRS operations since he began in 1969; for many years, Biber did two SRSs a day, three days a week. Another leading surgeon, Eugene Schrang, maintains a similar schedule, and together with a third leading surgeon, Toby Meltzer, presently performs a total of 400 to 500 SRS operations every year. Including the operations by other surgeons, this leads to an estimated total of 800 to 1,000 each year in the United States.

At, say, 1,000 operations per year, over forty years, there would be a total of 40,000 people in the United States who are postoperative male-to-female transsexuals. Dividing this 40,000 by the approximately 80,000,000 males in the age range from eighteen to sixty years, in which the surgeries occur, yields 1 in 2,000. The discrepancy between 1 in 2,000 and 1 in 50,000 makes all the difference in the world. The 1 in 2,000 figure is 50 times higher than the level consistent with a genetic defect. Furthermore, the figure of 1 in 2,000 pertains to postoperative transsexuals. Many are not counted in this statistic—preoperative and nonoperative transsexuals. Conway suggests that taking these additional people into account leads to a rate of male-to-female transsexualism greater than 1 in 500, implying that the oft-quoted medical consensus is "wrong by more than two orders of magnitude."[9]

Similar statistics for the United Kingdom reinforce this claim. In 1997–98, government services performed 44 male-to-female operations, and the private sector, 104 gender reassignment operations, for a total of about 150 per year.[10] Doing the math, 150 per year times 40 years yields about 6,000 people.[11] With about 18,000,000 males between ages 18 and 60, the number of postoperative transsexuals in the United Kingdom is about 1 in 3,000.[12]

Furthermore, the main government clinic where sex reassignment surgeries occur, the Gender Identity Clinic at Charing Cross Hospital, saw 470 new cases and, of these, performed 44 male-to-female operations, suggesting an approximate 10 to 1 ratio of people who self-identify medically as transsexual to people who qualify for the operation. So the number of transsexual-identified people may be 10 times as high as the number of postoperative transsexuals, which would bring the number of transsexuals in the United Kingdom to about 1 in 300, in approximate agreement with Conway's calculations for the United States.

Both the U.S. and U.K. estimates are back-of-the-envelope calculations and will surely be refined in the future. Still, the estimates are robust. Consider the alternative. If the fraction of transsexuals were 1 in 50,000, as required by the genetic defect criterion, then the U.S. population of 80,000,000 males would contain only 1,600 transsexuals. If these people all became postoperative during the forty years from age 20 to age 60, then the number of SRS operations would amount to only 40 per year. We know, though, that at least 400 SRS operations per year are carried out by just three of the many surgeons who perform the procedure. Thus, the claim that transsexuality is a genetic defect is directly falsified by data showing that ten times more SRS operations are performed each year than predicted by the hypothesis of genetic defect.

Moreover, the estimated percentage of transsexuals in the United States and the United Kingdom is in the same ball park as the percentage of *hijra* in India. The number of hijra is over a million people in a population of one billion, which works out to about 1 per 1,000 (see p. 341). Meanwhile, India features other transgender categories in addition to the hijra, so the number of transgendered people is greater than 1 in 1,000, although perhaps not as high as 1 in 500.

Exact estimates of the transgendered fraction of the population await a detailed statistical and demographic study. In the meantime, I'll accept 1 in 1,000 as a working number for the fraction of the population with transgender gender identity. A ratio of 1 in 1,000 for transsexuals reframes the discussion, removing transsexualism from the realm of genetic defects into the realm of natural, normal, though uncommon, forms of human variation. One in 1,000 is in the 99.9 percentile, a desirable score for college entrance exams, like an IQ of 130. Being transsexual is having a TGIQ, transgender-identity quotient, of 130.

If transsexualism is no longer scientifically tenable as a genetic defect, the question arises: What, if anything, could possibly be adaptive about being transsexual? In Western antiquity, and in non-Western cultures, male-to-female people enjoy special occupations in which it is appropriate and useful for a male-bodied person to live in female and private spaces. Part 3 of this book surveys these situations. Furthermore, in other vertebrate species, such as sunfish, feminine males have distinct social roles. Similarly, in other vertebrates, masculinity is used by females to regulate the degree of sexual interest by males and to control how often they are solicited. I don't know whether the function of such cross-gender expressions in other vertebrates is relevant to humans. Until now, no one has seriously considered the possibility that being transgendered is adaptive.

Everyone, even transgendered people, has been secure in the belief that transsexualism is a medical anomaly of some sort. As such, narratives about successful transgendered people are expected to relate how they achieved productive lives in spite of their "disability." These narratives may be appropriate for successful transgendered people surviving in present-day Western society, but might be inaccurate for other times and places, when being transgendered was inherently valuable. As future research investigates transgender expression from a positive perspective, more possibilities will surely emerge for how being transgendered could be adaptive in itself.

What about a genetic basis for transsexualism? Little is known about this, although the genetic and hormonal contributions are plausibly higher than for homosexuality because gender identity probably forms earlier in development, when genes and hormones play more of a role, than sexual orientation.

Putting these points together, transsexuality now seems too common to represent a mutation-selection equilibrium; speculative scenarios from comparative anthropology, history, and animal behavior suggest that transgender expression may be adaptive in special situations; and being transgendered is in itself neither painful nor disabling. So is transsexuality a genetic defect or a genetic disease? Probably not.

IS INTERSEXUALITY A GENETIC DEFECT?

The criterion for what counts as intersexuality is not well defined to begin with. Basically, if a doctor can't unambiguously assign a baby as male or female upon birth, then the baby is intersexed. Thereafter, the baby is run through a checklist to see what its "true" sex is, and a baby is "assigned" to one of the two sexes as a result. Based on the assignment, the genitals are often sculpted with a scalpel to match some expected norm for genitals. This procedure presupposes that the binary distinction between male and female applies to the whole body, which, as we've seen, is not necessarily true.

Let's see if intersexuality qualifies as a genetic defect or genetic disease. Most medical doctors would consider the answer self-evident and be annoyed at the question. Yet intersexuality fails to pass even the initial criterion of having "identifiable characteristics." Being intersexed is defined by the absence of characteristics, not their presence. As a result, the category of intersexuality includes hundreds of different genetic, biochemical, and anatomical states, a few of which are arguably genetic defects, and others not.

HYPOSPADIA

Of the many bodily states lumped under intersexuality, the most common is hypospadia—boys whose urethral opening on the penis vents below the tip. Although the assumed normal opening is at the tip, a study of 500 men revealed that only 55 percent had urethral openings there, whereas 45 percent had the opening somewhere below the tip.[13] If the opening is somewhere in the bulbous end of the penis (the glans penis), the hypospadia is considered minimal, and one in every two boys seems to have such mild hypospadia, although the point at which it is noticed at all varies among pediatricians. If the opening is along the shaft or below the penis on the body wall, it is called medium or severe, and the commonness drops to 1 in 1,725.[14]

Clearly, mild hypospadias should not be considered a defect because they are not painful, are apparently not deleterious in any way, and are far more common than expected with a mutation-selection equilibrium. Medical researchers are beginning to suggest that this variation in the location of the urethral vent is "normal" after all.[15]

CONGENITAL ADRENAL HYPERPLASIA

The next most common bodily state lumped under intersexuality is called congenital adrenal hyperplasia (CAH). There are a dozen or more subvarieties of CAH. A gene called CYP21 on an autosome (chromosome other than X or Y) makes a protein that catalyzes the conversion of progesterone to cortisol, a stress hormone, in the adrenal glands next to the kidneys. If this gene is absent or blocked, then progesterone accumulates, which is androgenic in itself and is also converted to other androgens, like testosterone, outside the adrenal glands. In females, the activity of this gene's protein influences the masculine/feminine balance of the body. The lower the activity of CYP21's product, the more masculine the body, and the higher the activity, the more feminine the body.

So-called nonclassic or late-onset CAH is the most common, and refers to CAH that arises anytime after the first five years of life. Medical attention is attracted when a girl has early puberty, thick hair in a masculine body pattern, possible male-pattern baldness, and menstrual irregularity. The trait varies from very common to rare: 1 in 27 for Ashkenazi Jews, 1 in 52 for Hispanics, 1 in 62 for Yugoslavs, 1 in 333 for Italians, and 1 in 100,000 for a mixed Caucasian population. A species-wide average has uncertain meaning given this enormous geographic variation, but one review places the overall commonness among humans at 1.5 percent, or 1 in 66 people.[16] In many girls, CAH simply produces a large clitoris, and in many XY-bodied people, it has no effect.

In contrast, classic CAH is observed at birth in females as ambiguous genitals that may include not only a large clitoris but also fused labia comprising a partial scrotum and a urethra contained in the clitoris, making a micropenis, together with a uterus.[17] In males, the genitals offer little clue. Two-thirds of classic CAH people also lose or "waste" salt because the adrenal glands don't produce an additional hormone needed for salt metabolism. This salt-losing or salt-wasting version (SL-CAH or SW-CAH) is contrasted with the version applicable to the remaining one-third of classic CAH people, called simple-virilizing (SV-CAH), which does not affect functional salt metabolism. Without being given cortisol and other hormones made by the adrenal glands, SL-CAH people are likely to die as infants. More males die of SL-CAH because their genitals do not suggest a diagnosis that salt metabolism is at risk, whereas the risk in females is more likely to be diagnosed because of intersex genitals.

Classic CAH is rarer than late-onset CAH, but also shows enormous geographic variation, ranging from 1 in 300 in Yupik Native Alaskans, 1 in 800 in other Native Alaskans, 1 in 3,000 on La Reunion Island, 1 in 5,000 in Switzerland, 1 in 7,000 in Brazil, 1 in 8,000 in the Arab population of Israel, 1 in 9,000 in Austria, to 1 in 40,000 in the United States. One review puts the worldwide average at 1 in 17,000.[18]

CAH reveals the microcosm of issues that intersexuality raises. A heterogeneous category of people is lumped together in a pathologizing narrative using power words like mutation and blocking. One tail of the distribution of CAH people does arguably suffer from a genetic disease: salt-losing CAH is genetic, painful, life-threatening, deleterious under all circumstances, and arguably rare enough to represent a mutation-selection equilibrium in the gene pool. People at the other tail of the distribution, however, are damned by association. Labeling people as CAH is stigmatizing. Nothing is wrong with a large clitoris. The nonpathologic side of CAH cannot qualify as a genetic defect because it's too common—thousands of times more frequent than a condition maintained at a mutation-selection equilibrium—and it's not deleterious under all conditions. A large clitoris may have been adaptive during our evolutionary history, seeing as the clitoris in many of our primate relatives is large and pendulous. We know little of the mating habits of early humans and can't rule out some positive function to a large clitoris in the past, or even today.

ANDROGEN INSENSITIVITY SYNDROME

A form of intersexuality called androgen insensitivity syndrome (AIS) pertains to people whose sex chromosomes are XY. The Y chromosome with its SRY gene helps the gonads to differentiate as testes and produce testosterone, while the X chromosome contains a gene that produces receptors for testosterone. This gene, called Xq11–12, has many alleles (150 to date),[19] which determine how much effect testosterone can have on the body. Thus, how much body masculinity is expressed by SRY on the Y chromosome depends on the outcome of its negotiation with Xq11–12 on the X chromosome. AIS is characterized by a very feminine body in an XY individual, as a result of receptors that don't bind strongly to testosterone so that the body's testosterone has little effect on the body's appearance.

AIS comes in three major classes. In complete AIS, the person has a fully typical female appearance with respect to external genitals, breasts, hair distribution, and voice, is raised as a girl, and is a girl as far as gender identity. Partial AIS involves a mixture of feminized and masculine features, leading to an ambiguous sex classification at birth. In mild AIS, the person is classified as male upon birth but later shows some feminine features, like body hair distributed in a female pattern and possibly impaired spermatogenesis.[20]

The commonness of complete AIS in births originally classified as male is reported to lie between 1 in 20,000 and 1 in 60,000. In births originally classified as female, complete AIS occurs in 1 in 8,000 births (detected as 1 to 2 percent of females showing an inguinal hernia). Combining the AIS reported for births classified as male with births classified as female leads to the statistic that 1 in 13,000 people overall is born with complete AIS. Partial and mild forms of AIS are reported to be one-tenth as common as complete AIS, but this may represent underreporting because the diagnosis would be based on less distinctive clues. Complete AIS is about as common as classic CAH, and it is in a sense the reverse of classic CAH, in that complete AIS leads to feminization of a body whose gonads are male, while classic CAH leads to the masculinizing of a body whose gonads are female.

Complete AIS does arguably qualify as a genetic disease. Although not necessarily painful, complete AIS is deleterious to fertility and rare enough to represent a mutation-selection equilibrium. Partial AIS, however, could simply intergrade with various nondiagnosable body types that are relatively androgynous and would be scored as normal. With 150 alleles already known at the Xq11–12 locus for the androgen receptor, probably quite a few are benign, and may even be beneficial in circumstances where less extreme styles of masculine body types are adaptive. Indeed, this locus might help modulate the degree of sexual dimorphism in our species. As with mild CAH, mild AIS isn't a disease and shouldn't be condemned by association with cases that do pose health or fertility risks.

CHROMOSOMAL VARIATION

Another pathway to intersexuality comes from sex chromosome configurations other than XX and XY. These people might be termed chromosomally intersexed, and some also have ambiguous external genitals. The most common are XXY at 1 in 1,000, XYY at 1 in 1,100, XXX at 1 in 2,000, a single X at 1 in 2,700, XXYY at 1 in 6,500, and XX males at 1 in 20,000. The people with XXY chromosomes show geographic variation ranging from 1 in 500 in Germany to 1 in 7,400 in Winnipeg, and people with a single X chromosome vary from 1 in 600 in Moscow to 1 in 9,500 in Edinburgh. Although some people with unusual chromosome counts suffer from health risks or from low or no fertility, many don't: "Many 47,XXX girls develop secondary sex characteristics at puberty, and are sometimes fertile."[21] Similarly, "many 47,XXY and 47,XYY males are undiagnosed because they present no symptoms which prompt a chromosomal analysis."[22] Sex chromosome configurations other than XX and XY are clearly quite common and cannot generally be called genetic diseases except in severe cases.

HERMAPHRODITISM

The rarest bodily state lumped under intersexuality is possessing both testicular and ovarian tissues simultaneously. About 1 in 85,000 people has this trait, averaged over our entire species. As with other intersex traits, though, large geographical variation exists. In southern Africa, one study showed that half of all the babies born with ambiguous genitalia were hermaphroditic, placing hermaphroditism on a par with more common pathways, such as classic CAH or AIS.[23] One developmental pathway to hermaphroditism is fusion of two embryos into one soon after conception, the reverse of how identical twins are produced.

As with other paths to intersexuality, the way hermaphroditism is expressed is also quite variable. One survey of 367 hermaphroditic people revealed that 30 percent had an ovary on one side and a testis on the other, 30 percent had an ovotestis (gonad with both ovarian and testicular tissue in it) on one side and an ovary on the other, 21 percent had an ovotestis on both sides, 11 percent had an ovotestis on one side and a testis on the other, and the remaining 8 percent had structures that were not classified or reported.[24] The structure of internal genital tubes and external genitals is similarly variable.

Medical consensus unquestioningly stigmatizes hermaphroditism as a genetic defect because of cancer risks and lower fertility. In addition, the rarity of this trait is consistent with a mutation-selection equilibrium. Still, one must recall that in some mammals ovotestes are the norm (see chapter 3). Hermaphroditic people should not be pathologized as violating some law of nature. They possess a trait that's rare in our species but common in others.

In summary, the descriptions of genetic and hormonal aspects of intersexuality are more extensive than for gender identity and sexual orientation because intersex bodily states form earlier in development than sexual orientation and gender identity. Some forms of intersexuality are too common to represent a mutation-selection equilibrium; counterparts to human intersexuality occur in some other species, where they are presumably adaptive; and many forms of intersexuality are neither painful nor disabling. The most common forms of intersexuality differ only cosmetically from nonintersexes—only the rarer forms are painful or deleterious. So, is intersexuality a genetic disease or a genetic defect? Usually not.

WHO NEEDS A "CURE"?

NOT GAYS AND LESBIANS

Even though there is no scientific basis for generalizing that LGBTI people are diseased, medical practice has for many decades tried to transform these people into the social norm of a heterosexist gender binary. Using whatever techniques are available or fashionable, medical practice has aimed to "cure" diseases that don't exist, thereby violating the Hippocratic Oath and abusing the human rights of a diverse people.

Specifically, therapists have tortured gay and lesbian people with a technique called aversion therapy.[25] The person, say a gay man, is brought to the clinic, exposed to erotic photographs of nude men, and then punished for any signs of arousal. In theory, the man is supposed to associate the erotic photograph with pain and learn somehow not to be aroused—much as a mouse is trained with rewards or punishment in operant conditioning. The punishments used can only be described as diabolical. In the 1960s the drug apomorphine was administered to induce vomiting (or hypnosis might be used to cause uncontrollable nausea); in the 1970s electric shock therapy was added, sessions sometimes lasting thirty minutes, repeated twenty or more times over several months. People were not only traumatized but physically burned. Even worse, electroconvulsive shock therapy (ECT), administered by either delivering shocks to the head or giving the drug metrazol, induced epileptic seizures with side effects of memory impairment and depression that could last for years.

After years of study, however, behavioral scientists have failed to come up with a theory or a cure for gayness; indeed, they have gradually thrown in the towel. In 1973 the American Psychiatric Association removed homosexuality from its list of mental disorders, but psychoanalysts persisted in describing homosexuality as a perversion well into the 1990s. Finally, in December 1998, the American Psychoanalytic Association, in its annual meeting in Manhattan, acknowledged its "own past homophobia," in part because of the coming out of a prominent Atlanta psychoanalyst.[26]

Behavioral scientists have also now gone on record that therapies attempting to convert gay, lesbian, and bisexual people to heterosexuality do not work and do more harm than good.[27] In Denver in 1998, the board of the American Psychiatric Association voted unanimously to reject therapy aimed solely at turning gays into heterosexuals. The American Psychological Association had made a similar decision the previous year.

But does this mean the spectre of a "cure" has disappeared? No. It's latest guise may be the promise of selective abortion of gay babies. A quotation from *Time* illustrates how claims of medical virtue can camouflage a social agenda: "Parents can use preimplantation genetic diagnosis to avoid having kids with attention-deficit disorder, say, or those predestined to be short or dull-witted or predisposed to homosexuality."[28] Notice the clever—and dangerous—juxtaposition of homosexuality with dull-wittedness and attention-deficit disorder. Hollywood, too, has taken up the issue of aborting a supposedly gay baby in the popular movie *Twilight of the Golds,* starring Faye Dunaway. Selecting babies to fit political specifications could fire competition among various biological constituencies, each with its own genetic agenda. If anti-gay groups breed gayness out of babies, pro-gay groups might breed gayness back in, thus conserving, or even expanding, the presence of gayness in the human gene pool. Thank goodness there isn't a simply gay gene.

Let's be clear: you can't cure homosexuality because there's no disease to cure. But I hesitate to become overconfident, assuming that the standing of our gay sisters and brothers as normal people has been permanently enfranchised by the vote of psychologists. What can be won by a vote can be lost by a vote. The value and naturalness of homosexuality must be as scientifically clear as the fact that the earth is round. Then the acceptance of homosexuality will not crumble when the political pendulum next swings.

NOT TRANSGENDERED PEOPLE EITHER

While homosexuality now basks in the glow of normalcy, transsexuality toils in the shade of stigmatizing pathology. After giving up homosexuality as a pathology, behavioral therapists turned their attention to gender variation as the new disease to cure. Transsexuals are presently listed by psychiatrists in their *Diagnostic and Statistical Manual of Mental Disorders* (DSM-IV) as having a mental illness called gender identity disorder (GID) and as suffering from "gender dysphoria."[29]

Various interventions have been tried to encourage children to assume gender-typical behavior. In the 1970s, for example, therapists aiming to "extinguish feminine behavior and to develop masculine behavior" in boys, used so-called social reinforcement, in which an adult in the child's playroom would notice, smile, and praise a child's gender-typical behavior but look away or pretend to read when gender-variant behavior occurred.[30] The children simply reverted to cross-gender behavior in the adult's absence or at home. In addition, the children didn't generalize to forms of behavior not presented to the adult. Another treatment used was "self-regulation," in which a boy was told through a "bug-in-ear" device to press a counter if he was playing with boy-toys. This big-brother-is-always-watching technique proved more effective in the short term than social reinforcement alone. At times self-regulation treatment has escalated into full-fledged incarceration accompanied by aversion therapy, although apparently not in the last ten years. Aversive conditioning and shaming are still used with cross-dressers, however. Today, attention is shifting to mind-altering drugs that influence serotonin uptake in the nervous system, like fluoxetine (Prozac) and clomipramine (Anafranil).

Do any of these tortures work? "Systematic information . . . is scanty," and "most authors have not found psychotherapy efficacious."[31] Critics ask, why continue then? Therapists, it seems, are not interested in critics. According to one set of therapists, "Most of the critics . . . are not clinicians. Those critics who are clinicians appear not to have had experience in the area. . . . In our clinical experience, we have found no compelling reason not to offer treatment to a child with gender identity disorder."[32]

What planet are these therapists on? Talk to trans people who came out during the 1970s and 1980s, and you will hear no end of horror stories. I recall seeing a woman who I felt somehow wasn't all there at a gathering of trans people in southern California in 1999. The organizer saw my hesitation and told me afterward, "Be patient, they did electroshock on her as a child." I'm grateful to have narrowly escaped such maiming by gender identity clinics.

In recent years, gender therapy has stopped trying to straighten out gender-variant people, as evidence mounts that "casts doubt on the view that transsexualism is a severe mental disorder."[33] Indeed, in a Scandinavian study that explicitly compared transsexual people with other groups, transsexuals were statistically indistinguishable from a reference group of healthy adults but differed significantly from a reference group of people with mental disorders.[34] Instead, gender therapy has come to mean helping people to accept themselves and coaching them on how to transition physically and socially from living in their birth-assigned gender to living in the gender they identify with. The purpose of the therapy isn't to cure transgenderedness as such, but to help people live the way they are.

In current practice, therapists "diagnose" the transgendered person as having gender identity disorder (GID), thereby setting the stage for the physician and surgeon to "cure" GID by prescribing hormones and surgery. The state of being transgendered is not itself changed or dissolved. Instead, the various health professionals facilitate the transition from a gender assigned at birth based on genitalia to a gender the person actually identifies with. In a sense, this system works. By acquiescing to being diagnosed with a disorder, a transgendered person gains access to the enabling medical technologies. Transgendered people often buy into this framework. Adopting a medical "explanation" helps transgendered people accept themselves, even though it is pathologizing.

Therapists serve as "gatekeepers" for those about to transition from one gender to another. I've heard therapists discuss the conflict of interest they feel as a result of being caught between judging and helping. Some therapists demand a stereotypical narrative from clients before recommending hormones and surgery. Transgender activist Patrick Califia recently commented, "None of the gender scientists seem to realize that they, themselves, are responsible for creating a situation where transsexual people must describe a fixed set of symptoms and recite a history that has been edited in clearly prescribed ways in order to get a doctor's approval for what should be their inalienable right."[35] Other transgendered people have found their therapists helpful. When beginning their transition, transgendered people may find that their therapist is their only friend. Therapists often convene support-group meetings that transgendered people find reassuring and validating. Unfortunately, these group meetings may spawn a subculture of dependency. Overall, the interaction between transgendered people and the gender therapy community is a mixed bag.

Since the present system has some good features, and matters are much better than they were twenty years ago, why not leave everything as is? Because the system is a lie: there simply isn't any disease to cure. Except for the stigma, how many transgendered people would wish they were straight? When gays were asked this years ago, they often wished to be "cured," but they say this much less today. As I reflect upon my own past, I don't feel I'd want to be any different than I am. I'm living a loving and richly interesting life.

Many have recognized that the present medical status of transsexuals is untenable. Furthermore, financial injustices in the present arrangement lead to unnecessary social costs. Transgendered people needing hormonal and surgical procedures must somehow find a way to pay for these themselves. In large cities, transgendered people may be seen working the streets, selling sex to raise money for hormones, often on the black market, and they are often unable to afford the more expensive procedures.

The representation of transgendered people by the behavioral health community is sloppy and unprofessional. The wording of the diagnostic criteria for GID is "ambiguous, conflicting, sexist, and overinclusive, as well as failing to acknowledge happy, well-adjusted transgendered people," according to psychologist Kate Wilson, who has carefully looked into this matter.[36] Still another group of psychologists, including primatologist Paul Vasey, find the DSM-IV self-contradictory in its treatment of gender identity disorder, charging that GID doesn't meet the DSM-IV's own criterion for a mental disorder, and that it "should not appear in future editions of the DSM."[37]

Pathologizing transgendered people indirectly marginalizes the few health professionals who do work with this group. These workers are often dedicated individuals who labor in isolation with little professional prestige or compensation. They are often self-taught, do their own research, and devise their own procedures on the job. Much of the best transgender practice has developed in this way and has yet to be synthesized into a curriculum widely available to medical students. The tendency of some health professionals to discuss transgendered people using exaggerated clinical jargon may reflect a fear of being marginalized by association with a stigmatized group. Just as a physician may catch an infectious disease from a patient, a therapist may catch an infectious stigma from a client. Both the self-interest of therapists and the welfare of transgendered people would be better served if behavioral health organizations reformed their transgender practice.

Being transgendered in today's society requires medical technology. A close analogy is being pregnant. A pregnant woman doesn't have a disease, but medical service is often needed. I believe being transgendered, like being pregnant, is best viewed as a normal human condition whose expression is aided by medical service. In practical terms, the existing drill for transgendered people of waiting periods, trial periods, and so forth prior to embarking on the medical aspects of transition does seem appropriate. I do not favor hormones on demand. I feel the dangers and significance of hormones and surgical procedures need to be weighed, and the liability of health professionals working with transgendered people respected. The issue is, rather, the pathologizing of diversity. Transgender procedures should be considered a medical service required for personal growth, not a therapy to cure a disease. Just as we should speak of "discovering" a woman to be pregnant, and not of "diagnosing" her, we should also speak of discovering transgender gender identity. Ascertaining this is a finding of fact, not a diagnosis.

Coming out as transgendered should be a source of joy and happiness to everyone, just like the birth of an eagerly awaited child. Someday, I'd love to see people celebrate someone's coming out as transgendered with a christening ceremony, much as occurred with two-spirited people in some Native American tribes (see chapter 18).

NOR INTERSEXES

With the loss of homosexuality as a pathology and the prospect of also losing transsexualism, medicine is making a last stand around intersex. Whereas homosexuality and transsexualism have belonged to mental health professionals, intersex belongs to surgery, endocrinology, and genetics. Although different professional associations are involved, the mind-set is the same: nature intends a heterosexual binary, and variation equals defect. Because of this mistaken premise, harm is done to helpless children hours to weeks after they are born.

The birth of an intersexed child sets off alarms. Here are representative descriptions: "The assessment of genital ambiguity in the newborn is a psychosocial emergency . . . all surgical, hormonal, and psychological therapy must be in concert and appropriate for the decision, and early and usually repeated reinforcement of the decision will be required."[38] "Abnormalities of ambiguous genitalia are considered a 'social emergency,' and a well-disciplined diagnostic and therapeutic team is enlisted to address the problem rapidly. . . . We emphasize the need for care in a specialized, multidisciplinary center where pediatric surgeons, pediatric urologists, pediatric endocrinologists, pediatric radiologists, geneticists, neonatologists, and pediatric anesthesiologists can bring their accumulated expertise."[39] This army of doctors is intimidating, and parents can't easily escape believing that something is terribly wrong with their child. Parental choice is negated by this throng of experts.

Doctors prefer to immediately assign the newborn as male or female and then hold firm through childhood and beyond. The criterion is penis size. In a newborn, "the size of the phallus is measured during a simulated erection. . . . A phallus less than 1.5 ± 0.7 cm is cause for grave concern and would lead one to recommend rearing as a female."[40] The male aspects of the genitals are trimmed off at the time, with female aspects to be sculpted during adolescence. "If, on the other hand, the phallus is of reasonable size and can respond to testosterone, then the child can be raised as a male." Thus penis size at birth is the primary criterion for forcing a gender assignment on the child, which then sets in motion how the child is raised and commits the child to an interminable sequence of visits for more and more surgeries during childhood and adolescence. Meanwhile, the parents are in the nearly impossible position of having to pretend that nothing's wrong, even though the birth of their child was greeted by a phalynx of doctors in a flurry of activity.

The Intersex Society of North America (ISNA) devised a ruler termed a "phallometer."[41] Penis length at birth in typical males ranges from 2.5 to 4.5 centimeters, and clitoral length ranges from 0.20 to 0.85 centimeters.[42] The phallometer is a ruler with 0.20 to 0.85 marked off as female and 2.5 to 4.5 marked off as male: hold the phallometer next to the phallus of the newborn and read off the child's sex. An XY child with a penis length of 1.5 is less than the conventional masculine range and is irreversibly assigned as female then and there. Simple. Too simple.

Today, clinicians seem to be focused not on whether their approach to intersex people must be rethought, but rather on how to improve technique. The surgeries might be getting better, but I'm not so sure. For hypospadias, over three hundred surgical "treatments" have been proposed, which may involve suturing and skin transplants in as many as three operations during the first two years of life, plus several more before puberty. As one reviewer summarized, "No consensus has formed about which technique consistently results in the lowest complication rates. . . . Every year dozens of new papers appear describing . . . surgery designed to repair previously failed surgeries."[43] Concern about the technical adequacy of penile reconstruction obscures the more fundamental issue of whether and when doctors should perform these procedures in the first place.[44]

For CAH, an astonishing intervention is used. Because external genitals start forming early in development, mothers carrying a child at risk for CAH take dexamethasone as early as four weeks after conception, although the CAH status of the fetus cannot be determined until the ninth week. Because of the way CAH is inherited, only one out of eight fetuses is affected, so the treatment is discontinued after the tenth week for seven out of eight fetuses. Seven mothers and seven fetuses suffer the side effects of the drug and DNA tests on behalf of the one child who does have CAH. Alternatively, for families with a history of CAH, in vitro fertilization may be recommended, with "preimplantation selection" of non-CAH embryos.[45]

After a CAH child is born, postnatal therapy begins. After therapy for any metabolic consequences from low cortisol, doctors proceed with so-called clitoral reduction, which shortens the clitoris by cutting out a piece in the middle and sewing the tip back on, and/or clitoral recession, which hides the clitoris under a labial hood.[46] Not long ago, the clitoris was removed altogether in the mistaken belief that female organism took place in the vagina and not in the clitoris. According to clinicians, such "surgical correction . . . should be started in infants aged between two and four months and continue, in stages, thereafter."[47]

For AIS patients, the immediate health risk is minimal. Nonetheless, the testes are removed from AIS infants because "they can become cancerous," although the risk of cancer doesn't appear until after puberty.[48] People with partial AIS are subjected to surgeries to sculpt their genitals and are assigned as girls. As a clinician writes, "The rationale is that for the purpose of sexual intercourse, it is easier to create a vagina than a penis."[49] Many partial AIS people do identify as female, but not all, and mistakes are made in sex assignment, with no end of trouble downstream. Doctors are now beginning to acknowledge "the frequency and vehemence of complaints registered in the newsletter of the AIS Support Group."[50]

One theme of intersex advocacy is that an infant's genitals should be left alone so that the child can elect later whatever plastic surgery he or she feels is needed. A second theme is that the child should be told the truth. Tales abound of people who were forced to visit the hospital during summer vacations for surgical procedures that were not explained; often they were directly lied to. As adults, some have resorted to legal measures to access their medical records and find out what was going on. Finally, intersex advocates stress that any procedure should benefit the child, not please the parents.

The medical party line is that the inability to say at birth whether " 'it's a boy' or 'it's a girl' could have profound negative effects on parents."[51] The child is supposedly helped by an early sex assignment, reassuring the parents that all has been fixed and allowing them to raise the child in a sex-typical fashion. But the encounter at birth with an army of specialist doctors precludes any semblance of normalcy. I've seen interviews with parents who simply wanted to be told the truth and loved their children as they were. I've personally met intersexed people who were spared childhood medical interventions because they lived overseas and who are perfectly happy. In contrast, people who were maimed as infants often have serious issues with medical practice.

This emphasis on treating the child to please the parents is typical of how LGBTI people are handled by the medical community. When children come out as gay, lesbian, or transgendered, some parents ask, "How can you do this to me?" and send the child to a therapist to be cured. If parents of a newborn intersexed child can't face up to their friends and say their baby is intersexed, and demand instead that the doctor fix their child, the responsibility of medicine is to say no. Children in a section of the Dominican Republic where one form of intersex is common grow up in a social holding pattern until they declare their sexual identity. I'm sure our society is strong enough to handle this situation too. The birth of an intersexed child is a great joy, just like the birth of any other child.

CURING THE OBSESSION TO "CURE"

One might comfortably assume that the obsession behind medicine's stamping out any deviance from a heterosexist binary has an impact only on LGBTI people. But no, medicine's obsession to cure nondiseases hurts each and every one of us. The medical profession has a long history of coining words that start with "dys-" (or "hys-") and end with "-ia." If you feel left out that you haven't been labeled with some dys-ia or hys-ia yet, don't despair, you probably already have been.

Merely being a woman in the 1860s was tantamount to having a medical condition. Women with a sensitive clitoris might awake to find it amputated by a doctor as a cure for "hysteria."[52] In a more benign approach doctors purchased vibrators to induce their patients to orgasm—hailing this method as a godsend compared with treating patients "with their own fingers." One researcher notes, "I'm sure the women felt much better afterwards, slept better, smiled more. Besides, hysteria was considered an incurable disease. The patient had to go to the doctor regularly, she didn't die. She was a cash cow."[53]

The fear in the nineteenth century was that women might become "oversexed." Today, not wanting sex all the time has become a disease. A study of sexual "dysfunction" reported that a third of the women surveyed did not want sex regularly, and 23 percent reported that sex was not pleasurable. Also, a third of the men said they climaxed too early, and 8 percent said they consistently derived no pleasure from sex. Overall, 43 percent of the women and 32 percent of the men reported one or more persistent problems with sex.[54] To cure the "disease" of sexual dysfunction, we now can purchase Viagra for men and Muse (an alprostadil cream) for women.[55] Typically, though, no disease exists. These drugs are aphrodisiacs.

If you're a man, don't think you're safe: you also have a congenital medical condition—your penis. In the 1960s, 95 percent of American-born boys were circumcised. Then, in the 1970s, the American Academy of Pediatrics declared there was no "medical indication" for circumcision. But in 1989 the academy reversed its decision, reporting "potential medical benefits." In 1999 the 55,000-member academy concluded the benefits are "not significant enough."[56] Imagine a species in which all males have a penis requiring surgical repair. Ridiculous.

By now everybody has been labeled with some horrible-sounding condition. Even being shy is a disease. Nineteen million Americans suffer from "social phobia"—7 percent of the population.[57] In 1998 sales of drugs to treat depression totaled: Prozac from Eli Lilly, $2.27 billion; Zoloft from Pfizer, $1.48 billion; and Paxil from SmithKline Beecham,

$1.16 billion. With figures like these, the drug companies have a staggeringly huge incentive to convince everyone that they're sick. Thus LGBTI people are not alone in needing to be "cured." Everyone needs to be "cured" of something involving their sex or personality.

A recent report from the U.S. surgeon general advocating a larger role for psychiatric and psychological therapy in mainstream health care[58] says one in five Americans experience a mental disorder in any given year, and half of all Americans have a mental disorder at some time in their lives. Who could possibly object to providing more counselors for people in stressful occupations?[59] One wonders, though, about quality control in a nationwide program of treatment for poorly defined behavioral conditions. The history of classifying people as mentally ill for political purposes is also hard to overlook. In the 1850s "drapetomania" was defined as the "disease causing slaves to run away."[60] Calling political dissidents mentally ill has been used to justify sending them to concentration camps.

Today's smorgasbord of personality-altering drugs is sometimes thought to legitimize mental diseases by implying a biological basis. On reflection, though, whether a behavior can be changed with a chemical is irrelevant. Behavior can always be changed with chemicals—one need only think of drinking alcohol. The important issue is how one goes about classifying a behavior as a disease. Our society is overmedicalized.[61] Too many conditions, both mental and physical, are branded as diseases without sufficient contextual research. How, then, should we care for people in pain? Part of the answer is for each of us as members of a human community to assume more responsibility for one another. We should know our neighbors, value them, and love them, and not pass the buck to health care professionals.

The time has come to take a stand, to say that we, in all our shapes and sizes, in all our gender expressions, sexual orientations, and body parts, are healthy. We are, all of us, descendants of those who rode together in the huge Ark, a vessel large enough to contain all the diversity of humanity plus all the diversity of creation. We are entitled to the presumption of health, unless proven otherwise. This inalienable right was ours long before the Bill of Rights gave us the presumption of innocence in judicial affairs.

Calling on Native American imagery, the novelist Paula Gunn Allen has written, "A society that believes the body is somehow diseased, painful, sinful or wrong is going to create social institutions that wreak destruction on the body of the earth herself."[62] By respecting ourselves, our natural health, we will live better lives, conserve the human rainbow, and protect our earth.

17

Genetic Engineering versus Diversity

Species need rainbows to survive, and today genetic engineering threatens our rainbow, as well as those of other species. Medicine threatens individuals, but genetic engineering threatens our entire species—our posterity. The damage from medicine is immediate, reflecting harm inflicted daily on people misclassified with fictional diseases, but harm from genetic engineering would reverberate through the future.

The threat of genetic engineering springs from an arrogant belief that we should manipulate our gene pool. At times genetic engineering proposes to redecorate a whole rainbow, or it may target specific colors, such as those for unusual expressions of gender and sexuality. All these threats begin with accepting the narrative that defective genes cause "genetic diseases." The empirical failure of this narrative, combined with the need to show a profitable return on investments, is propelling genetic engineering away from the unrealistic promise of delivering genetic cures toward the more attainable goal of delivering genetic weapons. The problems with genetic engineering all come back to a failure to understand and appreciate diversity, along with the inherently relational nature of life.

The most basic threat to our species' future is the belief that our entire rainbow is somehow muddied with dirty colors and must be cleansed. If Hitler had pruned humans into a "super-race," he would actually have produced a homogeneous species highly vulnerable to extinction from an epidemic. Most criticism of Nazism has focused on its cruelty to individual people, but it's important also to realize how disastrous Nazism's misguided scientific premise would have been from a population perspective. Nazi medicine diagnosed Jews as carrying genetic diseases and prescribed purifying the German race of Jews, the mentally ill, the handicapped, and homosexuals. German medicine provided an intellectual foundation for Nazi atrocities. The killings often took place in a cliniclike setting. One lesson from Nazi Germany is that medical consensus can be scientifically incorrect, morally wrong, and extremely dangerous. This lesson demands that we continually question medical consensus.[1]

CLONING WHOLE ORGANISMS

Cloning whole organisms, meaning that a child might be manufactured with nuclear genes identical to a parent, poses a potential threat to our rainbow. For this technique, the nucleus of an egg cell is made inactive, for example, by irradiation. Then a cell from an adult is fused to the egg cell, supplying a nucleus to replace the damaged one. The new nucleus takes over and starts to grow into an embryo. Sounds simple enough. A lamb named Dolly was cloned by Scottish scientists in February 1997 after 276 attempts.[2] Mice were cloned a few months later in Hawaii, and a year after that eight identical copies of a cow were cloned by Japanese scientists.[3]

Cloning is being heavily promoted today. One supposed benefit is to improve agricultural production by making "exact copies of animals who are superb producers of meat or milk."[4] Yet genetically identical cattle might be made more safely and cheaply simply by taking very young embryos and separating their cells into groups, more or less as happens naturally when identical twins develop. This "artificial twinning" would not require fusing an adult cell to an egg cell whose nucleus has been inactivated.

Another supposed benefit is to save endangered species: "On Jan. 8, 2001, scientists at Advanced Cell Technology, Inc. announced the birth of the first clone of an endangered animal, a baby bull gaur [a large wild ox from India and southeast Asia] named Noah. Although Noah died of an infection unrelated to the procedure, the experiment demonstrated that it is possible to save endangered species through cloning."[5] Reconstituting one specimen does not save a species. Hundreds of specimens are needed to make a viable population, and the specimens must be different from one another so as to reconstitute a genetic rainbow for the species.

Still another supposed benefit is the possibility of replacing dying or already dead pets. Researchers at Texas A&M cloned a domestic cat, producing a kitten called Cc starting from eighty-seven embryos.[6] The cloning experiments were funded by an eighty-one-year-old financier who wanted to charge wealthy pet owners to clone their animals. The financier said he would also like to clone socially useful animals, such as rescue dogs. (Meanwhile, a member of the cloning team envisioned giving the kittens AIDS, saying, "Cats have a feline AIDS that is a good model for studying human AIDS."[7])

Cloning animals is one thing, but cloning humans—will that really happen? Some scientists clearly intend to clone humans "sooner rather than later."[8] Already there have been unverified reports of cloned human embryos, first from South Korea and later from a privately held U.S. company, Advanced Cell Technology.[9] In May 2002 a former member of the University of Kentucky faculty, now an infertility treatment entrepreneur, testified before a U.S. House subcommittee that he had assembled a team to produce a human pregnancy within a year. He had lined up twelve couples around the world who wanted to conceive by cloning and claimed "the genie is already out of the bottle."[10] The suggestion is that cloning will eventually join the other services offered by fertility clinics.

Extensive cloning will endanger the cow as a species, just as extensive inbreeding of agricultural stocks does now. One might imagine substitutes being found for cattle, corn, or other stocks if they were challenged by an epidemic. But for us? How can we find substitutes for humans?

WHY CLONING DOESN'T WORK

Not to panic. By and large, cloning is not working for any species, including humans. In January 2002 the creators of Dolly, then five and a half years old, disclosed that she had arthritis in her left hind leg, hip, and knee. Not only had Dolly developed the disease at an unusually young age, but it had affected two joints not normally affected. In addition, it was noticed that the cells in Dolly's body had started to show signs of wear normally found in an older animal. According to one hypothesis, the genetic blueprint was "wearing out," as though a typewriter ribbon had run out of ink. Dolly died at age six, in February 2003, after developing a lung infection.[11]

"Failure far exceeds success" in attempts at cloning, as one science reporter indicates.[12] An investigator who made three hundred attempts to clone a monkey over three years, with no success, said, "We never obtained a single pregnancy." Most of the time, the process resulted in "grotesquely abnormal embryos containing cells without chromosomes . . . or cells with three or four nuclei, and one time even nine; or cells that looked more like cancer cells."[13] The scientist who cloned Dolly noted that teams who had succeeded with some species failed with others and that some species had not yet been cloned, such as rabbits, rats, dogs, and monkeys, in spite of extensive and well-funded efforts. And where cloning has worked, only 1 to 4 percent of the attempts are successful, the rest resulting in serious abnormalities.

When does cloning work? When lots of eggs are available: "Researchers get thousands of cow eggs from slaughterhouses . . . [which] makes it much easier to try cloning often enough, with enough slight variations in technique, that it eventually works." In contrast, with primates, "it will never be feasible. You would need a whole primate colony with thousands of animals. . . . If you want to make it into something that will have commercial value . . . the process has to be repeatable. Your success cannot be 1 or 2 percent. A 2 percent success rate is not a success; it's a biological accident."[14]

Nonetheless, genetic engineering companies continue trying to sell whole-animal cloning to the techno-gullible.[15] Advanced Cell Technology has accumulated twenty-four cows between one and four years old who are the result of thirty pregnancies. Their "success" rate of 80 percent is advertised as close to the 84 to 87 percent of conventional breeders. But this number is misleading, because the company doesn't report how many embryos were needed to generate the thirty pregnancies, thus omitting the stage where cloning encounters the most problems. Moreover, the scientist who cloned Dolly notes, "These results do not in any way eradicate the previous history of unusual deaths in animals cloned by essentially the same procedure." Still, Advanced Cell Technology attributes the "team's high success rate to six years of cloning practice and to their methods. They use actively dividing skin cells, for example, as opposed to the non-dividing cells from which Dolly was cloned. Care of the newborn calf in its first few days also makes a big difference." Such boasting doesn't convince scientists. One is quoted as saying, "Years of experience do seem to make a difference, but that only affects the percentage that survive, not whether they're normal or not. No one really knows why cloning has such hit-and-miss results."

A clue about what's going wrong with cloning comes from the admission that Dolly is "not quite a clone" after all.[16] Dolly contains nuclear genes from the donor that provided the nucleus plus mitochondrial genes from the egg cell that received the nucleus, so her genes come from *two* sources. To be a true clone, both the donor cell and the egg cell would have had to come from the same individual, which did not happen with Dolly. Furthermore, a mammalian egg cell is huge, containing perhaps a hundred thousand mitochondria. The adult cell used for Dolly, which was from a mammary gland, was much smaller, containing only two thousand to five thousand mitochondria, or 2 to 5 percent of the number from the egg cell. Although scientists therefore expected that 2 to 5 percent of the mitochondria in Dolly would be from the donor cell, with the rest from the egg cell, in fact 100 percent came from the egg cell. Evidently, the cytoplasm of the egg cell actively destroyed the 2 to 5 percent of foreign mitochondria. This suggests there is a genetic dynamic going on within the cytoplasm beyond the reach of the nucleus.

Molecular biologists are starting to realize that the basic concept of cloning is wrong. The nucleus doesn't unilaterally "control" the cell. The nucleus negotiates with the cytoplasm, and if the cytoplasm doesn't go along, the project aborts. The nucleus and the cytoplasm are partners in life.[17] New jargon is emerging to describe this reality: "Difficulties in 'developmental reprogramming' are thought to underlie clones' survival and health problems. When nuclei are transferred from a cell into an egg stripped of its nucleus, they must erase previous patterns of gene activity and start up new ones that drive embryo growth."[18] This techno-talk raises the subversive idea that the state of the cytoplasm is just as important as the genes are.

Computer scientists use the phrase "execution environment" to describe the state of a computer when a program is running. A logically correct computer program can fail to run, depending on how the computer is outfitted and what other programs are running at the same time. The cytoplasm provides the execution environment for the genes. Every computer scientist knows that focusing only on code while ignoring the execution environment is a mistake. Cloning biologists are making this mistake.[19]

CLONING "PARTS"

Partial-person cloning encompasses a potpourri of technologies that focus on specific tissues and genes instead of whole persons. Imagine, for example, that your parents conceived you in a fertility services laboratory and that you spent your earliest hours not in your mother's oviduct on the way to her uterus, but in a petri dish in the laboratory. Imagine that your parents instructed the technicians to separate you into two groups when you were only a few cells big. The technicians then implanted half of you into your mother's uterus and kept the other half in the laboratory. You then went on to develop in your mother's uterus, were born, and became an adult. Meanwhile, you have some cells in storage that were originally part of you. Because these cells—called stem cells—came from a time before your specialized tissues started to form, they can become cells of any tissue type: bone, nerve, kidney, liver, gonad, and so forth. How do you feel about this?

These cells sitting in a laboratory somewhere could have been implanted in your mother's uterus when you were. If they had been, you would have had an identical twin brother or sister to grow up with. Instead, your identical brother or sister is being grown to supply you with spare parts. If your kidney should start to fail, you could be given some stored stem cells to regenerate your kidney. This partial-person cloning is called "therapeutic cloning" to mask the reality that a potential person is being farmed to provide body parts for another.

Artificial twinning is one way to obtain stem cells for tissue regeneration. To my knowledge, no one is doing this. Instead, nuclear transplanting is under active development, with the adult who needs the kidney replacement supplying the cell to be cloned. As in whole-person cloning, a donor cell from the adult person is fused with a woman's egg cell that has had its nucleus removed. The new nucleus from the adult's donor cell will then supposedly command the cytoplasm remaining in the egg cell to grow into an embryo. But instead of implanting the embryo in a uterus, as in whole-person cloning, the embryo is held in storage for spare parts. Such partial-person cloning was carried out at the University of California, San Francisco, from 1999 to 2001, until lack of funding and the departure of the lead investigator shut down the program.[20]

The quality of organs from partial-person cloning would suffer from the same problems that affect organs in whole-person clones, but the cost-benefit picture is different. Someone whose kidney is already failing might gladly accept a slightly defective replacement that would not be acceptable for a newborn baby. The question of whether to clone embryos for spare parts is further complicated by promises of "improving" the embryonic cells with new genes. Cells sitting in a fertility services laboratory are available for manipulation. Perhaps some genes can be swapped out and others inserted in their place. Your new kidney then promises to be better than the one you started with, as a result of "tissue engineering." If your own body couldn't make insulin, then splicing in the gene that can sounds like a great way to cure diabetes.

The reasons offered for regenerating tissue are not always health-related. For athletes, regeneration could be the "perfect performance enhancer. You build muscle mass and strength even without exercise, and it is not detectable in the blood."[21] The tests for illegal drugs that are given to Olympic athletes would not reveal this gene therapy. Tissue regeneration is imagined to offer a kind of immortality, permitting defective tissues to be replaced with new tissues that never age. A gene called telomerase has been studied that enables tissue cells to divide more than fifty times in laboratory culture. When this gene is combined with undifferentiated embryonic germ and tissue stem cells, it will—researchers forecast—be possible to grow new body parts. A *New York Times* science reporter concludes, "The cell is a mechanism and, absent the gods' fury, it can one day be made to operate closer to our desire than evolution's uncaring design."[22] Such enthusiasm, to my mind, borders on hubris.

A SHOPPING MALL OF GENETIC PROJECTS

Theoretically, genes could be introduced into an adult organism either by a "friendly virus"—a previously dangerous virus rendered harmless by deletion of its damaging genes—or by directly injecting so-called naked DNA. With present techniques, however, the body tends to reject the virus, as though thinking it still unfriendly.[23] An eighteen-year-old died from a severe immune reaction to a soup of genes administered to him as part of a gene therapy experiment. As a result, the human gene therapy program was shut down at the University of Pennsylvania.[24]

Genetic engineering can also be used to make what *Time* calls "designer babies."[25] Drugs that stimulate egg maturation and release, such as those that led to the birth of octuplets in December 1998, could be used to make a population of tiny embryos.[26] The genetic profile of these embryos could then be scanned as "quickly as a supermarket scanner prices a load of groceries," and only those embryos that passed the test would be implanted. In other words, the desired embryos would be retained and the remainder discarded. "Instead of aborting a fetus, you're flushing down a bunch of sixteen-cell embryos—which to a lot of folks is a lot less of a problem."[27]

A less intrusive approach to "designing" a baby is simply to be very fussy about who the egg donor is. A sum of $50,000 was offered in 1999 for an egg donor who was at least 5 feet 10 inches, had scored at least 1400 on her combined SAT, and had no major family medical problems. More than two hundred women responded to the advertisement.[28]

Trans-species genetic introductions are widely used in agricultural genetic engineering, and they have now been suggested for animals as well, although sometimes in ways that seem frivolous. For example, at a meeting of Ars Electonica, a twenty-year-old international group that explores the intersection of science with aesthetics, an artist proposed to create a dog with fluorescent fur. The technique would involve extracting the gene for a protein in jellyfish and inserting it into the dog's genome, producing a transgenic animal whose fur would glow when green light was shined on it. This technique is actually used in cancer research, where tumors are made to fluoresce. The artist explained, "Society as a whole has not even become aware of the vocabulary of this new wave of research and technique. We cannot leave this vocabulary in the hands of the few, the politicians, the people from the business world, the scientists." The artist further claimed to be "increasing global biodiversity by inventing new life forms." An attendee at the conference criticized the idea, saying, "Initially, I was fascinated by his ideas, but I had this growing sense of unease at the arrogance of his proposal. It is one thing for an artist to experiment on a canvas, but it's entirely different to experiment on a living creature."[29]

A fluorescent green rabbit was in fact made in 1999—an albino rabbit with pigment genes from jellyfish spliced in. The creator (so to speak) wrote, "When the transgenic [animal] is sitting in your lap, looking into your eyes . . . we now have a different kind of otherness. . . . [Indeed,] racial traits are nothing compared with transgenic beings."[30] Although green dogs and rabbits may seem silly, other trans-species conceptions pose serious ethical questions.

Certain kinds of cruelty to animals are outlawed because animals can feel pain, can suffer. But what if animals could be engineered not to feel pain, say, for dogfights and cockfights? If an animal were engineered that didn't feel pain, would it be acceptable to be cruel to it? What would being cruel mean if the animal couldn't suffer anymore? What would the ethical status be regarding synthesized life forms compared with native life forms? Are synthesized life forms second-class on earth and before the eyes of God?

Conversely, what if human genes were transferred to animals, say, to grow an organ there prior to its being transplanted into a person? Is a pig with a human kidney more than just a pig? The *Wall Street Journal* has reported on calves that have been cloned to carry human genes for immunoglobulin, a blood substance involved in the human body's defense system: "Already, scientists have slipped dozens of different human genes into cows, sheep, goats, rabbits and mice in hopes of harvesting one or another human protein for use in treating disease. Each of these 'transgenic' animals makes only the single protein coded for by the particular gene. But the cloned calves' immunoglobulin genes can make antibodies to fight a huge array of ills, making the animals a potential living drug factory."[31] How many human genes does an animal need to become eligible for human rights?

What about protecting the gene pools of other species? Do we have a right to mold them to our wishes, as we have done since humans originally domesticated farm animals? Genetic engineering is more risk-prone than conventional animal breeding, and the potential for damage to our domesticated stocks is greater.

Conversely, do we have an obligation to endow other species with human virtues? Should we someday instill in other animals the capacity for language and moral reasoning, so that we could talk with them and engage in moral dialogue? And wouldn't that be useful? Animals could be employed as intelligent agents patrolling ecosystem services, killing poisonous snakes for us, while protecting endangered species. And wouldn't an evangelical sect want to religiously convert all these newly intelligent creatures? Obviously our responsibilities to other species are complex, and genetic engineering buffs are nowhere close to inviting public debate on the issues.

Scientists are trying to create life. Now that the entire genome of some organisms is known, those with the smallest genomes, such as the bacteria called mycoplasma, could be synthesized from scratch. These "minimal-genome organisms" could be manufactured by synthesizing their DNA and adding some artificially synthesized fat molecules and ribosomes. Perhaps surprisingly, a panel of ethicists drawn from Roman Catholic, Jewish, and Protestant faiths has concluded that "there is nothing in the research agenda for creating a minimal genome that is automatically prohibited by legitimate religious considerations."[32]

From my perspective, this new science of synthesizing life totally misses the point. Duplicates of one string of DNA make a population of one color—no rainbow at all, not life, merely a chemical. Being alive means having relatives, being a member of a community, being the product of an evolutionary process, and belonging to a multichromatic rainbow. A computer virus has more claim to life than does a minimal genome. Computer viruses replicate, mutate, and evolve, leading to a family tree. One might object that a computer virus is not self-sustaining because the virus depends on people to keep the host it lives in (computers) up and running. Still, someday computer viruses may domesticate people to serve them, by offering virtual sex as orchids do to bees, in return for being fed and nurtured. At that point, computer viruses would be as self-sustaining as any other species is in relation to its ecosystem.

Genetic engineering is being irresponsibly proposed for all manner of projects, most of which have nothing to do with health, and many of which are dangerous, and possibly cruel.

SOME RED FLAGS

Cloning whole organisms may pose a distant threat to our gene pool, as well as some immediate ethical issues, but genetic engineering is also dangerous right now and problematic. Should we be redecorating our rainbow, color by color? Like many others, I've benefited from medical technology and expect to in the future. But we need to talk. We need to discuss up front where this can go. Biomedical technology developers exude attitude, and they sound like boys playing with matches.

TRUST THESE GUYS?

James Watson, who won a Nobel Prize with Francis Crick for their 1953 discovery of the double-helix structure of DNA, writes in an essay for *Time,* "In 1948, biology was . . . near the bottom of science's totem pole, with physics at the top." Watson then approvingly claims that today molecular biologists are both revered and feared, just as nuclear physicists were in the 1940s. He asks, "Dare we be entrusted with improving upon the results of several million years of Darwinian natural selection?" and answers, "You should never put off doing something useful for fear of evil that may never arrive. . . . Moving forward will not be for the faint of heart."[33]

But before forging ahead, let's take a look at some of the ethics behind the scene. A spectacular ethical blindness became apparent in April 2002, when Craig Venter, leader of Celera, one of the two projects to sequence the human genome, admitted that his project had been sequencing his own genome.[34] He co-opted what was supposed to be an anonymous process for selecting the individuals to be sequenced and instead placed his own cells up for sequencing. The millions of dollars invested in finding out what the "human genome" is actually found out only what the "Venter genome" is.

One would expect that other molecular geneticists would be outraged. Not so. Watson says, "That doesn't surprise me; sounds like

Craig." Boys will be boys. The leader of the rival publicly funded Human Genome Project downplayed the event, saying only, "It doesn't have any great significance." Nonetheless, one of Celera's board of scientific advisers expressed reservations, stating, "Any genome intended to be a landmark should be kept anonymous. It should be a map of all of us, not of one, and I am disappointed if it is linked to a person." The technical utility of the entire project seems severely compromised by the link to a single person. More important, the integrity of the project is destroyed, for the selection clearly wasn't random. Moreover, Venter made use of the information for his private welfare. He discovered that he carried a gene, apoE4, disposing him to risk for abnormal fat metabolism and Alzheimer's. He then started taking fat-lowering drugs, showing, I suppose, that he at least believes his own work.

Genentech, one of the founding paragons of the biotech industry, in 1999 paid $200 million to settle a patent infringement lawsuit resulting from one of its scientists who "sneaked into his former laboratory at the University of California and smuggled out DNA samples." Also in 1999, Genentech paid $50 million to settle criminal charges that it had marketed a hormone for unapproved uses. Then, in 2002, Genentech was fined $300 million in compensatory damages because of improperly withholding royalty payments to the City of Hope Medical Center, and it faces additional punitive charges of $200 million because the company was found to have acted with malice or fraud.[35] In a different case, the former chief executive of another biotech company, ImClone Systems, was arrested in June 2002 for perjury and charges of insider trading after allegedly tipping off relatives to sell company stock when a drug developed by the company that was pending approval was not supported by enough data even to warrant a hearing.[36]

Even the information we're given may be suspect. The sequence from the Human Genome Project turns out to have a hundred large-scale errors that were not disclosed when initially published. Yet in self-congratulatory excess, the human genome science community commemorated the fiftieth anniversary of the Watson-Crick paper on DNA's chemical structure by declaring the genome complete as of 2003. The former president of the American Society of Human Genetics demurred, however, saying "to call it complete, . . . to match the 50th anniversary of the Watson-Crick paper, is a bit of a sham."[37]

WHAT GENE CHIPS TELL US

A tip-off that prospects for "improving upon the results of several million years of Darwinian natural selection" is more hype than reality comes from an important new technology called "gene chips," or "DNA microarrays." This technology looks at the expression of all the genes in an organism at the same time.

The technology was originally developed in yeast.[38] Just pick up a glass of wine or beer and a piece of bread and you're seeing the result of the two different states of yeast: metabolizing in the absence of air, which produces alcohol in wine and beer, and metabolizing in the presence of air, which produces CO_2 and causes bread to rise, making the batter light and spongy. What happens inside a yeast cell when it shifts from making alcohol to making CO_2? The pictures and graphs that come from gene chips dramatically portray whole banks of genes shutting down and others starting up, as the cell shifts from making alcohol to making CO_2. A yeast cell has about 6,400 genes, some of whose functions are known and others unknown. During the alcohol-CO_2 transition, approximately 710 genes increase their expression twofold while 1,030 genes decline by half. All told, about 27 percent of the total genome is involved in the transition. Of these, the big players are about 6 percent of the genome: 183 genes that increase fourfold and 203 genes that diminish fourfold. Half of the genes that respond during the alcohol-CO_2 transition have no known function and haven't even been named yet. Four hundred of these mysterious genes show no similarity to *any* previously known gene.

The main biochemical steps in synthesizing alcohol and CO_2 were discovered over fifty years ago, in some of the most inspiring work of early biochemistry. Probably no biochemical system in all of biology is better understood than yeast fermentation and respiration. The classic genes for steps in these pathways did increase and decrease as expected during the alcohol-CO_2 transition. Evidently, though, even in one of the best-studied systems of modern biochemistry, only half of the genetic story had been discerned. Imagine how much of the genetic story remains unknown for most human traits.

No single gene is responsible for making alcohol and no single gene for making CO_2. Many steps are required to synthesize alcohol or CO_2 from sugar. If a gene for any one of these steps is absent, the ability to make alcohol or CO_2 is destroyed. Each of these could be called a gene "for" alcohol or CO_2 because if it's removed, the capability for making alcohol or CO_2 disappears. We can't then plug a gene for alcohol production into coffee beans to get a coffee liqueur plant. We'd need to plug in at least hundreds of genes for each *known* biochemical step to get the trait of alcohol production. But that wouldn't work either, because the hundreds of mysterious genes would also have to be plugged in, and what they do is unknown. One-third of the entire genome is somehow involved. To get a qualitatively new feature inserted into an organism, most of its genome, not just a few genes, has to be redesigned. Let's get real here.

This technology is subtly subversive, destroying the one-gene-one-function mentality and arguing for genetic interdependency.[39] As with cloning, genetic engineering is selling promises it can't deliver on, at least in the foreseeable future. Genetic engineering will not endow organisms with qualitatively new capabilities they don't already have.

What, then, can genetic engineering do? In the foreseeable future, genetic engineering will plug holes in pathways that are mostly intact to begin with. If a single gene is missing in pathways for, say, blood clotting or insulin production, then yes, the missing gene could be inserted, restoring lost function. Similarly, a gene that alters an existing trait might be inserted. What seems realistic is more aptly termed "genetic tinkering" than genetic engineering. Genetic tinkering might do some good, although I believe its potential for serious harm far exceeds the benefits.

POLLUTING OUR GENE POOL

The Monsanto Company has genetically tinkered with corn to make it resistant to the herbicide Roundup: "Monsanto's Roundup herbicide is widely used around the world. . . . It provides effective control of weeds. . . . Monsanto has developed the 'Roundup Ready' gene to make valuable crop plants tolerate Roundup. This makes it possible for farmers to apply Roundup around and over the top of crops, effectively killing weeds without affecting the crop. . . . In Africa, where most weeding is done manually by women and children, the judicious use of herbicide tolerant crops can free millions of people from this task to engage in other productive activities."[40] In addition to corn, the Roundup

Ready gene has been placed in the genome of cotton and soybeans. The gene was added into these species using a plant virus that had been rendered friendly, the cauliflower mosaic virus (CMV).

In 2001 the gene pool of wild relatives of domesticated corn were discovered to be polluted with the signature of CMV, indicating that some gene transfer had taken place.[41] About 1 to 5 percent of the kernels on a wild corncob had signs of the Monsanto gene. How the Monsanto gene managed to get into the gene pools of wild relatives of domesticated corn wasn't clear. One suggestion was that that the friendly virus had run amok and was spreading throughout the genome of the wild relatives of domesticated corn. Follow-up studies claimed instead that simple cross-breeding between domesticated corn and its wild relatives had introduced the gene into the wild gene pool.[42] Either way, the gene pools of the wild ancestors of domesticated corn are now polluted with Monsanto genes.

For those concerned about genetically modified foods, so-called Frankenfoods, this news is bad enough. But could our own gene pool also become polluted from genetic tinkering? In January 2002 just such a scare was reported. A gene therapy trial was conducted by the biotech company Avigen, in which it attempted to insert a missing gene into the liver to cure hemophilia B, the rarer of the two kinds of hemophilia, which afflicts about 150 persons each year. The missing gene is for Factor IX, one of the proteins needed to make blood clot. First a gene for Factor IX was inserted into a friendly virus, originally related to a virus for the common cold. The virus was also outfitted with a gene that was supposed to make the Factor IX gene turn on only in liver cells, and not in cells of other tissues that the virus might infect. However, if the virus did enter germ cell tissue, the gene for Factor IX would be passed on to future generations, thus entering the human gene pool. In fact, the virus was detected in the seminal fluid of the patient, implying that the virus was being expressed in tissues outside the liver, although the virus was not detected in the sperm. Nonetheless, the patient was "required to wear a condom so as to avoid any chance of fathering a genetically altered child."[43]

Adding the gene for Factor IX to our gene pool might seem desirable, as it could cure the patient and his descendants of a rare form of hemophilia. However, the technique involves adding not only the gene for Factor IX, but also the whole virus used to transport the gene, including the gene that is supposed to make the virus operate only in liver cells. If the friendly virus doesn't behave as planned, the virus could insert at multiple spots in the genome, possibly disrupting genes elsewhere and becoming a source of genetic disease itself. The friendly virus probably can't jump to another person on its own, although I doubt that this possibility could be absolutely ruled out for any virus related to the common cold. However, genes could be passed into the gene pool as a result of someone not using a condom. This technology of using friendly viruses to transport supposedly desirable genes into the genome reminds me of biological control. In spite of the best assurances of agricultural scientists, introduced species have an agenda of their own and don't always behave as intended.

Restoring Factor IX in hemophiliac patients is the best-case scenario for gene therapy: the desired gene is undeniably good, the gene in the patient undeniably deleterious, and the side effects of the friendly virus perhaps not significant compared with the seriousness of the genetic disease being treated. In other, less clear-cut cases, though, genetic tinkering might inadvertently cause a genetic disease instead of curing one. The desired gene may not be as good as advertised, the patient's gene may not be as bad as claimed, and the friendly virus may not be very friendly.

If some pollution of our gene pool may come from an attempt at doing good that has gone astray, as in the near-mistake with Factor IX, other forms of genetic pollution may be more deliberate. To protect their product, genetic engineering companies will undoubtedly include trademark DNA with any human-synthesized genes added to our genome. The clients and their children will then be stamped with this mark for all posterity. One can also imagine people wanting personal genetic trademarks, like genetic coats of arms, to be carried in their descendants.

A successful technology of gene insertion will probably be followed by genetic hacking. Someone will find it great fun to introduce genes that manifest as color changes in skin or hair. A technology of gene insertion threatens to make our gene pool a public medium, a bulletin board for posterity. People could post messages like jpeg-encoded images in untranscribed regions of our genome, its "dead code," for all sorts of purposes—the Gettysburg Address, a pornographic photo, an embodied time capsule. What started as a technology promoted to improve agricultural productivity may become a technology that fundamentally changes the material of humanity.

You may think there's no reason to panic. None of this genetic hacking will happen tomorrow. As long as genetic engineering doesn't work, we'll neither obtain its advertised benefits nor suffer its unspoken dangers. But like cloning, someday genetic engineering will work, especially once a more sophisticated view of the gene-cytoplasm interaction is taken into account. We need to get ahead of the curve now because gene pool pollution is irreversible.

GENETIC BULLETS

The Human Genome Project and private companies intend to build databases of genes from people of various races—people from Africa, Asia, Europe, and pre-Columbian America.[44] Genetic engineers believe the genetic profiles will enable drug makers to tailor drugs to each individual's genetic composition, an approach called "pharmacogenetics."[45] Personalized medicine is a new field in genetic engineering, with significant economic potential.[46] It sounds wonderful, but genetic profiles of people could also be used to make genetically targeted poisons, genetic bullets. The potential for killing through genetic engineering is easier to realize than the potential to cure.

The genetic profiling envisioned by genetic engineers misrepresents natural genetic variation. Races are socially invented categories with little genetic distinctiveness between them and a lot of genetic heterogeneity within them. However, some genetic differences do exist among people from different places of origin. Species typically show "geographic variation"—location-specific genes. These genes allow geneticists to track early routes of human migration, as regularly reported in the press.[47] The genes get mixed up as migrations continue, but new local constellations of genes continually form in their place, so that at any time some location-specific genetic markers can usually be identified. So, while genetic profiling of races has limited biological validity, a temporary genetic profiling of place of origin is more plausible.

Warfare is often directed against peoples from different places, and geographic profiling of people threatens to aid this kind of war. Recall the friendly virus manufactured to operate only in liver cells: the virus detects some protein or gene made only in the liver. Let the virus instead detect some protein or gene made by people from a certain location. Upon such detection, let the virus spring to action and turn off some critical enzyme, causing disability or death. These genetic bullets could target an army from another place of origin.

People of different nationalities obviously differ in some genes. Height alone distinguishes the armies of some countries from those of others. A fusillade of genetic bullets shot by an army of tall people against an army of short people would accidentally make some mistakes. Some of the shortest people in the attacking army would be lost (friendly fire, collateral damage). And some of the tallest people in the attacked army would be missed. The infantry can then clean up. Genetic bullets would be no less precise than present technologies of warfare, and the first country to use them would win.

Alternatively, taking a cue from Monsanto's Roundup Ready gene, the soldiers of one army could be injected with a friendly virus that endows resistance to a general poison. Then the poison could be spread on the battlefield, weeding the enemy from caves and casbahs. Soldiers would volunteer for such missions. Any army that can recruit suicide bombers could undoubtedly also recruit soldiers who would agree to have their genetic makeup modified to include a gene protecting them from the lethal pesticide used to annihilate their opposition.

The former Soviet Union already attempted to use molecular biology to wage war. Kanatjan Alibekov, a senior defector from the Soviet germ warfare program, disclosed that Moscow had mastered the art of rearranging genes to make harmful microbes even more potent and harder to counteract. Anthrax was genetically altered to resist five kinds of antibiotics.[48] And researchers are working on anthrax right now in the United States. One of these is Craig Venter, whose original company, the Institute for Genomic Research (TIGR)—now managed by his wife and built to a staff of three hundred people with $40 million a year in research grants—has financing to sequence anthrax DNA. Venter acknowledges "severe pressure on me from the people who put up the money."[49]

Genetic bullets are easier to construct than genetic cures. Cloning isn't working, tissue regeneration isn't working, gene therapy isn't working. But it is possible to make anthrax more lethal. National defense (and offense) is where the money is. The venture capitalists who have under-written genetic engineering in the name of health are facing revenue chal-lenges,[50] so it seems logical that they will someday demand a return on their dollars. Weapons production is perhaps the only sector in which ge-netic engineering might be profitable in the foreseeable future.

Indeed, genetic engineering seems destined to become primarily a weapons industry and not a health industry, if present trends continue. The desire of molecular biologists for the power that nuclear physicists possessed during World War II has already been noted. A major govern-ment sponsor of the Human Genome Project is the Department of En-ergy, the U.S. agency responsible for managing nuclear power. Much re-search on genomics is taking place at the U.S. National Laboratories, which were active in developing nuclear weapons and had to shop around for new tasks after nuclear disarmament treaties were signed. The terrorist attacks of September 11, 2001, and the following anthrax mini-epidemic now provide the perfect justification for expanding the role of genetic engineering as a weapons technology.

An example of the extraordinary personal pressure being placed on biotechnologists to become weapons producers is captured in quotations from Sidney Drell, cofounder of Stanford University's Center for Inter-national Security and Cooperation (CISAC), deputy director emeritus of the Stanford Linear Accelerator Center (SLAC), senior fellow at Stan-ford's conservative Hoover Institution, and winner of the U.S. National Intelligence Distinguished Service Medal. In February 2002 Drell stated, "If anybody thought that the Cold War was over and the world was re-laxing, that there was less need for scientists to get involved, they ought to think again. . . . For biological terrorism, all you need is a small cadre of very sick minds that are intelligent, that have been trained with the lat-est medical knowledge, and they need a facility that's little more than a microbrewery." Drell goes on to say that "the scientific community has an obligation to society" to prevent biological terrorism, just as he cred-its science for stabilizing the threat from nuclear warfare with improved aerial reconnaissance technology.[51]

Genetic engineering seems poised to enter a new arms race. Genetic engineers now not only have a way of satisfying their fantasy of being feared and revered, like nuclear physicists during World War II, and a way of turning their financially disappointing industry into big profits; they also have a moral obligation to protect our free society from attacks by sick terrorists. Do we want a new military-industrial complex?

A DIFFERENT FUTURE

We have a choice about what we do with the knowledge we've gained. As this part of my book has shown, the diversity so evident on people's exteriors is found on the inside too, extending through every level of our bodies. Our species isn't divided into two classes, normal and different. Our species is naturally a rainbow of normalcies in every bodily detail. The distinction between male and female, as defined by gamete size, does not extend with similar clarity into genotypes, chromosomes, biochemistry, hormones, morphology, brains, mental capacities, gender identities, or sexual orientations. Apart from gamete size and associated plumbing, nearly every male trait is naturally possessed by some female, and nearly every female trait is naturally possessed by some male. Claims of a gender binary in humans based on small statistical differences against a background of great overlap amount to social myths.

Each person has an individual developmental narrative beginning with the gametes from which they fused, through embryonic life, infancy, childhood, adulthood—through an entire life history, including how their gender identity and sexuality have come to be expressed. This inherent human diversity becomes manifest through whatever social categories the local society supplies. Yet the truth of human biological diversity is overlooked in the accounts of human nature that biologists tell each other, propound in classes, and relate to the public. Biologists teach that people can be divided into all manner of binaries—male and female, gay and straight, normal and mutant, healthy and diseased, each with its defining template. Biologists teach that deviations from these templates can be corrected by isolating the genes that control development or by managing the environment in which children grow up. These biological teachings that pathologize what is a natural diversity are inaccurate and dangerous. They advertise an impossible fantasy of control. They intrude on the human rights of individuals and threaten the future of our species.

Instead of fearing diversity, let's reach out to other times in history and other cultures to see how the human rainbow has surfaced there. We might pick up some useful hints about how we too might better accommodate the diversity we increasingly recognize, for the institutions of a just society should promote ways for a biologically diverse people to live and prosper together.

PART THREE

CULTURAL RAINBOWS

18

Two-Spirits, Mahu, and Hijras

By looking at how the universal human rainbows of gender and sexuality fit into the social categories of other societies around the world and at other moments in history, we may glean some ideas about how our own institutions might function better. Perhaps we can avoid the lost time and needless expense of suppressing biological difference. As with animal diversity, the facts of cultural diversity in gender and sexuality are unexpected and engaging. Yet, like natural science, the social sciences of anthropology, sociology, and history, as well as theology, all discount the very diversity that their painstaking research and primary texts so clearly document. Instead, many are surprised to learn how widespread homosexual and transgender expressions are among the peoples of the world and throughout history. We've never been told.

This part of my book, "Cultural Rainbows," offers a worldwide historical survey of how gender and sexuality variation are manifested in human society. I experimented with different ways to organize this story. Should we simply go around the world—it's Tuesday, so this must be Tahiti? Or start with places where gender is a reflection more of occupation and social space than of body type, noting how in such cultures some male-bodied people are effectively women, and vice versa? What about emphasizing cultures that illustrate a collision between ancient, traditional social categories and modern, Western ones—between a view of gender- and sexuality-variant people as sacred and accounts that assign medical pathology to homosexuals and transsexuals? Why not move from cultures in which the social categories emphasize gender to cultures that stress sexual orientation? Or maybe we should contrast societies that expect a sex-reassignment type of body morphing with those that don't. Or we could emphasize the role of religion in the construction of gender categories and mores of sexual practice. The number of interesting angles is limitless. The organization I offer is somewhat arbitrary, but please keep an eye out for all of these aspects.

I've chosen stories that stand out to me as a transgendered woman. When writing about ecology and evolution, I wrote as a native about my hometown. With developmental biology, I wrote about the town next door. Here I write as a tourist in foreign academic lands, the last leg of my journey of discovery through academia. I apologize for my insensitivities to foreign academic traditions, but do not regret my criticisms. Social scientists frequently denounce scientists' pretense to objectivity. I find social scientists just as flawed. They too deny the human dignity of gender-variant people.

TWO-SPIRITED PEOPLE IN THE AMERICAS

Since settling in San Francisco, I've encountered many expressions of gender and sexuality I didn't know existed, distributed across countless ethnicities. People being as they are. Lovely. Unnamed and without words for themselves. We're just beginning to discover ourselves. Sometimes I think we know more about diversity in the deep sea than we do about ourselves. Yet long before San Francisco was founded as a Western city, the Native nations in the Americas offered a rich social environment for the people we now call transgendered, gay, and lesbian. Gender-variant people in Native America are often referred to as "two-spirited," with the details varying from tribe to tribe.[1]

Some tribes have held two-spirited people in exceptionally high regard, in part because of their religious role in ceremonies and beliefs about creation. Among the Zuni, for example, legend tells of a battle between agricultural and hunting spirits in which a two-spirited deity brought peace to the warring parties. Zunis reenacted this event ceremonially every four years, with a two-spirited person playing the role of the two-spirited deity.[2] Similarly, among the Navajo, the survival of humanity was believed to depend upon the inventiveness of two-spirited deities. Having two-spirited deities at the foundation of religion endowed two-spirited people with dignity and significance.

OSH-TISCH

The anthropologist Will Roscoe reports an account of how Hugh Scott, a retired army general, interviewed a famous two-spirited woman named Osh-Tisch from the Crow tribe in 1919.[3] In his first encounter with Osh-Tisch, General Scott "wandered into the huge buffalo-skin lodge of Iron Bull, head chief of the Crows." Iron Bull's lodge had been created by Osh-Tisch, who was also an artist, medicine woman, and shaman who had accumulated great prestige. Scott asked her why she wore women's clothes, although she was known to be physically male. "That is my road," she replied. How long had she been that way? She answered that since birth she "inclined to be a woman, never a man." What sort of work did she do? "All woman's work." Then, with great pride, she produced a dark blue woman's dress with abalone shell ornaments and a finely beaded buckskin dress with a woman's belt and leggings. Photographs of Osh-Tisch show a stately woman. Romantically, she was oriented to men.

Two-spirited people do not "pass" physically as members of the gender they identify with—their bodily state is known to everyone. A two-spirited woman is accepted as a woman, however, even though she is generally larger than a one-spirited woman and can't breastfeed. A two-spirited woman participates in women's domestic and economic activities and looks after the older children. She also carries out activities that take advantage of her height and strength, including, if necessary, fighting in battles. In fact, Osh-Tisch was distinguished for her valor. She also helped take care of wounded warriors. Even though fighting as a warrior was "man's work," Osh-Tisch was claimed by the other women as one of their own.

A young woman named Pretty Shield recalled the accomplishments of Osh-Tisch to a journalist: "Did the men ever tell you anything about a woman who fought?" "No." "Ahh, they do not like to tell of it, but I will tell you. . . . She looked like a man, and yet she wore woman's clothing; and she had the heart of a woman. Besides, she did a woman's work. She was not as strong as a man, and yet she was wiser than a woman. The men did not tell you this, but *I* have. I felt proud . . . because she was brave."

In the 1890s, an agent from the Bureau of Indian Affairs tried to interfere with Osh-Tisch, as well as other two-spirited people, by cutting off her hair and forcing her to wear men's clothing and do men's labor. The Crow people were so upset by this that the chief of the Crow Nation told the agent to leave.[4] This intervention by the chief of the Crow Nation on behalf of two-spirited people shows a remarkable depth of political support.

HASTÍÍN KLAH

The anthropologist Will Roscoe also describes Hastíín Klah, a famous two-spirited Navajo who was gay but not gender-variant.[5] Born in 1867, he showed an early interest in religion, learned his first ceremony at ten, and studied the healing power of native plants. In his early teens, he discovered a cave on a canyon ledge where a medicine man had left a ceremonial bundle. The walls of the cave were painted with images of Navajo gods, and Klah decided to become a medicine man. He became acknowledged as two-spirited. "He dressed in men's clothes and there was nothing feminine about him unless an indescribable gentleness be so called," but the Navajo considered him two-spirited because he wove blankets and was romantically interested in men rather than women.

As a two-spirited person, Klah was expected to assist his mother and sister in their weaving. Weaving was part of women's life cycle, offering a medium for expressing self-control and self-esteem, creativity and beauty. Weaving reflected a balance between the world of animals and plants—represented by animal fibers and plant dyes—and the world of humans, those who wear the cloth. Klah's artistic style was distinctive, using backgrounds of tan undyed wool from brown sheep and designs created with dyes from local plants, and it set a new standard of excellence for Navajo weaving. He pioneered the presentation of sand-painting images in tapestry. Previously, sand-painting images had been engraved only on the ground.

By his mid twenties, Klah had become recognized for his weaving. In 1893 the World's Columbian Exposition in Chicago sought a Native weaver to demonstrate this skill to the public. They wished to bring a man, but didn't realize that a male weaver was necessarily two-spirited. Klah spent the summer in Chicago working before crowds of sightseers.

At the exposition Klah met a wealthy Bostonian, Mary Cabot Wheelwright, who wrote, "I grew to respect and love him for his real goodness, generosity—and holiness, for there is no other word for it. He never had married. He helped at least eight of his nieces and nephews with money and goods. . . . He never kept anything for himself. It was hard to see him almost in rags at his ceremonies, but what was given him he seldom kept, passing it on to someone who needed it."[6] In the 1930s Mary Wheelwright began to consider her legacy, and collaborated with Klah in founding a museum now known as the Wheelwright Museum of the American Indian, in Santa Fe. Klah died at the age of seventy, just a few months before the museum was officially dedicated in 1937.

WOMEN WARRIORS

According to Roscoe, still other people were female-bodied and participated in manly pursuits. For example, Osh-Tisch shared her warrior days with another Crow woman. According to Pretty Shield, "The other woman was a *wild* one who had no man of her own. She was both bad and brave, this one. Her name was The-Other-Magpie; and she was pretty."[7] The-Other-Magpie is not reported to have been two-spirited, but her tale suggests that the envelope of Crow womanhood was wide enough to encompass traditional masculine behavior.

Together, Osh-Tisch and The-Other-Magpie saved the warrior Bull Snake, who had been wounded by a Lakota and fallen from his horse. Osh-Tisch "dashed up to him, got down from her horse, and stood over him shooting at the Lakota as rapidly as she could load her gun and fire." Meanwhile, The-Other-Magpie rode around, waving a stick and deflecting the attention to her with a war song. "Both these women expected death that day. . . . I felt proud of the two women," recalled Pretty Shield.[8]

Other female warriors were evidently transgender. Among the Cheyenne, two-spirited women "were often great warriors who even sat with the Chief Council and had an effective voice."[9] A Cheyenne artist depicted a bare-breasted woman firing a rifle; she was dressed like the male members of a special society who fought wearing only their breech-cloths. Two-spirited women were known for romantic relationships with one another. A photograph from 1890 shows a two-spirited woman from the Quechan. She wears a man's breechcloth, with male bow guards on her wrist, and stands with one hand on her hip in a charac-teristically male pose. She was reported to be married to a woman.[10] A recent study includes maps indicating the location of tribes with male-to-female, female-to-male, or both types of genders, together with a table about which combinations of relationships were socially acknowledged and approved.[11]

Being two-spirited primarily means being of a different spirit, march-ing to a different drummer, but not necessarily being gender-variant. As the narratives show, the two-spirit category spans people who in West-ern society probably would identify as lesbian, gay, or transgendered.

TRANSITION CEREMONIES

The anthropologist Walter Williams relates a Navajo coming-out cere-mony that provided the community with an opportunity to endorse and bless a young two-spirited person.[12] On the day of the ceremony, the youth was led into a circle. According to a Navajo shaman, "If the boy showed a willingness to remain standing in the circle, exposed to the public eye, it was almost certain he would go through with the ceremony. The singer, hidden from the crowd, began singing the songs. As soon as the sound reached the boy, he began to dance as women do." A youth who was not two-spirited would refuse to dance. But for a youth who was two-spirited, "the song goes right to his heart and he will dance with much intensity. He cannot help it. After the fourth song, he is pro-claimed." The youth was then bathed and received a woman's skirt. She was led back to the dance ground, dressed in feminine clothing, and an-nounced her new feminine name to the crowd. After that, she would re-sent being called by her old male name.

The Papago had a similar transition ceremony. A brush enclosure was constructed, with a bow and arrow and a basket placed inside. The youth was brought to the enclosure while the adults watched from outside. The youth was asked to go inside and then the brush was set afire. The youth had the opportunity to grab either the bow and arrow, or the basket, and then escape the fire. If a male youth grabbed the basket, she was accepted as two-spirited; otherwise he remained a boy.

Among two-spirited people, there is no tradition of body morphing to resemble the other sex. Males don't modify their genitals to resemble female genitals, nor do females bind their breasts to hide them. Two-spirited people are not necessarily comfortable with their bodies, though. A Mohave two-spirited woman was described as embarrassed when making love because "the penis sticks out between the loose fibers of the bark-skirt."[13] Another two-spirited woman was sensitive to teasing about her penis, preferring that it be referred to as a clitoris. Good-natured sexually explicit teasing was typical, but, as the lover of a two-spirited woman related, "I never dared touch the penis in erection, except during intercourse. You'd court death otherwise, because they would get violent if you play[ed] with their erect penis too much."[14] Even though two-spirited people may have felt dissatisfied with their bodies, genital morphing was not a condition for social acceptance, nor was it expected by their sexual partners.

THE CONQUEST MENTALITY

The Spanish conquistadors of the 1500s were brutal to two-spirited people.[15] In 1530 Nuño de Guzman said the last person he captured in battle who "fought most courageously, was a man in the habit of a woman, for which I caused him to be burned." While in Panama, Vasco Núñez de Balboa saw men dressed as women and threw them to his dogs to be eaten alive. Calancha, a Spanish official in Lima, later praised Balboa for the "fine action of an honorable and Catholic Spaniard."

Justifying the Spanish conquest of America turned on whether the natives were "rational," meaning possessing a combination of reason, intelligence, and morality, as defined by the Catholic Church. If the natives were rational, then conquering them was not just. If the natives were irrational, then conquering and Christianizing them was just, similar to the domestication of animals. Sex between men would be irrefutable evidence of irrationality. Thus the Spanish explorers had a vested interest in establishing that gender-variant people practiced same-sex sexuality, thereby justifying their conquest. Their conduct during the conquest went beyond domestication, though, because they were not limited by any moral opposition to cruelty to animals.

European repression has not annihilated two-spirits in America. The two-spirit tradition is ancient. Anthropologists have traced two-spirit imagery back 1,500 years in paintings on the walls of a kiva (a round ceremonial room) in New Mexico.[16] The two-spirit tradition lives today too, as Native American groups throughout the United States reclaim this heritage.

Anthropologists, however, have tended to dwell on differences rather than similarities between today's transgendered people and Native American two-spirits, often using prejudicial language. Anthropologists use gendered pronouns for two-spirited people, including Osh-Tisch, that are based on their genitals, rather than using the pronouns appropriate to their gender presentation, which were used by the Native people themselves.[17] These words erase successful gender crossing by two-spirited people. One anthropologist then refers to present-day transsexuals as "products of our culture" who pay the "heavy price" of "bodily mutilation" for "the ideology of biological determinism" and wind up feeling "no more comfortable as a woman than as a man," although this claim is unsupported by data. This anthropologist goes on to say a "gay identity is closer" to the two-spirited role than is a transgendered identity.[18] Converting the obvious transgender aspect of two-spirited people into a gay identity appropriates transgender experience.

Other anthropologists assign gender-variant two-spirited people to a third gender, neither man nor woman, denying that some two-spirited people actually did belong to their gender of identification and not to some third, intermediate zone. All women vary in height, strength, aptitude, and capacity for breastfeeding and reproduction. Were two-spirited women simply another type of woman, albeit taller and stronger than the others and without the ability to breastfeed? Or were physical differences used to split them into a different category? The narratives suggest that some two-spirited people were folded into the two major genders of man and woman without forming a distinct gender.

Overall, two-spirited people in Native American societies are a diverse group, spanning all the rainbows of gender and sexuality that we Westerners divide into the different social colors of gay, lesbian, and transgender. Polynesia, which we will examine next, shows an expression of gender and sexuality quite comparable to that of Native Americans. Polynesia was colonized much later than the Americas, and native institutions were not decimated to the same extent. However, native representations of gender and sexuality are now colliding with introduced Western ideas, leading to a conflict between the traditional and the modern.

THE MAHU IN POLYNESIA

The anthropologist Niko Besnier has related how French explorer Louis Antoine de Bougainville encountered the islands of Tahiti during a tour of the South Pacific in 1766–69. As his vessel approached the island, native canoes came out to meet it that were "full of females; who, for agreeable features, are not inferior to most European women. . . . The men . . . pressed us to choose a woman, and to come on shore with her; and their gestures . . . denoted in what manner we should form an acquaintance with her. It was very difficult, amidst such a sight, to keep at their work four hundred young French sailors, who had seen no women for six months . . . and the capstern was never hove with more alacrity than on this occasion."[19]

Alacrity soon gave way to condemnation. The London Missionary Society, which established an outpost in Tahiti, concluded that the island was "the filthy Sodom of the South Seas: In these islands all persons seem to think of scarcely anything but adultery and fornication. Little children hardly ever live to the age of seven ere they are deflowered. Children with children, often boys with boys . . . playing in wickedness together all the day long."[20] The missionaries were particularly bent out of shape by what the British captain William Bligh described as "a class of people called Mahoo: These people . . . are particularly selected when Boys and kept with the Women solely for the carnesses [sic] of the men. . . . The Women treat him as one of their Sex, and he observed every restriction that they do, and is equally respected and esteemed."[21] Captain Bligh had encountered the Tahitian version of two-spirited people, called *mahu*, which means "half-man half-woman." All the Polynesian islands have mahu, although they have different names on Samoa and Tonga. On Hawaii, like Tahiti, they are called mahu.

Captain Bligh states that mahu were included in the company of women. Other reports suggest that they were seen as women by the sailors, leading to surprising encounters. In 1789, one sailor wrote, "One of the gentlemen who accompanied me on shore took it into his head to be very much smitten with a dancing girl, as he thought her . . . and after he had been endeavouring to persuade her to go with him on board our ship, which she assented to, to find this supposed damsel, when stripped of her theatrical paraphernalia, a smart dapper lad."[22] The natives, for their part, followed along the beach laughing and enjoying the comedy. This passage shows that although the English may not have been able to tell mahu from nonmahu women, the Tahitians certainly could, raising the question of whether the mahu constituted a third gender or were merged into the gender of women.

Studies in recent times by Besnier show mahu working in women's occupations: cooking, cleaning house, gathering firewood, doing laundry, weaving mats, and making cloth. In urban settings, mahu are sought after as secretaries and domestic help. Socially, mahu live in women's space, "walking arm-in-arm, . . . gossiping and visiting with them" into old age. In appearance, mahu typically include some feminine characteristics and occasionally dress as women.[23] Mahu, it is said, adopt a "swishy gait" and are characterized by a "fast tempo, verbosity and the animated face, which contrasts with men's generally laconic and impassive demeanor." They are "coquettishly concerned with their physical appearance," wearing "flowers, garlands and perfume, and in urban contexts, heavy makeup." A picture emerges of mahu as being feminine, perhaps effeminate, but stopping short of a completely feminine presentation, while socializing and working in women's space.

Mahu are identified by their gender inclination as children, before the "awakening of sexual desires of any type."[24] Thereafter, mahu are likely to interact sexually with men, but a sexual orientation toward males is "neither a necessary nor sufficient criterion" for mahu status. Mahu do not consort sexually with other mahu. Also, a mahu may leave this status and become a man by marrying and fathering a child. Mahu are perceived as always available for sexual conquest by men. Mahu may sexually taunt men in a caricature of flirting. Mahu are often the target of harassment and physical violence from men, especially men who have had too much to drink.

Unlike two-spirits, mahu do not have access to male political power or prestige. They also differ from two-spirits by having a relatively low status in society. Mahu cannot aspire to the leadership roles that two-spirited people like Osh-Tisch did. Nowhere in Polynesia are mahu associated with religious life,[25] nor is there any public coming-out ceremony.

Yet the mahu, as half-man half-woman, share with the Native American two-spirits the characteristic of combining elements from both male and female genders. Also like two-spirits, mahu are accepted to some extent into women's space just as they are, without further bodily authentication. They therefore differ, as we will see, from Indian *hijra,* the eunuchs of the ancient world, and contemporary transsexuals, all of whom undergo a sex-reassigning body morphing. Nor are mahu thought of as neither man nor woman, or as lacking something, as are hijra, eunuchs, and transsexuals. Instead, mahu are half of each, like Native American two-spirits.

Although Polynesian society has been influenced by French colonial and missionary culture, Polynesian society remains largely intact, in contrast to the Native American cultures. Polynesian society, while decidedly non-Western, is not so different that the cultural gap is insurmountable. Thus Polynesia is an excellent site for further anthropological study of gender and sexuality.

The mahu have recently been shown by the anthropologist Deborah Elliston to include masculine women in addition to the feminine men that attracted the attention of the early explorers. One woman explained that a female-bodied person could be a mahu too, saying, "Mahu, that can be a man or woman because that's what it means, someone who's both."[26] Elliston reported initial difficulties in discerning the "the codes, cues, signs, and performances of female-bodied mahu." It became apparent, however, that occupations like truck driving and subsistence farming were coded as masculine, as well as certain gestures, clothing, and the wearing of short hair in a society where most women grow their hair very long.

Polynesians today are, on the whole, accepting of mahu, primarily because they view mahu as natural, as "being that way." Mahu make themselves known while still children by demonstrating transgendered styles of appearance or a preference for transgendered work. Boys with feminine inclinations and girls who are tomboys (in French, *garçon manqué*) are likely mahu. Gender identity is more important to mahu status than is sexual orientation. In fact, the sexuality of mahu varies. One study reported that male-bodied mahu usually had sex with men, especially young men, yet several had also had long-term relationships with women and were fathers, and still others were celibate. Female-bodied mahu usually had women as lovers, but many had had male lovers at some time, and others were celibate.[27] This emphasis on gender rather than sexuality resembles that of contemporary American trans people, who express all types of sexual orientation, including celibacy. Yet American trans people also may have relationships with one another, whereas mahu form relationships only with men or women, but not with other mahu.

Polynesians conceptualize people as being "mixtures" of male and female ingredients.[28] People differ from one another by having different ratios of male to female. The mixture of a male-bodied mahu consists of more femaleness than maleness in a male body, and vice versa for a female-bodied mahu. A male-bodied mahu who is attracted to males represents the attraction of the mahu's female ingredients to a male. Thus an elemental sexual binary is affirmed, but bodies are allowed to express different combinations.

According to Elliston, an especially interesting recent development in Tahiti, seen particularly in the capital city of Papeete, is the emergence of a Western transgender style called *raerae,* or *travesti* in French. Presently, these are exclusively male-to-female trans people who emulate a "specifically Eurocentric form of white femininity."[29] In public, most travesti wear revealing European women's clothes: miniskirts, skimpy shorts, halter tops, high heels—the kind of white femininity idealized in the mass media throughout French Polynesia. Most travesti work, at least part-time, as sex workers for male clients. Surprisingly, travesti say they have "chosen" to be as they are, meaning apparently that they choose to express their transgendered nature by this route rather than as mahu. Their path begins as males who have sex with other males, although without identifying as mahu. Later they transition into being travesti. Most take hormones, and many have had sex-reassignment surgery. Prior to transition, some fathered children.

Tahitians disparage the travesti because their dress is "over the top," their style is foreign, and they are thought not to be authentic, in contrast to the mahu, who "have always been that way." Here we are witnessing a collision between two different cultural transgender manifestations. How this collision plays out will make for some fascinating real-time anthropology. Similarly, the new categories of *homosexuel* and *lesbienne* have arrived in Tahiti, representing European gay and lesbian identities, which also don't map neatly onto the mahu category. Stay tuned.

THE HIJRAS IN INDIA

India's size guarantees that the aggregate number of transgendered people is huge, even if the fraction of the population that identifies as transgendered is small. With a population of more than one billion people, India has over one million transgendered people (one in one thousand) who belong to a group called the *hijras,* a combination religious sect and caste.[30] The hijras, who consist of male-to-female transgendered people, acquire members mostly from the lower and untouchable castes.

The religious aspect of hijra life focuses on devotion to the Mother Goddess, Bahuchara Mata, or Mata for short. The major hijra temple is located near Ahmedabad in Gujarat, north of Bombay in northeast India. The religion is principally Hinduism, with some elements of Islam.

Hijras perform celebrations for the birth of a male child, and at weddings they offer the blessings of Mata. With the Westernization of India, the demand for these ceremonies is declining, and hijras increasingly work in the sex trade or as beggars. Hijras are attempting to break out of this downward spiral, and some have recently been elected to public office. In January 2001 the new hijra mayor of Katni, a limestone mining town with a quarter million people, was featured in the *New York Times,* along with five hijras elected to other positions around India.[31] Another hijra political leader was covered three years earlier in the *Wall Street Journal.*[32]

According to anthropologist Serena Nanda, hijras are organized nationally into seven named houses. An elder from each house, called a *naik,* has jurisdiction over a geographic region, such as a medium-sized city or one section of a large city like Bombay. The naiks meet collectively as a *jamat,* or meeting of the elders, and function as a ruling board for the region. The jamat formally approves the admission of a candidate to the hijras. A candidate hijra is called a *chela,* or disciple, and is sponsored by and apprenticed to a guru, or teacher. To join the hijras, a candidate is taken under the wing of a guru, who then brings her to the jamat for induction. The chela gives the guru her earnings and submits to her authority. The guru is responsible for the welfare of her chela and for the initiation fees paid to the jamat. A guru usually lives with her chelas in a small commune, typically composed of five people. Occasionally a hijra marries and goes to live with her husband.

Hijra appearance ranges from passing as a nonhijra woman to mixed-gender appearance with gaudy clothes and a deep, booming voice. Hijras generally wear women's clothes, including a bra and jewelry, and have long hair in a woman's style. They pluck their facial hair to attain a smooth face. Hijras walk, sit, and stand as women do, and carry pots on their hips, which men don't. Hijras take women's names, and use feminine language, including feminine expressions and intonations. They request women's seating in public accommodations and sometimes demand to be counted as women in the national census.[33] Hijras may also exaggerate feminine dress and mannerisms to the point of caricature, use unfeminine coarse and abusive speech and gestures, and smoke cigarettes, which is normally a male "privilege."

Hijras are marginalized in Indian society and are not accepted as women by nonhijra women. They are forced to function outside the traditional two genders, instead forming a third gender. While Indians acknowledge gender variation, they do not accept the variation socially: "Don't make it sound like we're about to invite them in for a cup of tea."[34] Hindu society's attitude toward hijras is mixed. Their blessings at a wedding promise prosperity and fertility, but their curses may bring infertility or other misfortunes. A hijra may insult a family that does not meet her demands for money and gifts, starting with mild verbal abuse and ridicule, then moving on to stronger insults, and culminating in the most feared insult—lifting her dress to display her genital area.[35] Hijras' spiritual contribution is tempered with this element of extortion. Hijras are at once special sacred beings and objects of fear, abuse, ridicule, and sometimes pity.

CONTEMPORARY NARRATIVES

The range of personal styles among hijras can be seen in the people living together in a commune described by Serena Nanda.[36] The woman in charge was over six feet tall, with classically beautiful Indian features and extravagantly thick jet-black hair hanging below her waist. She wore chiffon saris and diamond earrings, with gold chains and wrist bangles. Another hijra, who managed a bathhouse, was enormously fat and masculine in appearance, with extremely hairy arms and a tattoo on her wrist. She wore no jewelry and was described as looking like a "gigantic Buddha." Still another was young, beautiful, and feminine, and lived with her husband at night.

Among the other hijras interviewed by Nanda was Kamladevi, age thirty-five, who spoke fluent English, Hindi, and Tamal, as she had gone to a Christian convent high school up to the eleventh grade. As a child, Kamladevi refused to wear pants and instead dressed only in *lungi,* a traditional skirtlike cloth of brightly colored silk or cotton. She wore eyebrow pencil and lipstick at school, which she removed before coming home. At age eleven, she had her first sexual experience with a boy and later had trysts with several male teachers in the school. Her parents tried to prevent her efforts to feminize. Her father, a police subinspector in the crime division, even assigned an orderly to watch after her. However, the hijras noticed Kamladevi and invited her to join them. She did, going off to Bombay, where she became a sex worker. Her fate was bleak, and she died soon after being interviewed.

Another of Nanda's interviewees—Meera—was a successful hijra guru at age forty-two. Meticulous in dress and conservative in demeanor, like a "middle-class housewife," she had a masculine face but an exceptionally intense feminine gender identity. At four or five years of age, she pretended she was a girl and walked with a sway. Her parents allowed her to wear a *bindi* (the colored dot Indian women place on their forehead) and to dress in girls' clothes. As an adult, she began taking female hormones to increase her weight: "Now I am nice and fat, like a woman."[37]

Meera had a husband, Ahmed. "If I feel dejected and there are tears in my eyes, then Ahmed will ask me, 'Why are you so depressed? What do you want? What has happened to you?' And if Ahmed is not well, even with just a headache, I will sit by him the whole night tending to him, massaging his head, his body. . . . He's guarded me so well. If anybody teases me or disturbs me, he gets very angry. . . . When he is not here the police and the urchins bother me, but when he is here everyone is silent. . . . If anybody troubles me, Ahmed will thrash them and send them away. God and he are one to me. . . . If Ahmed [went] away to another lady or another hijra, then I would shave my head and burn myself, like a widow who commits suttee." In a later interview, Meera showed a small baby she had adopted and was taking care of. After years of hormones, her breasts had fully developed. "Now my only wish is this, that my husband should be all right, my chelas should be all right, that God gives us enough money to sustain ourselves. God is great."

Meera's path to this point, though, was not direct. She had previously married a woman and fathered a daughter, later arranging a marriage for that daughter. Meera was evasive when asked about this part of her life. She knew, as Kamladevi put it, "To be a hijra, you should not have any relations with a woman."[38]

Sushila, who was interviewed at thirty-five, was born of a Tamil family in Malaysia. "From my earliest school days, I used to sit only with the girls," she recalled.[39] She became sexually active at thirteen with a fisherman who was married and lived with his wife, mother, and sister. When Sushila moved in with them, her parents never came to take her back because she wore a *bindi* and *kajal* (feminine makeup). "My family didn't like this," she explained; they found it "embarrassing." She returned home on her own initiative after a while, and one day met a hijra at the movies. The hijra invited her to join, saying, "You can always wear a sari and live" when you live with us. Sushila joined the hijras because "I hated my house so much." This began a back-and-forth with her family. "Come home," they would say. "I'll come like this only [dressed as a woman]." "No, we're such a big, honorable family, how can I let you come home this way?" "Then I won't come." Still, when her sister fell ill ten days later, she did return. "My father and brother both requested me to get inside one of the rooms and to change my dress into a *lungi* and shirt before people could see me. I told my father, 'If you people feel ashamed of me because I am wearing a sari, I don't want to embarrass you. Allow me to go away here and now.' " After two days, she returned permanently to the hijras, and at times worked as a sex worker.

Sushila took a husband, a Brahman who was a chauffeur for a large corporation. She spoke warmly of her husband and was concerned that she could not give him a child, which she felt was necessary for him to have "a normal family life." In a later interview, Sushila revealed that she had pulled off a coup. She had adopted her former husband as her son (!) and had arranged his marriage to a neighbor's sister, who was poor but respectable and quite pretty. The newlyweds had a son, making Sushila legally a grandmother.

Meanwhile, Sushila had found another husband. "What I find attractive in my man is the way . . . he likes to see me well dressed, tidy, with flowers in my hair, with a bindi, wearing new clothes, keeping the house clean, and not using bad language. . . . I have my husband's lunch ready by the time he comes home. I tend to his house. . . . You see how many people come and sit here with me to chat. . . . Now that I am respectable and talk to people well, people come and sit with me." Even though she was a former sex worker, she could say, "Now I have my husband and he's the only man for me. . . . Now I'm leading the life of a respectable woman with a husband, an adopted son, a daughter-in-law, and a grandson—and running a house."

The three hijras discussed so far were born and raised male but wished to live as women. Kamladevi and Sushila were prevented from dressing as women at home, Meera was allowed to, and all three joined the hijras to live, at least to some extent, as women. In contrast, Salima was born intersexed. Salima was interviewed while living on the street, sleeping on a tattered bedroll in Bombay. At this point she was not even a sex worker ("no customers are coming to me") because of her dishevelment, with three days' growth of beard and dirty hands, feet, and clothes.[40] She recalled, "My parents felt sad about my birth. . . . My mother tried taking me to doctors. . . . My father made vows at different places, but it was all futile. . . . My organ was very small. . . . The doctors said, 'No, it won't grow, your child is not a man and not a woman.' . . . If I was a girl they would have nurtured me and made me make a good marriage; if I was a boy they would have given me a good education. . . . But I have been of no use to them." Salima went on to explain, "From the beginning I only used to dress and behave as a girl. . . . I never thought of myself as a boy. . . . My parents had given me a boy's name. . . . I would give them [the teachers] a girl's name." At school, the teachers would not allow Salima to sit with the girls. "For this reason I stopped going to school."

When Salima met the hijras, her mother said, "Since you are born this way, do whatever you want to do, go wherever you want to go, do whatever makes you happy." So Salima joined the hijras, and "the pain in my heart was lessened." Salima was treated well among the hijras while she was protected by her guru. After her guru's death, however, she was ostracized. She found a husband for a time, but eventually died on the streets.

As can be seen in these narratives, hijras don't propose a new and distinct conception of gender. Hijras are a third gender by default, not by design. Denied entry into the gender they identify with, they wind up as a third gender. Many, perhaps most, hijras clearly wish for the life of a conventional nonhijra woman.

NIRVAN: GENITAL SURGERY

The word *hijra* is often translated to mean "not man, not woman." Hijras have a form of sex-reassignment surgery referred to as *nirvan,* or "the operation," which modifies the genital region to a state intermediate between male and female genitals. *Hijra* is also translated as "eunuch." Of the four individuals whose narratives we have been given, Kamladevi and Meera had the operation, Sushila was planning to have it, and Salima didn't "need" one.

The nirvan is an elaborate ceremony in which a person is separated from her male form, resides while convalescing in a liminal state, and is finally reborn as a "true" hijra and empowered as a disciple of Mata. The operation is performed by a *dai ma,* or midwife. Meera, who was qualified to perform a nirvan, did so many times. Specifically, the testicles and penis are removed with "two quick opposite diagonal cuts."[41] The mere mention of this highly symbolic action has probably made you uncomfortable.

Why would a hijra consent to a nirvan? Not only consent, but pay big money? Kamladevi paid Meera "so much . . . 27 saris, 20 petticoats, 27 blouses, 2 dance dresses, 1 big tin box, 9 stone nose rings, 200 rupees," revealing a strong motivation "no government, from the British to the Indian, has been able to erase."[42]

Understanding the nirvan is hindered by the pejorative descriptions of anthropologists. Nirvan has been called an "emasculation ritual" carried out as "part of a religious obligation." A man, it is said, offers his family jewels to a demanding goddess who devours, beheads, and castrates her consort. Supposedly, "identification with the Goddess through sacrifice of their genitals assures [hijras] of her life-giving presence, warding off death." Yet, the reports claim, instead of warding off death, the pathetic result is only "mutilated genitals."[43] Nirvan is construed as the irrational superstition of a primitive people.

According to their own narratives, hijras are not actually offering their genitals as a sacrifice. The genitals are not placed on an altar to Mata, but are quietly removed and buried in a pot at the base of a tree. If nirvan were a sacrifice, why would both penis and testicles be removed? Testicles alone would suffice if manhood were being yielded to Mata—that's what castration means. Construing hijra practice as an irrational devotion to a bizarre primitive deity denies dignity and agency to hijras and discounts the human diversity they represent.

So why do hijras undergo nirvan? Perhaps hijras are rational after all. Let's see if nirvan stands up to cost/benefit analysis, the benchmark of rational analysis. The costs are low. A hijra isn't giving up much when she cedes her male genitals to Mata. To a hijra, born male but identifying as female, male genitals are hardly family jewels. Kamladevi referred to her male organ as weak and useless, "not good for anything." Similarly, describing how she was before her operation, Lakshmi, a beautiful young hijra dancer, said, "I was born a man, but not a perfect man." And Neelam, who was waiting to have the operation, remarked, "I was born a man, but my male organ did not work properly." Preoperative hijras do not view their genitals as assets, so giving them up represents no cost at all. The operation itself is the major cost, both the sum paid to the *dai ma* and the pain of the six-week recovery process. The procedure itself is not painful, just "a small pinch" or "ant bite."[44]

The benefits are many:

1. *A feminine body.* The operation furthers the feminization already begun with women's dress, including padded bras, a

woman's style of long hair, a smooth face from plucking facial hair, feminine language, and change of name. Meera explained, "After the operation we become like women." Removing the testicles eliminates the main testosterone-producing glands, allowing a more feminine body contour to develop, and removing the penis allows the hijra to pass urine as a woman does. Hijras consider the operation's result to be beautiful, not a mutilation. Meera mentioned she had been ill and examined in a hospital. "The doctors were amazed at how excellently I had been 'made into a woman' by the operation. Only with this proof of their own eyes were they convinced of the power of the hijras to transform themselves from men into women." To the interviewer, she added, "You must take a picture of my operated area so that people in your country will also know the power and skill of the hijras."[45]

2. *Husbands' expectations.* Meera stated that her husband, Ahmed, said to her, "You're a man and I'm a man," and told her to have the operation. "So I went for the operation at that time." Similarly, Sushila said, "My husband wants me to get the operation done so I will look robust and nice, like the others."[46]

3. *Authenticity.* Nirvan provides proof that a person is a real hijra rather than a cross-dressing impostor. Peer pressure contributes too. Kamladevi admitted, "Having lived so many years, if I didn't get the operation done, it would be a great 'black mark' for me."[47]

4. *Power.* Nirvan endows a hijra with Mata's power. The operation ordains a hijra with the spiritual authority to bless in Mata's name. A cross-dressing male lacks this spiritual power. If he dances in a ceremony to bless a new baby or wedding and is discovered, he cannot claim a fee and is sent away embarrassed.[48] Furthermore, after nirvan, a hijra's threat to expose her genital area becomes credible, whereas before nirvan, any such threat would be a dangerous bluff.

Little wonder, then, that no government, from the British to the Indian, has been able to erase the hijra nirvan. The practice is rational in the local context, both now and in the past.

Any regrets? Sure, but not about the operation as such. After the operation, a hijra is repositioned in society's power structure, where Mata's blessings don't cut any ice. She's no longer a man and no longer enjoys the possibility of male power. She can't switch to guy mode anymore to get out of a jam. Kamladevi said, "Before the operation, even when we went out at night, we never had a fear. But now, suppose we see a drunkard, or a rowdy; now after the operation we get frightened. . . . The local rowdies and bullies come at night, knock at the door, wake us up, and forcefully have their way with us. But still, we must do it."[49] Welcome to a woman's world. Meera had no such regrets. She had Ahmed to "guard" her.

Thus the nirvan practiced by Indian hijras can be a rational choice, a way for a person with cross-gender identity to make a better life in local circumstances. Although nirvan is described as a religious obligation, this appeal to religion may be nothing more than a cover for the real reasons. Nontransgender people are rarely able to comprehend transgender motivation, and transgender people come to depend on social fictions. In the West, transsexualism is couched in medical fiction; in India, apparently religious fiction holds sway.

A COMPARISON OF HIJRAS AND TWO-SPIRITS

For two-spirited Native Americans, the cost/benefit table of surgery was not the same as it is for hijras. The costs were higher, as the technology wasn't available, and the pain, suffering, and danger were likely to be much greater than for hijras, who have perfected nirvan over hundreds of years. The benefits were much less too. Only the benefit of acquiring a more feminine body would seem to apply to two-spirits. A two-spirited woman didn't have a husband pestering her for an operation. Native Americans were easy about same-sex sexuality, and a two-spirit woman's partner knew what came with the turf. Nor was some bodily symbol necessary for authentication; a two-spirited person was authenticated by her transition ceremony. Finally, no one body was religiously correct. A two-spirited person was admired for her spirit, not her body. For these reasons, surgery was not a rational choice for Native Americans, and it wasn't done.

There are also other differences between Native American two-spirits and hijras. The two-spirit transition ceremony was held by the whole tribe and represented a person coming out into the society at large. A hijra nirvan is held within the hijra community, and the advance to full membership as a hijra is not witnessed, acknowledged, or endorsed by the wider society. Two-spirited people look outward to the whole tribe, where they can fulfill a role that benefits the greater good and aspire to succeed in the world at large. In contrast, a hijra must focus inward, and her existence depends on what she can extract from the larger community.

The concept of two-spirits is inclusive—a combination, union of man and woman, more than either alone. The concept of a hijra is exclusive—neither man nor woman, whatever's left over, intersection of man and woman, less than either alone. The two-spirited person is positioned to bridge gaps, to heal, construct, create. The hijra is positioned to threaten, to advertise loss, to demonstrate inability. India does not prosper from its hijras as much as Native America did from its two-spirits.

The two-spirited category is much broader than the hijra category. Hijras are limited to males and intersexed people who identify as female, but Indian society includes many other expressions of gender variance, which are poorly described and understood.

MORE TRANS PEOPLE IN INDIA

In southern India, the *jogappas* are similar to hijras in that they are male-bodied, wear feminine dress, take feminine names, wear their hair long in a woman's style, engage in bawdy bantering and flirting with men in public to solicit alms, and perform at marriages and the birth of a male child.[50] They follow the goddess Yellamma, considered a sister of the goddess Bahucharaji, whom the hijras follow. Unlike the hijras, though, the jogappas do not practice nirvan and are never referred to as eunuchs.

In northern India, the hijras coexist with groups referred to as *jankhas, kothi,* or *zenanas*.[51] Jankhas are male-bodied and seem to identify as men, but they dress as women on a regular basis. The group is heterogeneous. Some appear to be biding their time while waiting to apply to the hijras. Some compete with the hijras for money by playing at celebrations. Still another group, the kothi, are more complex, having a wife with children as well as a male lover.

The realm of masculine females is largely unexplored, although a historical study of lesbian expression from Sanskrit texts to the present has recently appeared.[52] I've been told of people called *mardana aurato,* or manly women, who have female partners.[53]

Now, with Westernization, people from the English-speaking upper classes are beginning to identify as lesbian, gay, and transgendered. The hijras are said to be uncomfortable with having their history appropriated and subsumed into Western categories, much as the mahu from Polynesia feel conflict over the introduction of Western categories for gender and sexuality.

19

Transgender in Historical Europe and the Middle East

ender variance was generally acknowledged by ancient writers in their descriptions of eunuchs, people similar to the *hijra*. We can find ancient eunuchs described in writings from the late Roman empire, from A.D. 100 to 400, as well as in the Bible and Islamic texts.

EUNUCHS IN THE ROMAN EMPIRE

As historian Mathew Kueffler recounts, the ancient Romans defined eunuchs as males who lacked functioning genitals.[1] The Roman lawyer Ulpian wrote, "The name of eunuch is a general one," and he enumerated three types. Eunuchs "by nature" were those whose genitals didn't continue developing at puberty. Such a person would have had sufficient genitals at birth to be classified as male. (In the early Roman empire, infants born with genitals too ambiguous to be classified initially as male or female were killed.) The second group of eunuchs had been castrated by a nonsurgical procedure—tying up the scrotum so that the testicles atrophied, or crushing the testicles, yielding genitals that continued to look male but didn't produce sperm or testosterone. Finally, there were eunuchs whose genitals had actually been removed surgically, leading to a genital area no longer male in appearance. Castration was illegal in the

Roman empire (but not outside, "across the border"), and by the third century, castrating a man against his will had become a crime deserving of capital punishment. Still, a Roman could make a living doing the procedures, and Plautianus, a government official, castrated not only "boys and youths," but "grown men as well, some of whom had wives."

The appearance of eunuchs depended on whether castration occurred before or after puberty. Most eunuchs were castrated before puberty, and their bodies assumed feminine characteristics. They retained the high voice and hairless body of their youth, and often grew breasts that, according to the writer Sidonius Apollinaris, "hang down like a mother's paps."[2] Fat also tended to deposit on the buttocks. The lack of both testosterone and estrogen, however, also led to longer limbs, curvature of the spine, osteoporosis, and sallow skin prone to premature wrinkling. The emperor Severus Alexander (A.D. 222–235) referred to eunuchs as a "third sex."

Many eunuchs were traded in the slave market after being sold initially by their parents. The demand for eunuchs exceeded the supply. Emperor Aurelian (A.D. 270–275) "limited the possession of eunuchs to those who had a senator's ranking, for the reason that they had reached inordinate prices."[3] Limiting demand by allowing only senators to purchase eunuchs was supposed to drive down their price. To increase supply, free-trade policies were encouraged. Constantine's law against castration forbade only the making of eunuchs "within the Roman Empire," allowing them to be castrated elsewhere. Although the Eastern emperor Leo I referred to the horror of "men of the Roman race, who have been made eunuchs in a barbarous country," he then granted permission "to all traders to buy or sell, wherever they please, eunuchs of barbarous nations who have been made outside the boundaries of Our Empire."[4]

Why were eunuchs so valuable? As domestic servants, eunuchs served as guardians for women and children. A noblewoman's eunuch slaves provided public transportation, acting as her porters and carrying her in sedan chairs whenever she traveled in public, allowing her to move about without male relatives. According to the writer Jerome, one noblewoman even brought her eunuchs with her into St. Peter's Basilica. In addition, eunuchs had long served in the royal administrations of the ancient Greek kingdoms of the eastern Mediterranean, working as mediators between men and women, and between servants and masters, both within the household and outside. The later Roman empire featured large numbers of eunuchs, so much so that writers often referred to them with phrases such as "a crowd of eunuchs, young and old," "crowds of eunuchs," "armies of eunuchs," and "troops of eunuchs."[5]

VARIATIONS AMONG EUNUCHS

Eunuchs encompassed people who were strongly female-identified as well as those who played on male turf. On the feminine side, Firmicus Maternus reported, somewhat disparagingly, on eunuchs who "feminized their faces, rubbed smooth their skin, and disgraced their manly sex by donning women's regalia. . . . They nurse their tresses and pretty them up woman-fashion; they dress in soft garments; they can hardly hold their heads erect on their languid necks. Next, being thus divorced from masculinity, they get intoxicated with the music of flutes."[6] Apuleius said such eunuchs renounced their previous masculine identities and called one another "girls" in private.

Some eunuchs were evidently marrying as women, prompting a ruling to outlaw this practice. In A.D. 342 the Christian emperors Constantius II and Constans imposed the death penalty "when a man married in the manner of a woman, as a woman wants to offer herself to men, where sex has lost its place, and where the offence is that which is not worth knowing, where Venus is changed into another form, where love is sought but not seen."[7] Presumably, the partners wishing to be married didn't see the relationship this way.

Other eunuchs were boyish and sought homosexual relations with older men. Roman men sometimes castrated a male slave so as to prolong his youthful beauty. This practice was discouraged for economic and moral reasons. The lawyer Paulus wrote, "He depreciates the value of a slave . . . who corrupts his morals or his body," remarking both on "the hurt done to the essential quality" of a slave used for sex and also an "overturning of the whole household."[8]

Still other eunuchs were successful in a male universe. Almost all fourth- and fifth-century emperors associated with powerful eunuch ministers. The office of grand chamberlain (something like head butler in the imperial court) was reserved for eunuchs. This position was accorded senatorial rank, with a status that increased over the years (from *clarissimus* to *illustris* to *eminentissimus*). The eunuch Eutropius was given a consulship, in part as a reward for leading a successful military campaign against the Huns in A.D. 398. Imperial politics were deadly, though, and Eutropius was executed a year later. Non-eunuchs resented the power of eunuchs and praised the emperor Severus Alexander for temporarily removing all eunuchs from court except for those who took care of women's baths, in the hope of confining eunuchs to the private space of women, away from the public space of men. Still, eunuchs continued to be appointed to high civil and military positions, partly because they had no family and thus posed no threat to imperial succession.

CYBELE PRIESTESSES

One of the most conspicuous occupations for a eunuch was priestess to the goddess Cybele, called the Mother of the Gods. Roman writers related her to Isis (Egypt), Asarte (Syria), Istar (Babylonia), Tannit (or Caelestis, from Carthage), as well as the Greek goddesses Rhea, Demeter, Aphrodite, and Hera, and the Roman goddesses Ceres, Venus, and Juno. Cybele was believed to control both agricultural and human fertility, including its underlying passions. Her own abundant fecundity aided in the birth of many of the gods (hence the name Mother of the Gods).

Cybele had a male consort, Attis, who was related to Osiris (Egypt), Tammuz (Syria), and Dumuzi (Babylonia), as well as the Greek gods Adonis and Dionysius and the Roman god Bacchus. (Perhaps the earliest existing transgender narrative is the Roman poet Catullus's recounting of the legend of Attis.[9]) The consort god rejects the mother god's love in favor of another. She castrates him in anger, and the consort dies of the wound. But because the mother goddess still loves him, she restores his life, minus the genitals.

Cybele could also be related to ancient Israel. At the time the Hebrew Bible was being written, the Hebrew mother goddess was called Asherah and her consort was Tammuz. These deities were later discarded, as worship consolidated around the god Yahweh, and monotheism replaced polytheism. Asherah was attended by eunuch priestesses, just like the other regional variants of the mother goddess. The Roman writer Jerome translated the Hebrew words for Ashreah, Tammuz, and the priestesses into corresponding words from his own Roman culture of a few hundred years later, suggesting a continuity in this religious tradition of as long as one thousand years.[10]

The Cybele priestesses were a stable and long-lasting transgender group. Recently, archaeologists have shown that a male-bodied corpse from Yorkshire, England, who was buried in female clothes and jewelry, was a fourth-century A.D. Cybele priestess. The religion was well established in the north of England: Hadrian's Wall at Corbridge contains an altar dedicated to Cybele.[11]

Spring was the time of year for rites lamenting death, followed by rites rejoicing at the restoration of life.[12] These rites, on March 24, were also when disciples of Cybele performed sex-reassignment surgery. With a sickle, priestesses-to-be severed their genitals in "an ecstatic frenzy." The ceremony used ornamented clamps, one of which was found in the Thames River near London Bridge and is now in the British Museum. After the operation, a Cybele priestess adopted women's clothing, including a veil and jewelry, and grew long hair. Particularly interesting is that the priestess placed her severed genitals on the doorstep of a house, and the women of that house gave her some clothing to start a wardrobe.

The whole operation was embedded in layers of symbolism. On a mythological level, the operation reenacts the castration of Attis at the hands of Cybele. In addition, the sickle used to cut the genitals is the same tool used in agriculture to cut wheat, which are seed-bearing stalks of grass. A pine tree, presumably with a conical shape, was felled too. The action of the priestesses was portrayed as a sacrifice of individual fertility to enhance the fertility of the community. Yet the writer Prudentius noticed the eagerness of the Cybele priestesses for the operation, and he wondered whether they could qualify as martyrs, because martyrs are unwilling victims of hardship beyond their control.[13]

COMPARING ANCIENT ROMAN EUNUCHS AND HIJRAS

The Cybelean myth is similar to how the Indian goddess Mata castrates her lover, as reenacted in the nirvan. As we saw, though, the descriptions of hijras do not indicate that the real-life nirvan has anything to do with mythology: they speak instead of the operation's feminizing and authenticating significance. The mythology behind the castration of Cybele priestesses may also have been a religious cover for a gender transition, judging from the eagerness of the priestesses for the procedure.

The Cybele priestess castration differs from the Indian nirvan in several respects, however. The Roman operation took place in a public setting and amounted to a gender-transition ceremony, like that of Native American two-spirits. The Indian nirvan, by contrast, is a private ceremony open only to other hijras, without the participation of nonhijra women. The Roman operation included a role for non-eunuch women, who endorsed the priestess's change of gender by giving her clothing. This social acceptance of the gender transition perhaps reflected the success of the Cybele priestesses in linking their gender change to the common good: permitting the agricultural season to start and culminate in a good harvest. Because Cybele priestesses didn't place their severed genitals on an altar to Cybele, but offered them instead to non-eunuch women, non-eunuch women were apparently necessary to validate the gender transition.

Both eunuch and non-eunuch priestesses of Cybele have been termed "cult-prostitutes" or "sacred prostitutes." Both eunuch and non-eunuch priestesses were apparently sexually active with worshippers as part of religious ceremonies to promote fertility and received alms for their temples in return.[14] Cybele priestesses were conspicuous when passed on the street, leading Augustine to complain of their corrupting influence, even though he admitted having "thoroughly enjoyed the most degrading spectacles" from them while still a youth.[15] This suggests a similarity to hijras, who also travel at times in groups and may also be described disapprovingly.

One would expect hostility from the early Christians toward the Cybele priestesses—after all, Cybele was a rival religion to Christianity. Indeed, the early Christian writer Lactantius described the public ceremonies as "insanity" and used today's trans-phobic language to decry the "mutilation" into "neither men nor women." Augustine ridiculed this "amputation of virility," in which "the sufferer was neither changed into a woman nor allowed to remain a man."[16] Yet the Bible itself takes a very different approach, as we shall now see.

TRANSGENDER IN THE BIBLE

Both the Hebrew (Old) and Christian (New) Testaments were written when many varieties of eunuchs were an obvious fact of life, an expression of humanity that did not fit the gender binary. Does the Bible attack and discourage eunuchs, and by extension, today's transgendered people? On the contrary. The Bible explicitly welcomes eunuchs.[17]

The Hebrew Testament starts with a passage warning Cybele priestesses not to set foot in the temple: "He whose testicles are crushed or whose male member is cut off shall not enter the assembly of the LORD" (Deut. 23:1 RSV). Later in the Hebrew Testament, however, the prophet Isaiah clarifies that eunuchs are indeed welcome in the temple if they honor the Sabbath. "For thus says the LORD: 'To the eunuchs who keep my sabbaths, who choose the things that please me and hold fast my covenant, I will give in my house and within my walls a monument and a name better than sons and daughters; I will give them an everlasting name which shall not be cut off'" (Isa. 56:3–5 RSV). In some editions, the LORD is translated as Yahweh, or Jehovah, the monotheistic Lord who emerged from the pantheon of gods available to the early Hebrews. Thus Yahweh promises eunuchs that they can have descendants through the house of God.

In the Christian Testament, Jesus himself speaks about eunuchs. On his way to Jerusalem, Jesus passed through Judea, where Pharisees questioned him about his views on marriage and divorce. In a back-and-forth exchange, Jesus discussed the situation of eunuchs who could not marry: "Not all men can receive this saying, but only those to whom it is given. For there are eunuchs who have been so from birth, and there are eunuchs who have been made eunuchs by men, and there are eunuchs who have made themselves eunuchs for the sake of the kingdom of heaven. He who is able to receive this, let him receive it" (Matt. 19:11–12, RSV).

Jesus thus acknowledges the multiple types of eunuchs: those who are intersexed (whose genitals don't develop at puberty), those castrated for administrative or domestic positions as slaves in imperial courts, and those who castrate themselves—the latter being for the "sake of the kingdom of heaven." This third type of eunuch that Jesus enumerates would include the people who might otherwise become Cybele priestesses. The kingdom of heaven is clearly open to all eunuchs, even those who have sought their own castration. The beginning and ending phrases, "not all men can receive this saying," and "he who is able to receive this, let him receive it," anticipate that understanding the topic is hard. Yet Jesus urges people to try to receive his teaching anyway.

PHILIP AND THE ETHIOPIAN EUNUCH

The apostle Philip actually put Jesus' teaching into practice—a eunuch was baptized and welcomed into the Christian church. Philip had gone to the city of Sama'ria to preach. "But an angel of the Lord said to Philip, 'Rise and go toward the south to the road that goes down from Jerusalem to Gaza.' This is a desert road. And he rose and went. And behold, an Ethiopian, a eunuch, a minister of the Can'dace, queen of the Ethiopians, in charge of all her treasure, had come to Jerusalem to worship and was returning; seated in his chariot, he was reading the prophet Isaiah. And the Spirit said to Philip, 'Go up and join this chariot.' So Philip ran to him, and heard him reading Isaiah the prophet, and asked, 'Do you understand what you are reading?' And he said, 'How can I, unless some one guides me?' And he invited Philip to come up and sit with him" (Acts 8:26–31 RSV).

Philip and the eunuch rode together as Philip explained the book of Isaiah. The specific passage Philip heard the eunuch reading was, "As a sheep led to the slaughter or a lamb before its shearer is dumb, so he opens not his mouth. In his humiliation justice was denied him" (Acts 8:32–33 RSV, which corresponds to Isa. 53:7–8). This passage urges people to speak out and not suffer humiliation, a remarkable call to activism in the face of injustice. And later in this same book of Isaiah comes the specific passage previously quoted (Isa. 56:3–5 RSV), which opens the Lord's house and walls to eunuchs. Thus Philip and the eunuch talked about specific passages in Isaiah affirming a place for eunuchs in the church. Moreover, the message was not to suffer humiliation in silence. In this way, "Philip opened his mouth, and beginning with this scripture he told him the good news of Jesus" (Acts 8:35 RSV).

"And as they went along the road they came to some water, and the eunuch said, 'See, here is water! What is to prevent my being baptized?'

And he commanded the chariot to stop, and they both went down into the water, Philip and the eunuch, and he baptized him" (Acts 8:36–40 RSV). This baptism welcomes into the church not only a eunuch, but a black-skinned foreigner as well. This definitive act of inclusiveness sets a high standard, which the Christian church has struggled to attain.

Collectively, the passages in Isaiah, Matthew, and Acts report the explicit teachings of the prophets Isaiah and Jesus, and convey an astonishing affirmation of diversity and inclusion. These passages are not ambiguous one-liners inviting misappropriation. Instead, they are clear, direct, and extensive. Both the Hebrew and Christian Testaments instruct full inclusion of gender-variant people in communities of worship.

GENDER-BENDING IN THE CHURCH

The Bible's clear endorsement of eunuchs led early Christians to wonder whether they should be eunuchs too. Because becoming a eunuch for the glory of God seemed a ticket to heaven, discussion unfolded on just how far one had to go to be considered a eunuch. It would be nice if the bar could be set low enough that celibacy alone could qualify one as a eunuch for the purpose of transportation to heaven.

But celibacy raised problems of its own. The Christian writer Ambrose promoted celibacy by saying that a celibate bishop was saving himself to be a "bride of Christ." Celibacy would be rewarded with sexual fulfillment in the afterlife. Ambrose continued, "Christ, beholding his Church . . . says, 'Behold, thou art fair. My love, behold thou art fair, thy eyes are like a dove.' " Ambrose claimed, "We kiss Christ . . . with the kiss of communion." "Open to me," Ambrose has Christ say to his bride, the Church, "and I will fill you."[18] The bishop Cyprian went on to assert that membership in the church entails submission to its bishop in the sense that a wife submits to her husband. This priestly gender-bending subordinates women and sets the stage for sexual abuse. As one climbs down the ladder from God to bishop, to priest, to confessor, each submits to the other in an alternating exchange of sexual identity. Homosexual abuse can masquerade as heterosexual submission.

Some early Christians did go all the way. Origen of Alexandria and a group of Christian men called Valesians practiced self-castration. Origen was criticized for taking the words " 'There are eunuchs who have made themselves eunuchs for the kingdom of heaven' in too literal and extreme a sense, thinking both to fulfill the words of the Savior and also, since although youthful in years, he discoursed on divine subjects with women as well as with men, to avoid all suspicion of shameful slander in the minds of unbelievers."[19] Origen was also one of the first to use the Bride of Christ metaphor extensively. I wonder if Origen wasn't in fact female-identified, as his association with women has been noted. In any case, he bought himself some recognition with the self-castration, and Jerome mixed grudging praise with criticism.

The eunuch category in Christianity was also populated with masculine women. Earliest Christianity features numerous stories of holy women who dressed and lived as men, the so-called "transvestite saints." The earliest, and perhaps best known, is Thecla, a companion of Paul. Thecla heard the preaching of Paul, converted to Christianity, and vowed to remain a virgin. She dressed as a man, traveled with Paul, and was baptized by him, also while dressed as a man. Similarly, Eugenia explained why she dressed as a man: "From the confidence I have in Christ, I did not want to be a woman. . . . I have acted manfully as men do, embracing boldly the virginity that is in Christ."[20]

Many such legends have been studied, and in each a link is made between pursuing holiness and renouncing a feminine identity, called being "clothed in Christ." Many of the women who dressed as men introduced themselves as eunuchs, possibly to explain their high voice, beardless face, and feminine body shape.[21] Jerome, however, condemned the women as "eunuchettes." Interestingly, some of the eunuchs and eunuchettes were said to travel with one another, an early precursor to today's occasional reciprocal trans couple.

There were critics, however, who refused to cut even real eunuchs any slack. As one historian summarizes, "Ecclesiastical sources frequently suggest that, in the struggle for ascetic virtue, eunuchs had 'cheated' and were not able to attain fully the celibate ideal. That is, celibacy was too easy for them because they did not have to struggle to attain it."[22]

The beginning of the end for real-life eunuchs came from those determined to substitute celibacy for castration. A monk, John Cassian, wrote, "The blessed Apostle is not forcing us by a cruel command to cut off our hands or our feet or our genitals. He desires, rather, that the body of sin, which indeed consists in members, be destroyed as quickly as possible by a zeal for perfect holiness." Cassian went on to found a particularly mean-spirited order of monks.[23] And Jerome stated, "Go then and so live in your monastery, free from all stain of defilement, that you may come forth to Christ's altar as a virgin steps from her bower."[24]

By the beginning of the fifth century, monasticism had become the new Christian masculine ideal. In this way, the Christian church several centuries after Christ totally appropriated the concept of a real-life gender-variant eunuch, the kind that Jesus and the prophets explicitly welcomed into the kingdom of God. The real-life street eunuch was replaced by the male monk, a make-believe eunuch. This loss of the eunuch category for gender-variant people forced gender variation underground. Famous gender-variant people surfaced now and then in Europe throughout the Middle Ages and into modern times, but only now is widespread natural human gender variation reemerging in Western society after a millennium of repression.

MUKHANNATHUN IN EARLY ISLAM

Like other ancient cultures, writings from early Islam record a transgender culture. People known as *mukhannathun* lived in the cities of Mecca and Medina (in present-day Saudi Arabia). Mukhannathun were "an identifiable group of men who publicly adopted feminine adornment . . . in clothing and jewelry." They are described in the *hadith,* which are accepted prophetic traditions, according to historian Everett Rowson.[25]

Hit was a mukhannathun who lived about the time of the Prophet Muhammad, around A.D. 630 in the Western calendar. Because women could be accompanied only by their children, female slaves, and mukhannathun, these last were well positioned to supply inside information about prospective brides to eager suitors. In describing a girl's charms to a potential suitor, a mukhannathun had to be discreet, however, and Hit earned condemnation for being too explicit, even crude in this regard.

According to one of the Prophet's wives, Hit told the Prophet that if he was victorious in taking a certain city, he should "go after Ghaylan's daughter; for she comes forward with four and goes away with eight." The reference was to the four belly wrinkles that wrapped around the sides of her body, so that eight could be seen from the back, a sensuous image for the time. The Prophet was outraged and said, "Do not admit these into your presence." According to Asian scholars, "The Prophet's words imply that the mukhannathath's awareness of what men found attractive in women was proof of his own sexual interest in them, and that this is the reason that he and others like him should be barred from the women's quarters." Hit was thus condemned not for "expressing his own appreciation of a woman's body, but [for] describing it for the benefit of another man." Granting a "license to be with women" was appropriate only for "one whose limbs are languid and whose tongue has a lisp by way of gentle natural constitution, and who has no desire for women and is not . . . in evil acts." Despite Hit's transgression, mukhannathun continued to play a significant role as matchmakers for eligible bachelors who, as a rule, had little opportunity to meet eligible women.

Tuways was a mukhannathun who was born in A.D. 632 and died in A.D. 711 at the age of eighty-two. He was distinguished as a musician who sang "art music" using measured rhyme. He was a musical innovator and trained the next generation of musicians, relying on a kind of tambourine called a duff. He was married and had fathered children. Tuways was a "leader of a group of male professional musicians who publicly adopted women's fashions and were appreciated for their wit and charm as well as their music, but were disapproved of by others who . . . saw their music and flippant style as immorality and irreligion." They were not matchmakers like Hit.

Al-Dalal was also a mukhannathun, less cultured than Tuways and, like Hit, best known for getting into trouble. Though physically beautiful and charming, al-Dalal's wit was crude and seriously irreligious. According to one story, "He farted during prayers and said, 'I praise Thee fore and aft!' " He was also a go-between who arranged assignations, and is depicted as encouraging immodesty and immorality in women. Al-Dalal was close to two scandalous women in Medina who were said to engage in "horse-racing, and while riding to have shown their ankle-bracelets." The women were killed, and when al-Dalal fled to Mecca, the women there viewed him as a threat: "After killing the women of Medina you have come to kill us."

Al-Dalal's sexual orientation was toward males. He "adored women and loved to be with them; but any demands [by them for his sexual favors] were in vain." In one story, a Syrian commander overheard his singing and invited him to visit. Al-Dalal refused to sing unless he was sold a beautiful servant boy. The commander meanwhile wished for a slave girl of a particular and very voluptuous description, which al-Dalal arranged. Another story relates that "after arranging a marriage, al-Dalal would convince the bride that her sexual excitement at the prospect of the wedding night was excessive and would only disgust her husband, and then he would offer to calm her down by having sexual intercourse with her first. He would then go to the groom, make the same point, and offer himself, passively, to cool him down as well." The outraged and "jealous" ruler Sulayman then ordered *all* mukhannathun castrated: "They are admitted to the women of Quraysh and corrupt them." Interestingly, even with explicit testimony about al-Dalal's homosexuality, which is condemned in the Quran, it was the corruption of women that was used to justified the punishment, not effeminacy or homosexuality.[26]

Although the castration of mukhannathun as punishment begins a repressive period for gender-variant people in Mecca and Medina, the supposed victims showed curious reactions:

Tuways: "This is simply a circumcision which we must undergo again."

al-Dalal: "Or rather the Greater Circumcision!"

Nasim al-Sahar: "With castration I have become a mukhannath in truth!"

Nawmat al-Duha: "Or rather we have become women in truth!"

Bard al-Fuad: "We have been spared the trouble of carrying around a spout for urine."

Zill al-Shajar: "What would we do with an unused weapon anyway?"

Reports of gender-variant entertainers resurface one hundred years later, in A.D. 813, again using the tambourine-like duff, together with a particular drum and a long-necked lute called a tunbar. Wit, more than music, now defined the presentation, described as "savage mockery, extravagant burlesque, and low sexual humor."

JEHANNE D'ARC, A MEDIEVAL ICON

Popular culture has cooperated in erasing the reality of gender-variant people. Joan of Arc, the famous heroine of movies, television specials, and books, is usually portrayed as a role model for young women, an icon of women's rights and militant feminism. But might Jehanne d'Arc, as "Joan" was called in medieval France, serve better as a hero for transgendered people? The trans activist and writer Leslie Feinberg argues that Jehanne d'Arc was a male-identified trans person killed specifically for his expression of gender identity. Feinberg and other researchers show there is more to the story of Jehanne than we've been told.[27]

Jehanne d'Arc was born in the province of Lorraine in France, around 1412. Fifty years before, the bubonic plague had killed one-third of the population of Europe. To make matters worse, France was at war with England. Marauding English armies were plundering the peasants of France, and French nobles were unable to oust them. Jehanne d'Arc, a peasant, emerged as the only military leader able to defeat the English.

At the age of seventeen, Jehanne d'Arc, dressed in men's clothing and with a group of followers, approached the heir to the French throne, Prince Charles, and offered to forge an army of peasants to drive out the English. Charles agreed and authorized Jehanne's command of a ten-thousand-strong peasant army. Jehanne d'Arc defeated the English, led by the duke of Bedford, at Orléans later that year, in 1429. Jehanne continued liberating towns occupied by English troops, making it possible for Charles to receive the crown. When Charles was crowned, Jehanne d'Arc stood beside him with a combat banner.

A year later, Jehanne was captured by the Burgundians, allies of England, who referred to Jehanne as *hommasse,* a slur meaning "man-woman," or masculine woman. The king of England, Henry VI, wrote to Pierre Cauchon, the bishop of Beauvais and leader of the Catholic Inquisition: "It is sufficiently notorious and well-known that for some time a woman calling herself Jeanne the Pucelle [the maid], leaving off the dress and clothing of the feminine sex, a thing contrary to divine law and abominable before God, and forbidden by all laws, wore clothing and armour such as is worn by men."[28] Jehanne was sold to the English by the Burgundians and then turned over to the French Inquisition, which charged Jehanne with cross-dressing.

In the tradition of the transvestite saints, who renounced sexuality (hence a maid) and affirmed masculinity, Jehanne claimed that dressing as a man was a religious duty compelled by voices spoken in visions. The verbatim court proceeding states, "You have said that, by God's command, you have continually worn man's dress . . . your hair short, cut *en rond* about your ears, with nothing left to show you to be a woman; and that on many occasions you received the Body of our Lord [Holy Communion] dressed in this fashion . . . and you have said that not for anything would you take an oath not to wear this dress." Therefore, the court concluded that "you condemn yourself in being unwilling to wear the customary clothing of your sex."[29] Thus Jehanne d'Arc was sentenced to die.

Jehanne d'Arc was burned alive at the stake in Rouen on May 30, 1431, at the age of nineteen. After the clothing burned off and Jehanne was presumed dead, the inquisitors raked back the coals to show the naked body, revealing "all the secrets . . . that belong to a woman, to take away any doubts from people's minds."[30] Jehanne d'Arc must have been convincingly masculine to require such extraordinary measures.

Leslie Feinberg writes, "Joan of Arc suffered the excruciating pain of being burned alive rather than renounce her identity. I know the kind of seething hatred that resulted in her murder—I've faced it. But I wish I'd been taught the truth about her life and her courage when I was a frightened, confused trans youth. What an inspirational role model—a brilliant transgender peasant teenager leading an army of laborers into battle."

20

Sexual Relations in Antiquity

S ame-sex sexuality was part of life in antiquity, although not as a category of personal identity. Whether one did or did not participate in same-sex sexuality no more defined who a person was, and who they thought themselves to be, than does, say, an appetite for French fries instead of potato chips. However, those who do eat French fries should avoid splashing ketchup on the table. Similarly, an appetite for same-sex sexuality required certain manners. The focus of social convention was not one's choice of sexual partner, but rather how the sexual practice was carried out.

ANCIENT GREECE

Plato, the ancient Greek philosopher (428–347 B.C.), extensively discussed same-sex male love in *The Symposium* and *Phaedrus*. Homosexuality was widespread in ancient Greece by the sixth century B.C. Plato's writings, together with a legal brief by the Athenian politician Timarkhos and the explicit art on many pieces of pottery, offer a glimpse into ancient love and sex, primarily from a male perspective.[1]

Ancient Greeks had a right way and wrong way of participating in male-male sexuality. In ancient Greece, male same-sex relationships were almost always between an older partner and a younger one. A post-puberty youth who had grown to full height assumed the passive role, and an older man the active role. Pre-puberty youths were not participants. A sexual encounter was initiated by the active party tickling the penis of the passive partner with one hand while caressing his cheek with the other. The active party's penis was erect, whereas that of the passive party remained flaccid. If the passive partner accepted the advance, the couple stood face to face and the active partner held the passive one around the chest, rested his head on or below the shoulder of the passive partner, bent his knees, and thrust his penis between the thighs of the passive partner, just below the scrotum. This position was called intercrural and differed from oral sex or anal sex in that thrusting between another's thighs was not considered penetration. This position might be thought of as the missionary position for gay male sex in ancient Greece.

The code for how a proper young man was to have male-male sex as the passive partner of an older man dictated that he never accept payment, that he refuse any unworthy active partner, that he avoid enjoyment, that he insist on an upright position, that he not meet the active partner's eye during orgasm, and that he avoid positions with penetration. The passive partner was understood to grant a favor to the active partner.

What about female-female sex? Women are often depicted with an ancient dildo, an artificial penis made of leather, called an *olisbo*. Women are shown using these in groups, and with one another, entering the mouth, vagina, or anus with them. Using the olisbo in anal intercourse among women suggests bringing sexual pleasure, and is not necessarily a gesture of subordination in a dominance interaction.

The positions in heterosexual encounters were distinct from those of same-sex affairs. The man began by patting the woman's genitals with one hand and caressing her cheek with the other. Thereafter, the woman might bend over, placing her hands on the ground, to allow anal intercourse. This position was common in paintings on vases, a position one historian conjectures was preferred because it eliminated the risk of unwanted conception. Alternatively, in a frontal position, the woman put her legs in the air and rested them on the man's shoulders. And in the "racehorse" position, the woman lowered herself on the man's erect penis while he sat on a chair. Thus penetration was reserved for the female partner in heterosexual encounters.

"CLEAN" VERSUS "UNCLEAN"

The intercrucal position for gay male sex was socially accepted and considered "clean." Alternative positions involving penetration were "unclean." This distinction is also apparent in Hebrew writings from about the same time. What happened when a man did break the code of conduct, becoming "unclean"? He was then not worthy of being a "citizen" of Athens and was expected to take himself out of activities reserved for citizens, such as addressing the assembly or holding government office. The penalty for an unworthy person who assumed the role of a citizen was death. The ancient Greeks took their honor code seriously.

Taking money for sex was perfectly legal, but not for a proper Athenian citizen. Boys and men who made a living from homosexual prostitution were predominantly non-Athenian, often slaves. Therefore, Athenians were not "denied the pursuit of their inclinations."[2] Yet the ancient Greeks had a strong sense of morality. Hubris is behavior in which one treats other people just as one pleases, with an arrogant confidence that one will escape paying any penalty for violating their rights. Homosexual or heterosexual rape, especially of visitors, was illegal, and was prosecuted as a crime of hubris.[3]

We think of clean and unclean today in terms of hygiene. The sense of unclean homosexual relations in ancient Greece had nothing to do with hygiene, but policed the gender binary. Men could, even should, have sex with one another, but by golly, men had better use different positions with one another than they did with women. By restricting the permissible positions, gay male sex in ancient Greece was kept masculine.

THE BIBLE

In view of the centuries-old belief that homosexuality is condemned by the Bible, one might expect to see in the Bible extensive and clear-cut statements that homosexuality is wrong. In fact, only a few biblical passages refer in any way to homosexuality, and none condemns homosexuality clearly and unambiguously. Lists of sins including adultery, theft, and lying omit any mention of homosexuality. Nor did Jesus ever speak of homosexuality. It seems homosexuality was hardly on the radar screen when the Bible was written. Why should the Bible be so explicit in affirming gender diversity, as seen in its marked approval of eunuchs, and largely silent about same-sex sexuality?[4]

RUTH AND NAOMI

One of the Bible's most loving passages pertains to a relationship between two women, Ruth and Naomi. Naomi went with her husband and two sons to live in the land of Moab. There her husband died. The two sons settled down and married Moabite women, Orpah and Ruth. After ten years, the two sons died without leaving any children. Naomi urged her two daughters-in-law to return to their families rather than stay to care of her. Orpah accepted and returned to her family. But Ruth refused to leave Naomi and composed one of the greatest pledges of love from one human being to another:

> Entreat me not to leave thee
> or to return from following after thee;
> For whither thou goest I will go,
> and where thou lodgest I will lodge.
> Thy people shall be my people,
> and thy God my God:
> Where thou diest, will I die,
> and there will I be buried.
> The Lord do so to me, and more also,
> if aught but death part thee and me. (Ruth 1:16–17)

In this statement Ruth says not only that she will leave her people to go with Naomi, but also that she will join Naomi's faith. To a Bedouin, family and faith are the highest values, and Ruth is offering to give up both to follow Naomi. Because Naomi is older than she, Ruth assumes Naomi will die first, and she wishes to buried next to her when she dies.

Naomi and Ruth set out for Bethlehem, Naomi's place of origin. Once there, Naomi found a husband for Ruth. Just as Ruth cared for Naomi, she in turn would need a child to take care of her in her old age. Ruth conceived a son with her new husband, and the women of Bethlehem said to Naomi, "Blessed be the Lord. . . . [H]e shall be unto thee a restorer of thy life, and a nourisher of thine old age; for thy daughter-in-law, which loveth thee, which is better to thee than seven sons, hath borne him" (Ruth 4:14–15). This relationship between two women remains a precious model of a loving partnership even today.

JONATHAN AND DAVID

Another loving relationship concerns two men, Jonathan and David. David, the son of Jesse, was a musician, "ruddy, and withal of a beautiful countenance, and goodly to look to" (1 Sam. 16:12). David came to the court of King Saul of the Israelites, where he met Jonathan, the king's son, who had already distinguished himself as a hero in a major battle against their enemies, the Philistines. Thereafter "the soul of Jonathan was knit with the soul of David, and Jonathan loved him as his own soul. . . . Then Jonathan and David made a covenant, because he loved him as his own soul. And Jonathan stripped himself of the robe that was upon him, and gave it to David, and his garments, even to his sword, and to his bow, and to his girdle" (1 Sam. 18:1–4).

After a political intrigue, Saul decided to send David away. In a tantrum leading to David's dismissal, Saul insulted his own son, Jonathan, by disparaging Jonathan's relationship with David: "Do I not know that you have chosen the son of Jesse to your own shame?" (1 Sam. 20:30). In ancient Greek, the passage is: "Do I not know that you are an intimate companion to the son of Jesse?" Saul's insult was intended to break up a homosexual relationship that he found threatening.

At their parting, "David . . . prostrated himself with his face to the ground. He bowed three times, and they kissed each other, and wept with each other; David wept the more. Then Jonathan said to David, 'Go in peace, since both of us have sworn in the name of the Lord, saying, "The Lord shall be between me and you, and between my descendants and your descendants, forever." ' He got up and left; and Jonathan went into the city" (1 Sam. 20:41–42). Jonathan remained with his wife and father, and eventually died in battle. In a eulogy, David wailed, "O Jonathan, thou wast slain in thine high places. I am distressed for thee, my brother Jonathan: Very pleasant has thou been unto me: thy love to me was wonderful, passing the love of women" (2 Sam. 1:26).

Thus two of the most beautiful biblical passages about love pertain to same-sex relationships. In contrast, three other biblical passages comprise the main artillery aimed at gay and lesbian people.

SODOM AND GIBEAH

The story of Sodom begins with a recent Bedouin immigrant, Lot. Two angels approached Sodom in disguise, and in Bedouin fashion, Lot extended his hospitality by inviting them into his home for the night. They accepted. But while Lot prepared food for them, news of their arrival spread across the town. Before Lot and his guests could retire for the evening, "the men of the city . . . called unto Lot, and said unto him, 'Where are the men which came in to thee this night? Bring them out unto us that we might know them.' And Lot went out at the door unto them, and shut the door after him, and said, 'I pray you brethren, do not so wickedly. Behold now, I have two daughters which have not known man; let me, I pray you, bring them out unto you, and do ye to them as is good in your eyes: only unto these men do nothing; for therefore came they under the shadow of my roof.' And they said stand back" (Gen. 19:4–9a). Then they began to break down the door. At this point, the two angels pulled Lot in and bolted the door, but not until they had blinded everybody outside. Next they instructed Lot and his family to leave the city immediately, for at any moment it would be totally destroyed, and an earthquake came and gobbled up Sodom.

The men of Sodom intended to homosexually rape the two visitors, unaware that they happened to be angels. Where is the wrong: in the homosexuality, the rape, or the rape of two visitors? All of these, or just some of these? Which of these possible wrongs was the cause of Sodom's destruction? The sin is in raping visitors. The homosexuality is irrelevant. That is clear from another biblical episode.

Consider the similar destruction of another town, Gibeah (Judg. 19:22–30). A Levite traveling with his servant and concubine came to the town of Gibeah. No one offered hospitality, except a foreigner living there. When they were inside, the men of the town assaulted the house and wanted to rape the visitor. Like Lot, the host offered his daughter instead, but the Levite pushed his concubine out, saving the host's daughter from rape. The concubine was raped during the night and found dead in the morning. The Levite carved the body of his concubine into twelve pieces and sent one piece to each tribe in Israel, so they'd really get the message about what happened. The tribes of Israel then collected an army and destroyed Gibeah.

In this case, heterosexual rape was consummated and homosexual rape avoided, but Gibeah was still destroyed. Thus whether the sexuality is heterosexual or homosexual is overshadowed by the greater evil of rape and the transgression of the hospitality extended to a visitor. The sin of Sodom has nothing to do with homosexuality as such; quoting this passage of the Bible as part of an anti-gay agenda is simply mistaken.

LEVITICUS

There is a famous one-liner in Leviticus addressed to men, called "The Holiness Code": "You shall not lie with a male as with a woman; it is an abomination" (Lev. 18:22). The word "abomination" means doing an "unclean" act. This passage specifically condemns male sex in which one male partner penetrates the other, particularly anal sex. Nothing is said about other male homosexual positions, nor anything at all about female homosexuality. The passage says nothing against homosexual relationships—the love, trust, and shared living of a committed homosexual partnership. This passage outlaws a particular sexual position, not homosexuality. Males may obey this command by abstaining from sexual penetration, for instance, by using the intercrural "gay male missionary position" of ancient Greece.

The Israelites had a long list of what was considered clean and unclean. Pigs, camels, lobsters, and shrimp were unclean and were not to be eaten. Sowing a field with two different kinds of seed at the same time, or weaving a cloth from two kinds of thread, was unclean. Menstruation in women, seminal emission in a man, attending a burial, and giving birth made one unclean for a certain length of time. Since the time of the Hebrew (or Old) Testament, many branches of Christians and Jews have updated the list of what's clean and unclean. Jesus, for example, tried to move people beyond a laundry list of do's and don'ts. "Listen and understand," he said. "Out of the heart come evil intentions, murder, adultery, fornication, theft, false witness, slander. These are what defile a person, but to eat with unwashed hands does not defile" (Matt. 15:10,18–20). Same-sex sexuality isn't mentioned.

PAUL'S LETTER TO THE ROMANS

The third passage from the Bible cited against gay and lesbian people is from the Christian (or New) Testament. The passage is from the first chapter of Paul's letter to the Romans. Paul begins by criticizing people who worship deities in human or animal form, such as the deities in ancient Egyptian art: "They exchanged the glory of the immortal God for images resembling a mortal human being or birds or four-footed animals or reptiles" (Rom. 1:23). Continuing, Paul says, "Therefore, God gave them up to degrading passions. Their women exchanged natural intercourse for unnatural, and in the same way also the men, giving up natural intercourse with women, were consumed with passion for one another. Men committed shameless acts with men and received in their own persons the due penalty for their error" (Rom. 1:26–27).

As in Leviticus, Paul's focus is on homosexual acts, not on homosexual relationships. This passage implies that same-sex sexuality is unnatural, that female same-sex sexuality is equivalent to male same-sex sexuality, and that people who participate in same-sex sexuality will receive just penalties. Let's see what these points mean, one by one.

Is same-sex sexuality unnatural? This claim has been interpreted in two ways. Followers of the Stoics, the ancient counterpart of scientists, state that nature functions according to "laws" that human reason can discern, and virtue consists of living by reason, not by emotions or feelings. Morally, one should discern the laws of nature, and then follow them. With regard to sexuality, the Stoics further asserted that the "natural" purpose of sex was to procreate, and that therefore nonprocreative sex, including same-sex sexuality as well as sex during menstruation, was unnatural.[5] The difficulty with the Stoic interpretation of "natural" is that the Bible becomes open to falsification as science advances and nonprocreative, yet still natural, functions of sex are discovered.

Alternatively, unnatural has been interpreted to mean "out of character."[6] Someone with a healthy appetite who stops eating is acting "unnaturally," suggesting that something is amiss. The sense of Paul's letter is that if someone who has been worshipping God gives this up to follow other deities, their behavior will become unnatural for them—they will start behaving out of character, having lost their sense of moral direction. One might even wake up in the midst of a homosexual orgy. For someone who is primarily gay, the opposite is unnatural. For people who are gay to force themselves into heterosexual sex may be unnatural for them, and unfair to their partner too. Someone who has lost faith in God's love might wake up in the clutches of people who want to change their sexual orientation. By this interpretation, which I prefer, the Bible is making a moral statement rather than a scientific statement.

The passage also shows that by Roman times, female same-sex sexuality had acquired the same status as male same-sex sexuality.[7] The Hebrews didn't bother mentioning female same-sex sexuality in Leviticus because female-female "rubbing" need not involve penetration and therefore wasn't "sex."[8] By lumping in sex between women with sex between men, Paul's letter moves beyond concern for who "penetrates" whom, focusing instead on the context and motivation for sexuality.

Finally, Paul's letter to the Romans speaks of a penalty "received in their own persons" from wanton sexual behavior. Undoubtedly, sexually transmitted diseases were contracted during orgies. This penalty is not specifically because of same-sex sexuality, but because any sex in orgies invites unsanitary habits.

Thus the Bible contains two extended passages portraying the love between same-sex partners, Naomi and Ruth and Jonathan and David. Then, in the stories of Sodom and Gibeah, rape and inhospitality to visitors are condemned. In Leviticus, male same-sex sexuality involving penetration is required to be socially clean. And in Paul's letter to the Romans, out-of-character sexual excess is condemned as an indication of losing one's moral compass, female homosexual acts are equated with male homosexual acts (thereby discouraging any focus on who's penetrating whom), and all are warned of the negative consequences of wanton sex. The passages urge responsible approaches to sexual expression for both homosexual and heterosexual people. Heavy stuff, to be sure, but totally consistent with affirming full participation of gay and lesbian people in all aspects of Christian religious life.

The realization that biblical passages cannot justify excluding gay and lesbian people is not new. As Nancy Wilson, a member of the clergy, writes, "The vast majority of Christians . . . still believe outdated, erroneous, homophobic biblical interpretations. The Church leadership refuses to teach what it knows. The violence and hatred perpetuated against gays and lesbians in our culture is silently—and sometimes not so silently—co-signed by the Church. Church leadership knows that teaching the truth about homosexuality and the Bible will be controversial, difficult and, at first, costly. The fear of controversy, of loss of money, of criticism from the radical right keeps the truth locked up."[9]

Beyond the issue of religious inclusion for gay and lesbian people lies the question of why the Bible deals with gender variation explicitly but with same-sex sexuality only incidentally. The answer I suggest is that social categories of identity for gender variation and sexual orientation did not form at the same time. When the Bible was written and perhaps deep in our prerecorded past, eunuchs were recognized as a distinct category. Homosexuality wasn't. Homosexuality as a social category of identity first emerged in Germany during the late 1800s, making it a rather recent social category.[10] The Bible explicitly recognizes eunuchs for religious inclusion, while remaining relatively silent about homosexuals because, like everyone else, they are covered by the general moral dictum of "love thy neighbor as thyself."

21

Tomboi, Vestidas, and Guevedoche

T his chapter leaves the Middle Ages and moves on to three examples of how human variations in sexuality, gender presentation, and bodies are being accommodated within contemporary societies.

GENDER EXPRESSIONS IN INDONESIA

When I was about ten years old, I lived in the town of Bogor, in the hills of Java, Indonesia. I remember the wildlife, the flocks of fruit bats descending into the trees at sunset, the rain squalls, the steam rising from the road, the yummy rice cakes, the beautiful batik cloth, the red ants, the flowers—yes, the tropics at its best. I don't remember anything at all about sexuality in Indonesia. Yet lots was happening in that realm too. Anthropologists working in both Sumatra and Java have detailed the surprises that result when Western concepts of lesbian and gay are applied to local expressions of gender and sexuality.

An American anthropologist, Evelyn Blackwood, writes, "The term tomboi is used for a female acting in the manner of men. Through my relationship with a tomboi in West Sumatra, I learned of the ways in which my concept of 'lesbian' was not the same as my partner's, even though we were both, I thought, women-loving women."[1] Blackwood developed a romantic relationship with Dayan, a woman in her mid twenties who appeared boyish, in T-shirt and shorts and short hair, but otherwise not particularly masculine or tough. Dayan, however, thought of himself as a man. "I finally had to admit to myself," Blackwood reports, "that tombois were not the Indonesian version of butch. They were men."

Tombois pride themselves on doing things like a man: they play *koa* (a pokerlike card game), smoke, go out alone at night, drive motorcycles with their partner in back, and move in and out of their partners' houses. Their partners are women no different from others, and these women sometimes leave them to take up with a non-tomboi man.

A Dutch anthropologist, Saskia Wieringa, investigated women's communities on Java, where she found a well-developed butch/femme (b/f) culture that she felt was obsolete after having been "socialized in the Dutch women's movement where earlier b/f culture was rejected as 'old' lesbian. . . . The butches tried to teach me to be one of them and the femmes made clear what they expected from me in the way of chivalry and lovemaking. . . . The Jakarta butches voiced their astonishment at my preference for reciprocity. 'Isn't that confusing?' "[2] Butches were expected to have a decent job, not only to survive but to provide for their girlfriends, and were subject to a dress code—pants, shirts, and underwear bought in men's clothing stores, bandages to flatten the breasts, and a performance style—a little swagger, head up defiantly, and cigarette in hand, plus gendered language. Femmes passed as ordinary women, though they often dressed exaggeratedly, with ribbons, frills, heavy makeup, and high heels. Femmes worked as secretaries, and some were in sex work as well.

"I was indeed confused . . . ," Wieringa confesses. "I had never doubted androgyny as the major characteristic of 'new-style' lesbians. . . . We were feminists. . . . Roles, we announced, were derived from heteropatriarchy. We were proud to be liberated." When the Java butches were asked why they were not proud of their women's bodies, they answered that their bodies did not matter much to them. They wanted to love women and noticed that persons with male bodies had much less trouble finding women partners. But however much the butches conformed to male gender behavior, they didn't define themselves as male; at times they defined themselves as a third sex. The butches discussed a friend who was undergoing a sex change operation. They considered this option, but decided not to follow. When asked why, all mentioned health risks and costs. "None stated they preferred their own bodies."

Thus, gender expression in these two lesbian communities in Indonesia doesn't seem to coalesce around a single androgynous center, but divides into trans man and femme poles in Sumatra, and into butch and femme poles in Java. Why?

Women in the butch/femme cultures of the major cities of Europe and the United States during the 1940s, 1950s, and 1960s viewed themselves as expressing their innate desires.[3] Feminists criticized them as copycats of conventional masculinity and femininity, as not being true to themselves, as perpetuating an oppressive social order, and not being radical or courageous enough. Now opinions are starting to change. Butch lesbians and trans men are subjecting themselves to enormous prejudice from other women and from the straight world at large. Many increasingly feel this path requires more courage than melding into an unremarkable middle. The integrity of masculine identity in women is increasingly being appreciated as its own form of "self-determination," as Wieringa recently concluded.[4]

VESTIDAS IN MEXICO CITY

Although transgendered people today are not likely to be burned at the stake as Jehanne d'Arc was, the social options for transgendered people in many parts of the world are not much better. Let's consider the contemporary situation in Mexico City, where the most conspicuous transgender expression is the street queen, a transvestite sex worker, or *vestida*.

MEMA'S HOUSE

Neza (Ciudad Nezahualcóyotl) is a suburb within Mexico City with a bad reputation: dirty, dangerous, and poor, it is a place where middle- and upper-class citizens do not venture. As the Swedish sociologist Annick Prieur reports, Mema's house is a gathering place for youth in Neza.[5] Mema is a sex worker, hairdresser, cook, clerk, vendor, and AIDS educator who focuses on trans people. Usually ten to twenty persons stop by every day, mostly in the evening, for family-style meals. Mema's house is a sanctuary. The people staying with Mema are youngsters run out of their homes because of being feminine and gay. Their femininity makes it difficult to keep their sexuality private, so the families harass them to the point of running away.

After beatings from her stepfather and being made to sleep in the yard under the washstand like a dog, Pancha ran away from home at age eight, living in the streets and then with Mema. Pancha recounts that at age ten, her mother "asked me whether I was a man or a woman." She regrets that her mother "doesn't accept my wearing makeup at home. . . . She did not like me to wear tight pants and women's blouses, and to have long hair that I dyed. They always cut it. But now they let me let my hair and nails grow."[6] This acceptance is partly because Pancha contributes some of her earnings from sex work to the household. Wages from sex work often lead to reintegration into the family.

Marta, another transgender woman, recalls, "I liked dolls, I adored them. For the Holy Kings' night, they gave me a present, a car or a truck. And then I would play with my cars for a while. But I was more interested in my little sister's dolls. I played with them, I asked her to let me borrow them. And I went off to play with the neighbor girls."[7] She continues, "I was fascinated by grown-up men in the bathroom. . . . And I don't regret it, I like it. . . . I was six. A neighbor . . . talked to me, he seduced me with ice-cream . . . and I was delighted. . . . He went to his bed and started to undress. . . . I was tempted, so I got close out of curiosity to touch. . . . And then it continued, he kept on giving me ice-cream . . . and I continued to be his lover until I was nine. . . . For sex I was wide awake from an early age." How did this come to be? "I think I was born like that. . . . I said that to the doctor who treated me, who injected male hormones into me. . . . Since I was six or five years old, I was attracted to men. And that's not something you do if you don't like it." The doctor's treatment made Marta's legs become hairy. Marta was teased at school for having long eyelashes and was expelled at age twelve in spite of good marks. Then she was beaten by her parents and driven from her home.

At this time Marta met other vestidas. "I though they were women, but somebody told me no, they are men dressed like women. I didn't believe it, but I said if they are men I can join them, I want to be like that. I want to look like a woman. So I got to know them, and they supported me. Mema helped me, thank God he helped me. And he bought me shoes, and clothes. I started to make myself up like a woman, in his hairdressing parlor. I made up my eyelashes, I painted my nails. I let them grow." At fourteen, Marta was introduced to making money from sex, which until then had been freely given. "There comes a moment when you have to decide for yourself. And I felt locked in by men's clothes; there came a moment when I said 'away, away all men's clothes. I don't want it anymore.' And I put on women's clothes. I felt like Cinderella, I shed the old clothes and put on the new ones. What I wanted to be."

Marta wants to change sex surgically. "When I pee, I say 'Ai, this penis isn't mine'. . . and I would like . . . to cut it off." Still, Marta says, "I have a lot of pride. I'm homosexual. I'm homosexual, but I have come close to a woman. I mean physically, with everything, with my face and my body. I am a woman, isn't that so? That doesn't mean that in order to be a woman, I stop being a homosexual . . . inside myself I'm proud that I as a homosexual have managed to look like a woman. And that people can see that a gay can get where he wants to. Because I have heard that many homosexuals have been important people through history, isn't it true? Like writers, painters, a lot of things, and in the whole world. So one can feel pride." A remarkable narrative combining transgender elements with gay pride, situated in a homosexual world.

Homosexuality and femininity are completely intertwined in this group, so much so that feminine presentation—streetwalker style—and homosexuality are nearly synonymous. Vestidas dress and act provocatively, talk a lot about sex, joke about sex incessantly, and spend a lot of time and energy on sex. Teenagers. No positive role models are available for these young people, and they have been excluded from education. Often teachers notice their feminine inclinations and initiate sexual relations themselves. All the vestidas interviewed by Prieur were bullied at school, and most stopped attending soon after elementary school, although some were hoping to finish high school, and one even wished to be a teacher. All in all, the vestidas did not talk much about the future.

Other than hairdressing, the main occupation of vestidas is sex work. In the streets, vestidas keep to areas where the sex workers are known to be transvestites and clients know what they're shopping for. Vestidas are often picked up by police, paying bribes to obtain release. Life on the wild side can include parties, drinking, drugs, thefts, quarrels, detention, and violence. They often steal from their clients. They take risks in behavior and dress, and seem to invite trouble. But with no belief in a future, why not? This life probably also describes nontransgendered teenage sex workers in this economic situation.

A REFUSAL TO LISTEN

Sociologists often violate the primary narratives they record. For example, Marta says she was "born that way," but that is apparently a forbidden premise in sociology. Prieur writes, "Some readers might be led to think, 'Marta is a transsexual.' This is . . . contrary to my constructivist approach." Prieur goes on to claim that Marta was "put into a homosexual role long before he had become conscious about his own sexual desires."[8] No. Marta directly states her sexual interest in men preceded her first homosexual encounter and insists that she consented thereafter.

Indeed, are we even allowed to believe Marta's own account? According to Prieur, Marta and others "are colored by the time that has passed, by the common interpretations of homosexuality and effeminacy that they have learned later, and presumably also by their own wishes to present relatively coherent stories about themselves. And I believe all these factors lead jotas [a pejorative term for an effeminate homosexual] toward emphasizing the early determination of homosexuality and effeminacy in their accounts. These men have become what they think they were born to become."[9] Time and again, social scientists feel entitled to ignore the primary narratives of the individuals they study and to substitute their own views. But perhaps the vestidas can at least be permitted their own opinions about whether they feel beautiful. No. According to Prieur, "Male domination has structured their modes of perception and appreciation in ways that have made them perceive their choice as a positive one."[10]

Prieur also doesn't like the way vestidas look, suggesting that their "attire signals fuckability." Underneath the short skirts and body-hugging fabrics, Prieur points out that "penis and testicles must be kept hidden between the legs with tight-fitting briefs or even adhesive tape."[11]

She indicates that her own strong reactions to these "bodily transformations" prompted her to take a look at her own attitudes, and she notes how Scandinavian "people seem strongly provoked by . . . transformations Mexican homosexual men are forced by the macho society around them to make," especially the way these men become "effeminate, instead of remaining naturally masculine."[12]

The vestidas don't agree. When Prieur challenged Gata about her femininity, Gata replied that as a teenager she felt very hurt when men turned her away. But "by being more feminine, more like a woman . . . the tables were turned. Men started to beg me for sex, they kissed me, and I liked it." Gata then challenged Prieur by claiming, "My boobs are bigger than yours." Prieur admits, "I defended myself by saying that mine at any rate were natural. Gata retorted, 'That doesn't matter . . . it is an achievement, . . . a thing you have been able to provide yourself. As if you wanted a house.'" Prieur is thus forced to acknowledge that, for vestidas, "their shaped and fashioned bodies are symbols of social standing, obtained through hard work and privation. At the same time, the body is an investment which may ensure their earnings as prostitutes." She summarizes, "The question is not whether the femininity is genuine or false, but whether it works. And indeed it does."[13]

Prieur also doesn't like the way vestidas act, having expected to find "a woman's soul trapped in a male body." Instead, vestidas have "more of a manly than a womanly attitude." Here, Prieur clarifies: "According to my standards of femininity," a real woman "looks like a woman," "resembles a woman emotionally," is "warm," "cares for others," enjoys "helping others," "pleasing them," and overtly "expresses her feelings, both joy and sadness."[14] This stereotype of women is evidently not one the vestidas generally observe. But what if the comparison had been made to tough-girl street gangs or to nontransgendered street sex workers? How many of these people would meet a Scandinavian academic's middle-class standard of femininity?

Throughout her multiyear study, Prieur refers to vestidas only as effeminate homosexual men, suggesting that vestidas are female impostors and denying their identity as the transgendered girls and women some obviously are. Vestidas have little chance to integrate into the life of women (although it would have been interesting to interview the women clients of the vestidas who worked as hairdressers). For this reason, neither dress nor deportment can develop in line with local women's values. A vestida has no mom mentoring her as a woman. Popular media glamour comes to define feminine presentation. Vestida sex workers receive positive reinforcement from their male clients for morphing themselves into fetishistic bodies. This fate befalls nontransgendered sex workers as well. Finally, vestidas are socially young for their chronological age. They've come into their social femininity a decade or more later than nontransgendered girls of the same chronological age, and their look connotes immaturity. Thus the imperfect presentation of vestidas doesn't necessarily discredit the authenticity of their feminine identity, as Prieur argues.

Too many sociologists don't accept transgendered people at their word, perhaps because doing so would admit that there is some truth to the biological account. Instead, these sociologists cling to the belief that vestidas and other transsexuals have "chosen" to live as a different sex. Prieur writes, "Transsexuals . . . may be the only persons in the world who actually have chosen their sex, yet they are the last ones to claim that sex is founded on choice."[15] Perhaps transgendered people are correct. Transgendered people don't choose their sex, or gender, any more or less than nontransgendered people do.

The prejudicial investigation of transgendered expression by sociologists joins the flawed analysis from biology, anthropology, and theology. In my opinion, social scientists who cannot avoid being so judgmental about the subjects they study should find another occupation. The gritty and determined refusal to acknowledge, accept, and affirm transgendered people is an academic counterpart to burning Jehanne d'Arc at the stake—an attempt to deny and erase a valid aspect of humanity.

GUEVEDOCHE IN THE DOMINICAN REPUBLIC

Enough intersex people once lived in three rural villages of the Dominican Republic that a special third-sex social category flourished there, until stifled by recent medical interventions. Called *guevedoche,* or penis at twelve, these intersex people are usually raised as girls but mature into making sperm, and so become biologically male. Guevedoche are born with unfused, labialike scrotal tissues, an absent or clitorislike penis, and undescended testes. Some guevedoche are identified as such as birth, others are classified as female, but in either case they are raised as girls, not boys. Until age twelve or so. Then the voice deepens, the muscles develop, testes descend, the phallus grows, erections occur, and semen with sperm is produced that is vented below the phallus.

Of the eighteen subjects on whom the anthropologist Gilbert Herdt gathered data, two had died, one lived as an asexual hermit, one continued to live as a woman and was married to a man, one had an ambiguous gender identity—dressing as a woman but considering himself to be male—and thirteen had transitioned to male.[16] Most of these thirteen married women and took male occupations as farmers and woodsmen, while their wives were homemakers or gardeners. Thus a large fraction, thirteen out of eighteen, did transition from female to male. The transitions occurred between the ages of fourteen and twenty-four, with an average age of sixteen (not twelve, as the name *guevedoche* implies), some time after puberty's testosterone splurge.

No information is available about how the guevedoche children felt before their transition. Did most think they were girls and wake up gradually to the realization they were boys? Or did they refuse to buy into the idea they were girls to begin with, and feel relieved when their genitals developed to confirm their feelings? Or did most feel like a third gender, not identifying with either male or female until their developing genitals gave them a clear sign of identity? Or did they not care which gender they were but simply decide that being a man offered the better deal? And what about the person who stayed female, and the person who became gender-ambiguous? No one knows. What is clear is that this form of intersex was agreeably accommodated into the social structure of these villages.

Social scientists have been interested in whether the guevedoche comprise an instance of a third-sex social category. Perhaps these villages show a society in which three *body types*—male, female and guevedoche—are accepted as equals: three types of sexed bodies, not simply three behavioral templates. The guevedoche category is, however, a placeholder, a temporary location for a child while the anatomy develops to reveal the person's "true" sex. The villagers really see only two sexes, plus a third category for those waiting to mature.

This situation can't be studied further, because medical doctors told the villagers that guevedoche were males and shouldn't be raised as girls at all. They gave the villagers technology to tell a guevedoche from a female at birth, so the social category of guevedoche is now extinct. All guevedoche are simply raised as boys from the start. No data are available about whether this medical intervention has proved worthwhile.

A tribe in New Guinea offers another case where intersex people are common enough to become a social category. Here again, though, the tribal people discern two real sexes plus a third category for temporarily holding children who are still physically maturing. Thus comparative anthropology hasn't found any societies containing three sexes, in the sense of three equal body-type categories. Many societies acknowledge substantial variation in behavioral templates, but body types remain sorted into only two primary categories, male and female.

From a biological standpoint, it seems inevitable that some society will someday devise three or more body-type categories that can't be sorted solely into the male and female binary. Biologists will continue to acknowledge the two gamete sizes corresponding to male and female function. But biology has many precedents for multiple body types with different mixtures of male and female function. The biotechnology of tissue engineering from stem cells will probably allow people someday to choose whether they make sperm or eggs at the same or different periods of life. This ability may complete the reproductive potential of people born intersex or may satisfy a yearning to be both father and mother sometime during life. Such people would attain a capability already well developed among other vertebrates; they would be truly transsexual and constitute additional body types beyond those that are solely male or female for their entire lives. Those now considered transsexual have changed gender, using the genitals as bodily markers for gender identity. Changing gonadal function from producing eggs to producing sperm, or vice versa, would actually change sex, a conceivable prospect in the future.

22

Trans Politics in the United States

A cross-cultural survey of gender expression and sexuality would seem incomplete if the present-day United States were omitted. What's happening here, where I write from, today? I believe what's interesting here is that, all around us, new social categories are emerging to hold the people who formerly lived invisibly in the closet. This birth comes with pains and leaves stretch marks. The pain comes from the extraordinary threat of violence that transgendered people face just living their daily lives. The stretch marks come from the efforts to bend existing categories to encompass people whose reality is grudgingly being acknowledged.

VIOLENCE AND THE GAY-TRANSGENDER RELATIONSHIP

Trans people launched the U.S. gay rights movement with the famous Stonewall riots in Greenwich Village in 1969. The New York police had harassed drag queens and other transgendered people in a gay bar to a point where violent rebellion broke out and spilled over into the streets.[1] Yet, in the following decades, gay political advocacy groups formed that did not include transgendered people in their mission statements. Soon the sheer numbers of gay and lesbian people crowded out transgendered people and their issues. Organizations such as the Human Rights Campaign (HRC), a powerful gay lobby, sponsored legislation covering job discrimination and hate crimes that didn't mention gender identity, only sexual orientation.[2] Time and again, the initiative and experience of trans people have been ignored or appropriated by gay organizations. Nonetheless, violence against transgendered people continues to demonstrate that the political futures of gay and transgendered people are intimately intertwined.

Soon after the terrible murder of a gay student, Matthew Shepard, in Wyoming, a trans girl was killed in Austin, Texas, under similar circumstances.[3] Eighteen-year-old Lauryn Paige, born as Donald Fuller, was found murdered in a wooded area of southeast Austin. Lauryn was dressed as a woman. Lauryn's father said, "He's been that way all his life. We always knew he was a little different, and we pretty much accepted it, but we didn't allow it around the house. We just knew he wasn't happy unless he dressed up." Lauryn had often been seen walking along South Congress Avenue, where police regularly conduct prostitution raids. Police Commander Gary Olfers said, "We are dealing with sadistic killers. There was more than one [stab] wound, and they were brutal in the application of those wounds." The autopsy revealed a cut across Lauryn's throat 9 inches long and 3 inches wide.

"The police description of this murder is heartbreakingly familiar," said leading transgender activist Riki Anne Wilchins of GenderPAC. "Sadistic killers, multiple stab wounds, bludgeoned and/or shot repeatedly . . . it's a familiar litany of brutally violent acts done to gender-different people: Chanelle Pickett, Brandon Teena, Christian Paige, Deborah Forte, Vianna Faye Williams, Jamaica Green, Jessy and Peggy Santiago, Tasha Dunn . . . and the list goes on." Yet Lauryn's murder received hardly any mention in the press, and no gay organizations followed up, even though the events parallel the well-publicized murder of Matthew Shepard.

Lauryn's death was not an isolated incident. The transgender activist and columnist Gwendolyn Smith maintains a website for a project called "Remembering Our Dead" that contains the names of the transgendered people killed every day. Each year, a moving ceremony of remembrance is held in cities worldwide.[4]

Another recent example is the tragic murder of Barry Winchell, a twenty-one-year-old private at Fort Campbell on the Tennessee-Kentucky border in the summer of 1999. Winchell's death was used by gay advocacy groups to force the U.S. government to reevaluate its don't-ask-don't-tell policy in the military. Yet, as the *New York Times Magazine* eventually reported in a cover story, "The fact is that Winchell, killed for being gay, wasn't gay."[5] He was straight. He had dated only women in the past, nontransgendered women. At the time, he was in love with a beautiful transgendered woman, Calpernia Addams. The Nashville-based Lesbian and Gay Coalition for Justice paid Addams a visit and suggested that "for the sake of clarity" she should tell reporters that she was really a he, because "how can you say he [Winchell] was gay-bashed if he was dating a woman, you know?" Addams agreed, and in subsequent news accounts, she was Winchell's "boyfriend" or "cross-dressing friend." This devastating lie erased Addams's existence and the basis of her relationship with Winchell. A Nashville gay activist concluded, "We don't have a vocabulary for dealing with these issues."

On June 2, 2001, newspapers carried reports that more people were dismissed from the military in 2000 than in any other year since the don't-ask-don't-tell policy was initiated. Nearly half of the discharges were from Fort Campbell, Kentucky, home of the 101st Airborne Division, where Winchell was beaten to death with a baseball bat in 1999. The *San Francisco Chronicle* described Winchell as someone who was "thought to have been gay," thus recognizing that he wasn't gay. This description wrongly suggests, however, that the attack was simply a case of mistaken identity. Winchell was dating Calpernia Addams, who was known to be transgendered and sometimes worked as a performer in a drag bar. The murderer, who frequented the bar himself, was feeling genuine homophobia misdirected toward a straight fellow soldier.

The implications of this instance of homophobia for the military are still unclear. After the September 11, 2001, attack, the Pentagon issued an order suspending discharge proceedings against service members who disclosed their homosexuality. A similar order was issued during the Persian Gulf War. Apparently gay and lesbian troops are just fine in time of war.[6]

Just as trans women have been converted into cross-dressing gay men, trans men have been converted into cross-dressing lesbians. Billy Tipton was a jazz musician who married a woman, and adopted and raised children with her. Upon his death in 1989, he was discovered to have female genital anatomy. Yet the lesbian community refers to him as "her." Similarly, Brandon Teena, a young trans man who was raped and murdered in Nebraska in 1993 when he was discovered to be transgendered, is referred to in the lesbian press as a cross-dressing lesbian who passed as a straight man. This validates what the rapists were trying to show: he could be raped like a woman, so he was a woman. In fact, Brandon Teena went to great pains to be taken seriously as a man and referred to himself using masculine pronouns. He wasn't passing as a straight man, he was one.[7]

A recent transgender killing occurred in, of all places, the San Francisco Bay Area. A seventeen-year-old transgendered girl, Gwen Araujo, was killed during the night of October 3–4, 2002, although her body was not found until mid October. Born Edward Araujo, Gwen was living as a girl and was romantically involved with several guys. Two of the boys she had been intimate with began to suspect she was born male, and plotted to ascertain for sure what genitals Gwen had. They planned to punish her if she was discovered to have male genitals: "I swear, if it's a fucking man, I'm gonna kill him. If it's a man, she ain't gonna leave."[8] At a party on October 3, 2002, in Newark, a girl pulled up Gwen's skirt and outed her to the crowd: "It's a fucking man." One of the boys with whom Gwen had been intimate then cried out, "I can't be fucking gay." The girl who had outed Gwen then tried to console the boy, saying, "It's not your fault. I went to high school with you, and you were on the football team. Any woman that knows you after this, it's not going to matter. Just let her go."[9] Nonetheless, the four boys proceeded to beat Gwen into a bloody pulp, strangled her with a rope, put her in a truck for a four-hour drive to a spot in El Dorado County near the Sierra Nevada mountains, dumped the body in a shallow grave, and covered it with heavy rocks, dirt, and a tree trunk. Then the four boys got back into their vehicle, drove to a hamburger stand, and ordered breakfast from the drive-up window. The murder went unreported for days.

Many nontransgendered people seem to have trouble acknowledging that this type of violence occurs regularly in their backyards. These crimes are not ordinary crimes, but genuine hate crimes. They show that being outed as gay or transgendered is not merely discourteous and insulting, but seriously compromises personal safety. The crimes also show how closely intertwined trans and gay violence is—criminals don't draw clever distinctions between gender identity and sexual orientation, they go for the jugular and don't ask why.

COALITION-BUILDING

After decades of turbulence, the political landscape within the lesbian-gay-bi-trans-intersex community is starting to stabilize. Since the 1969 Stonewall riots, the gay political stance has usually been that the public isn't ready for transgendered people: best to establish rights for gays, then move on to the transgendered. I believe this political analysis is wrong. The general public doesn't want to be bothered with a new category of people to protect each year. In excluding transgendered people, gay advocacy organizations lose the moral high ground. They can be attacked for representing only "special rights" rather than human rights. The Human Rights Campaign seems to think that transgendered people are too few in number to contribute financially or at the ballot box and that politically they are a liability. The trans position is that we were there at the start, and we're there for the heavy lifting. Much of the violence passed off as gay-bashing is really violence against gender-variant people. Legislation to protect sexual orientation without mentioning gender identity misses the point. Gay people are often gender-variant too—it's their gender variance that places them at risk more than their sexual orientation, because their sexual activity is carried out in private, whereas their gender variance is publicly visible.

But the political situation is improving. On March 10, 2001, the HRC finally added transgender rights to its mission statement, although it hasn't yet included gender identity in its legislative proposals. Many gay organizations have added a "T" to their name, to signify the inclusion of transgendered people. In San Francisco, the Harvey Milk Democratic Club included transgendered people a few years ago, followed by the Alice B. Toklas Democratic Club—a rare unity among these rival progressive and centrist political organizations. More important, Mark Leno, a courageous political leader who represents the largely gay Castro district in San Francisco, sponsored legislation to include transgender medical needs in the city health insurance plan. On April 30, 2001, the Board of Supervisors passed the proposal with the necessary nine out of eleven votes, accompanied by thoughtful and articulate public debate. National news media covered the event, which was the result of unified action by gay, lesbian, and transgendered people on behalf of transgender issues, one of the few since Stonewall. The action set a new legislative standard in the United States for inclusion of the full human rainbow in gender expression and sexuality.[10] Mark Leno has now moved on to the California Assembly, where he has become one of the most effective legislators in the state.

HOW MANY LETTERS IN THE ALPHABET?

The confusion over who belongs to the lesbian-gay-bi-trans-intersex community no longer seems a major issue. For a while, each year seemed to bring some new group needing recognition: first gay men, then lesbians, then bisexuals, then transgendered people, and finally intersexed people. People have wondered whether this alphabet-soup approach to including sexual and gender minorities would ever stop. Do we have to amend our laws every year, as a new political constituency clamors for recognition and protection?

I believe the present list is now coherent and complete, and the game of gender scrabble may be over. Theoretically, gays and lesbians affirm traditional binary distinctions in gender identity and body, but flip the directions of sexual orientation. Bisexuals challenge the binary in sexuality, trans people challenge the gender binary, and intersexed people challenge the body binary. Collectively, all these identity categories seem to span the body-gender-sexuality space, and if any one category is omitted, the remaining group shows a gap. Perhaps we've finally attained a body-gender-sexuality community where anyone, I hope, can now find their spot by combining the elements they need from all these categories.

I believe all of us in the body-gender-sexuality community have more in common than the rhetoric of our identity politics sometimes suggests. First, each of us "comes out." We come to terms with ourselves, our family, school or employer, friends and colleagues, and society. We accept, often after years of denial, the stigma and danger of being ourselves. We differ in detail: a gay man realizes his sexual attraction to another man, a lesbian woman realizes she's not just a tomboy, a trans woman braves an obstacle course of hecklers on her first appearance in public, and an intersexed person forgives. Traumatic moments. Compare: A girl tells Dad she won't go into engineering. A boy tells Mom he won't be a doctor. An unmarried woman tells her family she's pregnant. A man tells his family he's marrying outside of their race or religion. These too are serious moments of self-definition, but they rarely equal what someone who's queer goes through when coming out. Second, each of us is told we're impossible. We're not supposed to exist, our reality is denied by science, religion, and custom. We're theoretically problematic. Yet we do exist. And we're good.

HOW MANY GENDERS?

The lesbian-gay-bi-trans-intersex community may seem complete, but there is a larger question to be addressed. Trans people must continually locate themselves within the traditional binary distinction between man and woman. Many trans people affirm traditional gender norms in their personal lives, while not wanting to impose such norms on others. Many transgendered women identify simply as straight women and lead lives little different from other women of similar age and occupation, regardless of their unusual history. Similarly for transgendered men. Short profiles of successful transgendered people, usually conforming to gender norms, have been assembled by Lynn Conway.[11]

Other trans people transgress gender norms. The most outspoken is the transgendered author and activist Kate Bornstein, who writes, "I know I'm not a man . . . and I've come to the conclusion that I'm probably not a woman, either. . . . The trouble is, we're living in a world that insists we be one or the other."[12] Some people feel they inhabit a space between man and woman—a third gender.

Because third-gender spaces exist in other cultures, many wonder whether U.S. culture is too rigid to allow for a third (or fourth) gender—forcing people to locate in one or the other of the two main genders—or whether people actually choose to identify with the main genders. The biggest difficulty with affirming a third-gender identity is knowing what that means. Those transitioning from one traditional binary gender to the other have a clear sense of where they want to end up and a clear unhappiness with where they started. Moving to a third gender requires lots of exploring and trying different combinations, some harmonious blends of both genders, others glaring and provocative declarations of resistance.

One experimental genre is called gender-queer, or gender-fuck. One of the first groups to pioneer a gender-fuck presentation was the Sisters of Perpetual Indulgence, a group of men who dress in nuns' habits while sporting beards or mustaches.[13] Founded in 1979, they have established chapters in many cities throughout the world, and are active in charity. They are outrageous and eagerly sought after for parades and as master of ceremonies at public events. They successfully taunt many religious leaders, who can't seem to resist swallowing the bait.

Young butch lesbians and young trans men are exploring interesting and appealing new combinations of the masculine and the feminine as full-fledged lifestyles. These new models of gender suggest that a third gender may become more of a real option in coming years.[14]

THE EVOLUTION OF HUMAN RAINBOWS

To repeat: lesbians, gays, bisexuals, transgendered people, and inter-sexed people all exist. But why do we exist? Can we theorize about how gender and sexuality diversity evolved in humans, as we did earlier for animals? I don't think so. Methodologically, one can't study human evolution the same way one studies animal evolution. Humans cover the globe. Animals can usually be pinpointed to one type of environment and traits tied to a particular function in that environment. Instead, our species has evolved in response to all the physical and social environments that our gene pool encounters. The social bonds built through same-sex sexuality might keep one alive in ancient Greece but cause death during the Catholic Inquisition. Our species' evolution reflects both positive and negative pressures.

The relatively short history of same-sex sexuality as an identity category may spring from how common same-sex sexuality is. Policing same-sex sexuality as a distinctive category in the face of this commonness, and in the absence of any visible phenotypic markers, takes society's constant energy. Just witness how much time and money is wasted on legislative and legal activities to repress gays, as though they will just disappear someday if enough is invested into stamping them out. Transgender categories seem to have a longer history, perhaps because transgendered people are both visibly distinctive and relatively uncommon, and therefore require less energy to maintain in descriptive boxes. I sometimes wonder if the identity categories of gay and lesbian will simply dissolve someday because no one wants to bother with the distinctions anymore, whereas transgender categories may persist longer.

Transgender expressions are seemingly tied to occupations, and the earnings from the occupation can benefit either the transgendered person directly or their extended family. In antiquity, being a eunuch qualified one to work in the private space of men and women, out of the public sphere of male-male competition. The parents of eunuch slaves presumably benefited from the sale of their child into slavery, and perhaps this enabled them to raise additional children. Similarly, Native American two-spirits sometimes directly helped raise relatives, and some Mexican vestidas give money to their families. These family benefits may be significant for evolution through kin selection, especially with an already-high reproductive skew in the population. Because people with transgendered identities were valuable in particular occupations, they may have helped perpetuate their presence in the human gene pool by benefiting their families. Furthermore, throughout history, transgendered people have often produced children of their own. Their occupation and temperament may be directly advantageous in some circumstances, and they may be sought as mates accordingly. Overall, the evolution of transgender expression, like same-sex sexuality, reflects both prosperity in positive times and repression in negative times.

We'll probably never know why any particular color occurs in human gender and sexuality rainbows. Nonetheless, our species, like others, clearly does contain natural rainbows of gender expression and sexuality. These rainbows emanate from our gene pool, our shared humanity. Society carves these rainbows into categories like a cookie cutter carves a marbled cookie dough into cookies. We shape our cookies through our policies on human rights. Should we have just two very broad categories, man and woman, accommodating same-sex sexuality and gender crossing—two huge cookies filled with chocolate chips, raisins, nuts, colored sprinkles, and more? But would two large categories still allow discrimination against those who wish a third gender? Or should we have lots of tiny cookies, each with special flavors—the M&Ms approach, a proliferation of identity politics? Or maybe some big cookies plus some little ones? I don't know. I do know that what won't work is stuffing our species into two small categories of gender and sexuality.

I believe the rainbow always has more colors than society has categories, and that society is always trying to cram humanity's rainbow into the few categories it does have. Social scientists have the opposite perspective; they think diversity results from society producing difference among people who are biologically the same. I don't agree. The biology I know tells of endless variation, not of a few universals. This endless biological variation is always poking through social categories, spilling over the borders, fudging the edges.

Still, the rainbow isn't static. When we modify society, its institutions and categories, our species' substance slowly changes in response to new forces of natural selection that now reside within society, leading to new rainbows that then flow back into society again, a glacially slow cultural-biological back-and-forth.

A MORAL IMPERATIVE

Until now, I have focused on empirical grounds for affirming the full human rainbow of diversity. I now turn to a moral imperative for embracing diversity taken from one religious tradition. The Bible doesn't approach diversity by affirming selected categories one by one. It's true that today we might wish the Bible were more direct in affirming homosexual people. But in another thousand years, we may discern more categories of identity that we don't presently recognize, whereas others will have coalesced. What will the Bible have to say then? The same as it says now. The Bible affirms all of biological diversity—even unnamed or rearranged categories—in the story of Noah's Ark, a story that spans three chapters of Genesis.

Let's look again at this story: "The earth was filled with violence. And God said to Noah 'Make yourself an ark.' " Noah was told "Of every living thing of all flesh, you shall bring two of every sort into the ark. . . . Of the birds according to their kinds, and of the animals according to their kinds, of every creeping thing of the ground according to its kind, two of every sort shall come in to you, to keep them alive. . . . Take with you seven pairs of all clean animals, the male and his mate; and a pair of the animals that are not clean, the male and his mate; and seven pairs of the birds of the air also, male and female, to keep their kind alive upon the face of all the earth."

Genesis continues: "And Noah and his sons and his wife and his sons' wives with him went into the ark, to escape the waters of the flood. Of clean animals, and of animals that are not clean, and of birds, and of everything that creeps on the ground, two and two, male and female, went into the ark with Noah, . . . every beast according to its kind, and all the cattle according to their kinds, and every creeping thing that creeps on the earth according to its kind, and every bird according to its kind, every bird of every sort. They went into the ark with Noah, two and two of all flesh in which there was the breath of life."

After forty days of flooding, the waters receded. "And God said, 'Bring forth with you every living thing that is with you of all flesh— birds and animals and every creeping thing that creeps on the earth—that they may breed abundantly on the earth, and be fruitful and multiply upon the earth' " (Gen. 6–9, RSV).

These passages make clear that all organisms belong on the Ark, both the "clean" and the "unclean," and each "according to its kind." Yet the Ark is usually depicted as having only one male and one female from each species. "Kind" means more than simply a species. "Kind" includes the varieties within a species. "All the cattle according to their kinds" means all varieties of cattle. All cattle belong to the same species because they all interbreed. So "all the cattle according to their kinds" indicates that the Ark contained all the varieties of cattle, and by extension, all the variants within every species. God didn't tell Noah to pick and choose, including some varieties and excluding others. Therefore, the Ark would have harbored full rainbows of gender expression and sexuality, as well as all other dimensions of biological diversity.

In the story of Noah's Ark, the Bible gives a single overarching protection for all biological diversity. The message is comprehensive in its inclusion, and without qualification. We should not look to the Bible for affirmation of each new category of diversity that we distinguish. The

Ark covers all, now and forever. The message of Noah's Ark is to conserve all biological kinds.

THE TRANS AGENDA

I went to dinner once with a public speaker who, upon learning that I was transgendered, showed annoyance at my existence. "What do you people want?" he insisted. One would have thought the answer obvious: to enjoy the rights everyone else has. But that answer wasn't specific enough. In response, and to conclude this chapter, here is my list of "what we want." My trans agenda consists of six points:

1. We want to be cherished as a normal part of human diversity.

2. We demand the freedom to offer our own unfiltered narratives—we demand our own voices.

3. We want to be treated with courtesy and dignity. We don't want to lift our skirts to show we're female or drop our trousers to show we're male. We want to be respected as people, not bodies.

4. We demand that the killing of transgendered people stop. We support extending existing anti-hate crime legislation to include gender identity as a protected category.

5. We want equal participation in public social institutions, including employment, education, housing, marriage, adoption, military service, and religious life. We support extending any existing antidiscrimination legislation to include gender identity. We support legislation to allow any two people to enter into marriage. We support rescinding the don't-ask-don't-tell policy in the U.S. military. We support baptism and ordination, regardless of sexual orientation or gender identity.

6. We want full-service health plans to cover gender-transition medical services, similar to the coverage of pregnancy benefits. Although some health plans cover only catastrophic illness, others cover many procedures, from acupuncture to physical therapy, and these comprehensive plans should not exclude transgender benefits.

Of these six points, the first is the most important, and the rest follow from it. I feel we have earned these rights by our collective contributions to family and society, and I know we can be even more productive if we were not laboring under threats to our personal safety, or diminished by stigma. I also feel that a contemporary society must grant these six points if it wishes to be considered moral and civilized.

Policy Recommendations

This book has presented information about diversity, gender, and sexuality that is not widely known or appreciated. What's next? Do we sit and ponder, and then gradually forget? Or are we compelled to take action? I believe some action is warranted. The facts in Part 1 about diversity in vertebrates calls for some educational reform at premed, medical school, and continuing education stages. Part 2, which reveals how genetic engineering and medicine attempt to cure nondiseases, develop bio-weapons, and tamper with our gene pools, implies a need for explicit standards of professional ethics and for public involvement in the supervision of biotechnology. The survey in Part 3 of how gender and sexuality are manifested in various cultures calls for a public affirmation of our human rainbow. Here then are some specific recommendations for actions we might take.

EDUCATION

PREMEDICAL CURRICULUM

I recommend that the undergraduate curriculum for premedical students require instruction in biological diversity, particularly in gender and sexuality, and that medical schools enforce this requirement as a condition for admission. Presently, the U.S. premedical curriculum consists of courses in organic chemistry, biochemistry, genetics, and cell biology, with perhaps some physiology thrown in. As a result, premeds are professionally acquainted with only seven species: a bacterium, a worm, a fruit fly, a chicken, a rabbit, a mouse, and the human being—the species on which most medical laboratory studies have been done. Limiting the undergraduate curriculum to these seven species leaves doctors clueless about natural diversity, which translates into pathologizing anything surprising. Imagine a young doctor's alarm when first encountering a baby with unusual genital plumbing, a woman with masculine anatomy, a man with feminine anatomy, or a person who introduces their same-sex partner at a cocktail party? Yet this is all so ho-hum in light of comparative vertebrate genital morphology, multiple gender expressions that include feminine males and masculine females, and over three hundred vertebrate societies that feature same-sex courtship and mating.

MEDICAL CURRICULUM

Patients and their doctors obviously need to talk about many matters pertaining to sex, gender, and sexuality. Yet Stanford University Medical School, for example, doesn't offer a single course on human sexuality, although the medical students have organized one on their own, relying on outside speakers. I feel this curricular lacuna raises questions about the adequacy of the medical school curriculum, assuming Stanford is representative. Therefore, I recommend that medical schools provide education in human sexuality as a condition of being accredited to offer an M.D. degree.

A difficulty in implementing this recommendation lies in finding distinguished faculty in human sexuality. Human sexuality is one of the weakest academic subjects in all of biomedical science. To remedy the lack of outstanding scholarship in human sexuality, I recommend that the National Institutes of Health establish special grants to support junior and midcareer scientists from other biomedical disciplines to switch their study to human sexuality. Furthermore, I recommend that continuing education for medical doctors include training in both gender and sexuality diversity, as well as human sexuality.

PSYCHOLOGY CURRICULUM

I recommend that the psychology curriculum include a yearlong core course on the principles of biology, covering three areas:

Classification: Psychologists are rather casual from a biological standpoint in their willingness to classify people into various "types" and "subtypes." Biologists, on the other hand, are cautious even when classifying organisms into species and would never dream of setting up formally diagnosed categories for the phenotypic variation within a species. In the last two decades, biology has developed sophisticated statistical methods for classifying, and these principles of modern classification should be applied in psychology.

Evolutionary biology: Psychologists are increasingly misusing evolutionary theory to concoct evolutionary stories that are supposed to "explain" human behavior. Psychologists need a better understanding of the standards and methods for proposing and testing hypotheses in evolutionary biology.

Molecular genetics and endocrinology: Psychologists frequently speak about genes and hormones in terms that are naive and decades out of date. Psychology needs more contemporary information on gene and hormone action to support an improved and less dichotomous account of the nature/nurture distinction.

MEDICAL PRACTICE

FDA-CERTIFIED LIST OF DISEASES

The U.S. Food and Drug Administration (FDA) should take responsibility for maintaining the official list of diseases used by the health professions as the basis for diagnosis. Presently, lists of diseases are maintained by numerous professional societies and specialist groups. The Merck Manual is the guide for physical ailments, the DSM-IV for behavioral conditions, and various treatments have their own "standards of care." In addition, the professional societies take stands on particular procedures, like circumcision. Meanwhile, the major health insurers and health maintenance organizations have their own lists of covered diagnoses, allowed prescriptions, formularies for appropriate drugs, and so forth. I suggest instead that the FDA be charged with developing an official list of conditions that are considered diseases. The process by which the list is created should include input from medical professionals, health insurers and health maintenance organizations, and patient advocacy groups.

FDA-APPROVED MEDICAL PROCEDURES

The FDA should regulate surgical and behavioral therapies in the same way it does pharmaceutical therapies. The standard for a cure by the scalpel or on the couch should be no different than that from a bottle. To be certified by the FDA, a drug must pass tests about effectiveness, mode of action, and side effects. The same should be required of surgical procedures and behavioral therapies. To be certified by the FDA, a surgical procedure or behavioral therapy should be shown to work, why the procedure or therapy works should be understood, and the side effects should be quantified. Furthermore, guaranteed follow-up study should be provided. These steps could eliminate or greatly reduce the spurious curing of nondiseases.

GENETIC ENGINEERING AND BIOTECHNOLOGY

OATH OF PROFESSIONAL ETHICS

Biotechnologists should publicly affirm a professional standard of ethics. Medical doctors have long affirmed some version of the ancient Hippocratic Oath, which commits them, above all, to do no harm.[1] Admission to a master's and Ph.D. program in the biotechnology disciplines, including molecular biology, biochemistry, and genetic engineering, should also include an oath. This oath should be: *I promise to protect the human gene pool. I promise to use biotechnology for peace.* At the very least such an oath would make the moral imperative clear.

One might object that this oath does not go far enough. I think it's too early to require an oath to protect the gene pools of other species, even though we might in the future. Until that time, we should require an oath for protecting the human species.

PROFESSIONAL BIOTECHNOLOGIST LICENSE

Like medical doctors and engineers, biotechnologists should be licensed. The examinations leading to certification should test knowledge of public safety, including epidemiological and public health implications. The test should also examine knowledge of the ethical dimensions of their practice and include a reaffirmation of their oath to protect the human gene pool and pursue peace. Thereafter, only a licensed biotechnologist should be eligible to be a principal investigator on a government grant to conduct research affecting the human genome, just as only licensed medical doctors are eligible to conduct research on human subjects with government funds.

CORPORATE POLICY ON ETHICS

Biotechnology companies should be required to make a corporate commitment similar to the oath that individual biotechnicians take. Companies should affirm corporate policy to protect the human gene pool and pursue peace. I recommend that institutional investors divest their stock portfolios from biotech companies that do not make this commitment, and that private investors not purchase stock in such companies. I recommend that companies not making this investment not be eligible for government grants and contracts.

EPIDEMIOLOGICAL IMPACT REPORT

Turning to the products of biotechnology, the U.S. Food and Drug Administration should authorize therapies only after a satisfactory epidemiological impact report has been produced. This report would be the medical equivalent of the environmental impact report required for construction projects that impact the common good. The epidemiological impact report would require that ecological homework be carried out before the therapy is approved. For infectious diseases, the impact report would detail how to administer the drug to minimize the development of drug resistance in pathogens. For genetic diseases, the impact report would explain why the human rainbow needed repair. The report would clarify whether the proposed gene therapy really did remedy a genetic defect, or was solely gene cosmetics. The report would project the implications of the proposed genetic engineering product for the human gene pool.

An epidemiological impact report for genetically engineered therapies is the equivalent of impact reports demanded by consumer and environmental groups for genetically engineered crops.[2] The concerns raised about Frankenfoods offer a mild foretaste of protests that will emerge about genetic engineering of the human rainbow once the aspirations of the biotech industries become better known.

A COMMON CODE FOR ECOLOGICAL AND ENVIRONMENTAL IMPACT REPORTS

At present, the Environmental Protection Agency (EPA) regulates plants engineered to produce their own pesticides, and the U.S. Department of Agriculture (USDA) is responsible for assuring the ecological safety of genetically engineered plants. Their regulatory processes have been strongly criticized. Hence this final recommendation: A common code or standard for ecological and epidemiological impact reports should be developed for the FDA, EPA, and USDA.

A PUBLIC SYMBOL

I propose we construct a "Statue of Diversity" that would be to the West Coast what the Statue of Liberty is to the East Coast: a beacon of welcome and a fundamental statement of American values and way of life. In the November 2000 election, I ran for the office of supervisor in the South of Market district of San Francisco. I proposed that a Statue of Diversity be constructed in San Francisco harbor.[3]

The statue could be built on Treasure Island, an island in the harbor, near the Bay Bridge, which was the site of the 1939–40 World's Fair. The statue would be passed by vessels en route to docks along the Oakland and San Francisco shores, would be readily visible from both cities, and could be easily accessed by a five-minute ride from the Ferry Terminal at Market Street in San Francisco and by a slightly longer ride from Jack London Square in Oakland.

I imagine that the design would result from an open competition and could contain symbols affirming biological and cultural diversity from many spiritual traditions. The facility could also house a plaza and public space for recreation, the arts, and nature.

The statue would offer hope to those who have suffered discrimination, encouraging people to look beyond the broken promise of meritocracy, a promise so often made but rarely kept. This Statue of Diversity would be a bold statement of America's moral leadership.

Notes

INTRODUCTION: DIVERSITY DENIED

1. See: C. Yoon, 2000, Scientist at work: Joan Roughgarden, a theorist with personal experience of the divide between the sexes, *New York Times,* Oct. 17, pp. D1–D2; also a fifty-minute interview from January 22, 2001, on GenderTalk Web-Radio, program 294, with Nancy Nangeroni and Gordene MacKenzie, available at http://www.gendertalk.com/real/251/gt294.shtml.

2. H. Adams, L. Wright, and B. Lohr, 1996, Is homophobia associated with homosexual arousal? *Psychological Review* 103:320–35. Psychologists define "homophobia" as the dread of being in close quarters with homosexual people and having irrational fear, hatred, and intolerance of homosexual people.

3. J. Roughgarden, 1991, The evolution of sex, *Amer. Natur.* 138:934–53.

4. J. Roughgarden, 1998, *Primer of Ecological Theory,* Prentice Hall; J. Roughgarden, 1995, *Anolis Lizards of the Caribbean: Ecology, Evolution, and Plate Tectonics,* Oxford University Press; J. Roughgarden, R. May, and S. Levin, eds., 1989, *Perspectives in Ecological Theory,* Princeton University Press; P. Ehrlich and J. Roughgarden, 1987, *The Science of Ecology,* Macmillan; J. Roughgarden, 1979, *Theory of Population Genetics and Evolutionary Ecology: An Introduction,* Macmillan.

1: SEX AND DIVERSITY

1. C. Darwin, 1962 [1860], *The Voyage of the Beagle,* Anchor Books, esp. pp. 393, 394, 398.

2. E. Mayr, 1964 [1942], *Systematics and the Origin of Species,* Dover Publications; E. Mayr, 1963, *Animal Species and Evolution,* Harvard University Press.

3. See: T. Dobzhansky, 1954, Evolution as a creative process, *Proc. 9th Int. Cong. Genet.,* in *Caryologia,* pp. 435–49.

4. H. Muller, 1950, Our load of mutations, *Amer. J. Hum. Genet.* 2:111–76; N. Morton, J. Crow, and H. Muller, 1956, An estimate of the mutational damage in man from data on consanguineous marriages, *Proc. Nat. Acad. Sci. (USA)* 42:855–63.

5. See quote from H. Muller, p. 166 in M. Kimura and T. Ohta, 1971, *Theoretical Aspects of Population Genetics,* Princeton University Press; cited in R. Lewontin, 1974, *The Genetic Basis of Evolutionary Change,* Columbia University Press, p. 30.

6. K. Petren and T. Case, 1998, Habitat structure determines competition intensity and invasion success in gecko lizards, *Proc. Nat. Acad. Sci. (USA)* 95: 11739–44.

7. On geckoes, see: R. Radtkey, S. Donnellan, R. Fisher, C. Moritz, K. Hanley, and T. Case, 1995, When species collide: The origin and spread of an asexual species of gecko, *Proc. R. Soc. Lond.,* ser. B, 259:145–52. On whiptail lizards, see: C. Cole, 1975, Evolution of parthenogenetic species of reptiles, in R. Reinboth, ed., *Intersexuality in the Animal Kingdom,* Springer Verlag; O. Cuellar, 1979, On the ecology of coexistence in parthenogenetic and bisexual lizards of the genus *Cnemidophorus, Amer. Zool.* 19:773–86; O. Cuellar, 1977, Animal parthenogenesis, *Science* 197:837–43. See also: L. D. Densmore, C. Moritz, J. W. Wright, and W. M. Brown, 1989, Mitochondrial-DNA analysis and the origin and relative age of parthenogenetic lizards (Genus *Cnemidophorus*): IV. Nine *semilineatus*-group unisexuals, *Evolution* 43:969–83. On all-female fish, see: R. C. Vrijenhoek, 1984, The evolution of clonal diversity in *Poeciliopsis,* pp. 399–429 in B. J. Turner, ed., *Evolutionary Genetics of Fishes,* Plenum Press. See also, on vertebrates generally: R. C. Vrijenhoek, R. M. Dawley, C. J. Cole, and J. P. Bogart, 1989, A list of the known unisexual vertebrates, pp. 19–23 in R. M. Dawley and J. P. Bogart, eds., *Evolution and Ecology of Unisexual Vertebrates,* New York State Museum.

8. O. Cuellar, 1974, On the origin of parthenogenesis in vertebrates: The cytogenetic factors, *Amer. Natur.* 108:625–48.

9. H. Carson, 1967, Selection for parthenogenesis in *Drosophila mercatorium, Genetics* 55:157–71.

10. M. Olsen, 1965, Twelve year summary of selection for parthenogenesis in the Beltsville small white turkey, *Brit. Poultry Sci.* 6:1–6; M. W. Olsen, S. P. Wilson, and H. L. Marks, 1968, Genetic control of parthenogenesis in chickens, *J. Hered.* 59:41–42.

11. G. L. Stebbins, Jr., 1950, *Variation and Evolution in Plants,* Columbia University Press; O. P. Judson and B. Normark, 1996, Ancient asexual scandals, *Trends Ecol. Evol.* 11:41–46; and R. Butlin, 2002, The costs and benefits of sex: New insights from old asexual lineages, *Nature Reviews: Genetics* 3:311–17.

12. G. C. Williams and J. B. Mitton, 1973, Why reproduce sexually? *J. Theor. Biol.* 39:545–54; G. C. Williams, 1975, *Sex and Evolution,* Princeton University Press.

13. O. Solbrig, 1971, The population biology of dandelions, *American Scientist,* 59:686–94.

14. D. Tilman, 1990, Constraints and tradeoffs: Toward a predictive theory of competition and succession, *Oikos* 58:3–15.

15. See: R. Lewontin, 1974, *The Genetic Basis of Evolutionary Change,* Columbia University Press; J. Gillespie, 1994, *The Causes of Molecular Evolution,* Oxford University Press; H. Muller, 1932, Some genetic aspects of sex, *Amer. Natur.*

66:118–38; special issue on the evolution of sex in *Nature Reviews: Genetics,* vol. 3, April 2002.

16. G. Kolata, 1998, Scientists see a mysterious similarity in a pair of deadly plagues, *New York Times,* May 26, p. B9.

17. A. Cullum, 2000, Phenotypic variability of physiological traits in populations of sexual and asexual whiptail lizards (genus *Cnemidophorus*), *Evol. Ecol. Research* 2:841–55.

18. This claim has theorem-like generality. Consider a locus with two alleles in a random-mating diploid population. Suppose the fitnesses of the three genotypes at time t, $W_{11,t}$, $W_{12,t}$, and $W_{22,t}$, are independent identically distributed random variables. The average fitness at time t is the expectation of the W_{ij} with respect to the genotype frequencies at time t. The genotype frequencies in an asexual species are unconstrained, whereas the genotype frequencies in a sexual species are reset each generation to Hardy-Weinberg ratios. Therefore, the variance through time of the average fitness in an asexual population is greater than in a sexual population. Hence, the geometric mean of the average fitnesses through time is necessarily lower in an asexual population than in a sexual population, implying that a sexual population inevitably outlives an asexual population. For a computer simulation, see my paper: J. Roughgarden, 1991, The evolution of sex, *Amer. Natur.* 138:934–53. The simulation shows how a sexually reproducing species gradually outlasts a clonal species in a fluctuating environment, even if both start out with equally diverse rainbows. See also: W. D. Hamilton, 1980, Sex versus non-sex, *Oikos* 35:282–90; and W. Hamilton, P. Henderson, and N. Moran, 1981, Fluctuations of environment and coevolved antagonist polymorphism as factors in the maintenance of sex, pp. 363–81 in R. Alexander and D. Tinkle, eds., *Natural Selection and Social Behavior: Recent Research and Theory,* Chiron Press; R. May and R. Anderson, 1983, Epidemiology and genetics in the coevolution of parasites and hosts, *Proc. R. Soc. Lond.,* ser. B, 219:281–331; L. Nunney, 1989, The maintenance of sex by group selection, *Evolution* 43:245–57; R. Michod and B. Levin, eds., 1988, *The Evolution of Sex: An Examination of Current Ideas,* Sinauer; and the April 2002 issue of *Nature Reviews: Genetics.*

2: SEX VERSUS GENDER

1. See: Y. Iwasa and A. Sasaki, 1987, Evolution of the number of sexes, *Evolution* 41:49–65.

2. Having one gamete size is called isogamy. Having more than one gamete size is called anisogamy. See: G. Bell, 1982, *The Masterpiece of Nature,* University of California Press; R. Hoekstra, 1987, The evolution of sexes, pp. 59–91 in S. Sterns, ed., *The Evolution of Sex and Its Consequences,* Birkhäuser.

3. V. A. Dogiel, 1965, *General Protozoology,* Clarendon Press.

4. C. Bressac, A. Fleury, and D. Lachaise, 1994, Another way of being anisogamous in *Drosophila* subgenus species: Giant sperm, one-to-one gamete ratio, and high zygote provisioning, *Proc. Nat. Acad. Sci. (USA)* 91:10399–402; S. Pitnick and T. A. Markow, 1994, Male gametic strategies: Sperm size, testes size, and the allocation of ejaculate among successive mates by the sperm-limited fly *Drosophila pachea* and its relatives, *Amer. Natur.* 143:785–819; S. Pitnick, G. S. Spicer, and T. A. Markow, 1995, How long is a giant sperm? *Nature* 375:109.

5. R. R. Snook, T. A. Markow, and T. L. Karr, 1994, Functional nonequivalence of sperm in *Drosophila pseudoobscura, Proc. Nat. Acad. Sci. (USA)* 91:11222–26; C. Bressac and E. Hauschteck-Jungen, 1996, *Drosophila subobscura* females preferentially select long sperm for storage and use, *J. Insect Physiol.* 42:323–28; R. R. Snook, 1997, Is the production of multiple sperm types adaptive? *Evolution* 51:797–808. See also: P. Lee and A. Wilkes, 1965, Polymorphic spermatozoa in the hymenopterous wasp, *Dahlbominus, Science* 147:1445–46; R. Silberglied, J. Shepherd, and J. Dickinson, 1984, Eunuchs: The role of apyrene sperm in lepidoptera? *Amer. Natur.* 123:255–65; P. Cook, I. Harvey, and G. Parker, 1997, Predicting variation in sperm precedence, *Phil. Trans. R. Soc. Lond.,* ser. B, 352:771–80; M. Watanabe, M. Bon'no, and A. Hachisuka, 2000, Eupyrene sperm migrates to spermatheca after apyrene sperm in the swallowtail butterfly, *Papilio xuthus L.* (Lepidoptera: Papilionidae), *J. Ethol.* 18:91–99.

6. The state where both mating types, A and B, have the same gamete size has the interesting property of being an ESS (evolutionary stable strategy) that is dynamically unstable. If the optimal size for a zygote is say 2 mg, then the optimal size for each gamete is 1 mg, so they sum to 2 mg upon fusing. Therefore, conditional on type A having a gamete size of 1 mg, then the optimal gamete size for type B is also 1 mg, and a mutation within B that deviates from this optimal size of 1 mg will not increase when rare. Conversely, conditional on type B having a gamete size of 1 mg, the optimal gamete size for type A is also 1 mg, and any mutation within A that deviates from this size will not increase when rare. Thus the state where both type A and type B have a gamete size of 1 mg is an ESS. But this state is not dynamically stable to perturbation. If the gamete size of type A decreases somewhat, then selection favors increasing the gamete size within type B to compensate, in an escalating progression leading to increasingly divergent gamete sizes, culminating in one being as small as possible and the other as big as needed to fully provision the zygote. See: G. Parker, R. Baker, and V. Smith, 1972, The origin and evolution of gamete dimorphism and the male-female phenomenon, *J. Theor. Biol.* 36:529–53; N. Knowlton, 1974, A note on the evolution of gamete dimorphism, *J. Theor. Biol.* 46:283–85; G. Bell, 1978, The evolution of anisogamy, *J. Theor. Biol.* 73:247–70; J. Maynard Smith, 1978, *The Evolution of Sex,* Cambridge University Press; R. Hoekstra, 1980, Why do organisms produce gametes of only two different sizes? Some theoretical aspects of the evolution of anisogamy, *J. Theor. Biol.* 87:785–93; G. Parker, 1982, Why so many tiny sperm? The maintenance of two sexes with internal fertilization, *J. Theor. Biol.* 96:281–94; H. Matsuda and P. Abrams, 1999, Why are equally sized gametes so rare? The instability of isogamy and the cost of anisogamy, *Evol. Ecol. Research* 1:769–84; I. Eshel and E. Akin, 1983, Coevolutionary instability of mixed Nash solutions, *J. Math. Biol.* 18:123–34; J. Madsen and D. M. Waller, 1983, A note on the evolution of gamete dimorphism in algae, *Amer. Natur.* 121:443–47.

7. J. Butler, 1990, *Gender Trouble,* Routledge, rpt. on pp. 80–88 in C. Gould, ed., 1997, *Key Concepts in Critical Theory: Gender,* Humanities Press; S. Kessler and W. McKenna, 1978, *Gender: An Ethnomethodological Approach,* University of Chicago Press.

8. C. Francis, E. L. P. Anthony, J. Brunton, and T. H. Kunz, 1994, Lactation in male fruit bats, *Nature* 367:691–92.

3: SEX WITHIN BODIES

1. D. Policansky, 1982, Sex change in plants and animals, *Ann. Rev. Ecol. Syst.* 13:471–95.

2. R. Warner, 1984, Mating behavior and hermaphrodism in coral reef fishes, *American Scientist* 72:128–36; G. Mead, E. Bertelson, and D. M. Cohen, 1964, Reproduction among deep-sea fishes, *Deep Sea Research* 11:569–96.

3. D. Robertson and R. Warner, 1978, Sexual patterns in the labroid fishes of the western Caribbean: II. The parrotfishes (Scaridae), *Smithsonian Contributions to Zoology* 255:1–26.

4. R. Warner and S. Hoffman, 1980, Local population size as a determinant of a mating system and sexual composition in two tropical reef fishes (*Thalassoma* spp.), *Evolution* 34:508–18.

5. D. Robertson, 1972, Social control of sex reversal in a coral reef fish, *Science* 1977:1007–9; J. Godwin, D. Crews, and R. Warner, 1996, Behavioral sex change in the absence of gonads in a coral reef fish, *Proc. R. Soc. Lond.,* ser. B, 263:1683–88; J. Godwin, R. Sawby, R. Warner, D. Crews, and M. Grober, 1999, Hypothalamic arginine vasotocin mRNA abundance variation across the sexes and with sex change in a coral reef fish (unpublished manuscript).

6. H. Fricke and S. Fricke, 1977, Monogamy and sex change by aggressive dominance in coral reef fish, *Nature* 266:830–32; J. Moyer and A. Nakazono, 1978, Protandrous hermaphrodism in six species of the amenonefish genus *Amphiprion* in Japan, *Japan. J. Ichthyology* 25:101–6. See also: J. Moyer and A. Nakazono, 1978, Population structure, reproductive behavior and protogynous hermaphrodism in the angelfish *Centropyge interruptus* at Miyake-jima, Japan, *J. Ichthyology* 25:25–39.

7. A test of this idea may be possible by comparing with clown fish, such as those from Batavia Bay in Indonesia, who do live in anemones large enough to support more than one pair of adults and in whom polygamy may occur instead of monogamy. A table reviewing these species appears in J. Roughgarden, 1975, Evolution of marine symbiosis—a simple cost-benefit model, *Ecology* 56:1201–8.

8. E. Fischer, 1980, The relationship between mating system and simultaneous hermaphrodism in the coral reef *Hypoplectrus nigricans* (Seranidae), *Anim. Behav.* 28:620–33; P. Pressley, 1981, Pair formation and joint territoriality in a simultaneous hermaphrodite: The coral reef fish *Serranus tigrius, Z. Tierpsychol.* 56:33–46.

9. G. Mead, E. Bertelson, and D. M. Cohen, 1964, Reproduction among deep-sea fishes, *Deep Sea Research* 11:569–96.

10. Discussion based on T. Kuamura, Y. Nakashima, and Y. Yogo, 1994, Sex change in either direction by growth-rate advantage in the monogamous coral goby, *Paragobiodon echinocephalus, Behav. Ecol.* 5:434–38.

11. Ibid.

12. Discussion based on P. Munday, M. Caley, and G. Jones, 1998, Bi-directional sex change in a coral-dwelling goby, *Behav. Ecol. Sociobiol.* 43:371–77.

13. A population with both females and hermaphrodites is called gynodioecious, and a population with both males and hermaphrodites is called androdioecious. See: M. Geber, T. Dawson, and L. Delph, eds., 1998, *Gender and Sexual Dimorphism in Flowering Plants,* Springer Verlag; D. Charlesworth and M. Morgan, 1991, Alloca-

tion of resources to sex functions in flowering plants, *Phil. Trans. R. Soc. Lond.*, ser. B, 332:91–102; J. Pannell, 1997, The maintenance of gynodioecy and androdioecy in a metapopulation, *Evolution* 51: 10–20; A. Liston, L. H. Rieseberg, and T. Elias, 1990, *Datisca glomerata* is functionally androdioecious, *Nature* 343:641–42.

14. Qing-Jun Li, and Zai-Fu Xu, W. John Kress, Yong-Mei Xia, Ling Zhang, Xiao-Bao Deng, Jiang-Yun Gao, and Zhi-Lin Bai, 2001, Flexible style that encourages outcrossing, *Nature* 410:432; S. Barrett, 2002, The evolution of plant sexual diversity, *Nature Reviews: Genetics* 3:274–84.

15. Medical doctors have introduced the terms "true hermaphrodism" for the gonadally intersexed and "pseudohermaphroditism" for the genitally intersexed. These terms are useless. Nothing "true" or "pseudo" is involved, and these terms layer a value judgment on what should be simply a matter of description.

16. On white-tailed deer, see: C. Crispens Jr. and J. Doutt, 1973, Sex cromatin in antlered female deer, *J. Wildlife Management* 37:422–23; J. C. Donaldson and J. Doutt, 1965, Antlers in female white-tailed deer: A four-year study, *J. Wildlife Management* 29:699–705; J. Doutt, and J. Donaldson, 1959, An antlered doe with possible masculinizing tumor, *J. Mammalogy* 40:230–36; D. Taylor, J. Thomas, and R. Marburger, 1964, Abnormal antler growth associated with hypogonadism in white-tailed deer of Texas, *Amer. J. Veterinary Research* 25:179–85; J. Thomas, R. Robinson, and R. Marburger, 1970, *Studies in Hypogonadism in White-tailed Deer of the Central Mineral Region of Texas,* Texas Parks and Wildlife Department Technical Series No. 5, Texas Parks and Wildlife Department; J. Thomas, R. Robinson, and R. Marburger, 1965, Social behavior in a white-tailed deer herd containing hypogonadal males, *J. Mammalogy* 46:314–27; J. Thomas, R. Robinson, and R. Marburger, 1964, Hypogonadism in white-tailed deer in the central mineral region of Texas, in J. B. Trefethen, ed., *Transactions of the North American Wildlife and Natural Resources Conference* 29:225–36, Wildlife Management Institute; W. Wishart, 1985, Frequency of antlered white-tailed does in Camp Wainright, Alberta, *J. Mammalogy* 35:486–88; G. Wislocki, 1954, Antlers in female deer, with a report on three cases in *Odocoileus, J. Mammalogy* 35:486–95; G. Wislocki, 1956, Further notes on antlers in female deer of the genus *Odocoileus, J. Mammalogy* 37:231–35.

17. On black-tailed deer, see: I. McT. Cowan, 1946, Antlered doe mule deer, *Canadian Field-Naturalist* 60:11–12; B. Wong and K. Parker, 1988, Estrus in black-tailed deer, *J. Mammalogy* 69:168–71. On other deer, see: D. Chapman, N. Chapman, M. Horwood, and E. Masters, 1984, Observations on hypogonadism in a perruque silka deer *(Cervus nippon), J. Zoology, London* 204:579–84; G. Lincoln, R. Youngson, and R. Short, 1970, The social and sexual behavior of the red deer stag, *J. Reproduction and Fertility* (Suppl.) 11:71–103; D. Wurster-Hill, K. Benirschke, and D. Chapman, 1983, Abnormalities of the X chromosome in mammals, pp. 283–300 in S. Sandberg, ed., *Cytogenetics of the Mammalian X Chromosome* (Part B), Alan Liss. On moose, see: A. Bubenik, G. Bubenik, and D. Larsen, 1990, Velericorn antlers on a mature male moose *(Alces a. gigas), Alces* 26:115–28; O. Murie, 1928, Abnormal growth of moose antlers, *J. Mammalogy* 9:65; W. Wishard, 1990, Velvet-antlered female moose *(Alces alces), Alces* 26:64–65.

18. G. Sharman, R. Hughes, and D. Cooper, 1990, The chromosomal basis of sex differentiation in marsupials, *Australian J. Zoology* 37:451–66.

19. G. Pfaffenberger, F. Weckerly, and T. Best, 1986, Male pseudohermaphroditism in a population of kangaroo rats, *Dipodomys ordii, Southwestern Naturalist* 31:124–26.

20. J. Baker, 1925, On sex-intergrade pigs: Their anatomy, genetics, and developmental physiology, *Brit. J. Experimental Biology* 2:247–63; J. Baker, 1928, Notes on New Hebridean customs, with special reference to the intersex pig, *Man* 28:113–18; J. Baker, 1928, A new type of mammalian intersexuality, *Brit. J. Experimental Biology* 6:56–64; W. Rodman, 1996, The boars of Bali Ha'i: Pigs in paradise, pp. 158–67 in J. Bonnemaison, C. Kaufmann, K. Huffman, and D. Tryon, eds., *Arts of Vanuatu,* University of Hawaii Press.

21. B. Bagemihl, 1999, *Biological Exuberance,* St. Martin's Press, p. 234.

22. M. Cattet, 1988, Abnormal differentiation in black bears *(Ursus americanus)* and brown bears *(Ursus arctos), J Mammalogy* 69:849–52.

23. L. Frank, 1996, Female masculinization in the spotted hyena: Endocrinology, behavioral ecology, and evolution, pp. 78–131 in J. Gittleman, ed., *Carnivore Behavior and Evolution,* vol. 2, Cornell University Press; L. Frank, 1997, Evolution of genital masculinization: Why do female hyaenas have such a large 'penis'? *Trends Ecol. Evol.* 12:58–62.

24. M. Harrison, 1939, Reproduction in the spotted hyena, *Crocuta crocuta* (Erxleben.), *Phil. Trans. Roy. Soc. Lond.,* ser. B, 230:1–78.

25. L. Frank and S. Glickman, 1994, Giving birth through a penile clitoris: Parturition and dystocia in the spotted hyena *(Crocuta crocuta), J. Zoology London* 206:525–31; L. Frank, M. Weldele, and S. Glickman, 1995, Masculinization costs in hyenas, *Nature* 377:584–85.

26. M. East, H. Hofer, and W. Wickler, 1993, The erect "penis" is a flag of submission in a female-dominated society: Greetings in Serengeti spotted hyenas, *Behav. Ecol. Sociobiol.* 33:355–70; H. Hofer and M. East, 1995, Virilized sexual genitalia as adaptations of female spotted hyenas, *Revue Suisse de Zoologie* 102:895–906.

27. Frank, 1996, Female masculinization in the spotted hyena.

28. Bagemihl, 1999, *Biological Exuberance,* pp. 336–37; H. Butler, 1967, The oestrus cycle of the senegal bush baby *(Galago senagalensis senegalensis)* in the Sudan, *J. Zoology London* 151:143–62; G. Doyle, 1974, Behavior of Prosimians, *Behavior of Nonhuman Primates* 5:154–353; R. Martin, G. Doyle, and A. Walker, eds., 1974, *Prosimian Biology,* University of Pittsburgh Press; D. Lipshitz, 1996, Male copulatory patterns in the Lesser Bushbaby *(Galago moholi)* in captivity, *Int. J. Primatology* 17:987–1000.

29. D. Dewsbury and J. Pierce, Jr., 1989, Copulatory patterns of primates as viewed in broad mammalian perspective, *Amer. J. Primatology* 17:51–72.

30. See the photograph of the elongated clitoris of a white-bellied spider monkey *Ateles belzebuth* on p. 112 in N. Rowe, 1996, *The Pictorial Guide to the Living Primates,* Pogonias Press.

31. See the photograph of male and female woolly monkeys, *Lagothrix lagotricha,* including an elongated and pink-tipped clitoris on p. 116 in ibid.

32. Ibid.

33. For the respective cetaceans discussed in this paragraph, see: M. Nishiwaki, 1953, Hermaphroditism in a dolphin *(Prodelphinus caeruleo-albus), Scientific Re-*

ports of the Whales Research Institute 8:215–18; R. Tarpley, G. Jarrell, J. George, J. Cubbage, and G. Stott, 1995, Male pseudohermaphroditism in the bowhead whale, *Balaena mysticetus, J. Mammalogy* 76:1267–75; J. Bannister, 1963, An intersexual fin whale *Balaenoptera physalus* (L.) from South Georgia, *Proc. Zool. Soc. London* 141:811–22; S. De Guise, A. Lagacé, and P. Béland, 1994, True hermaphroditism in a St. Lawrence beluga whale *(Delphinapterus leucas), J. Wildlife Diseases* 30:287–90.

34. R. Jiménez, M. Burgos, A. Sánchez, A. Sinclair, F. Alarcón, J. Marin, E. Ortega, and R. D. de la Guardia, 1993, Fertile females of the mole *Talpa occidentalis* are phenotypic intersexes with ovotestes, *Development* 118:1303–11; A. Sánchez, M. Mullejos, M. Burgos, et al., 1996, Females of four mole species of genus *Talpa* (Insectivora, Mammalia) are true hermaphrodites with ovotestes, *Mol. Reprod. Dev.* 44:289–94.

4: SEX ROLES

1. E. Bertelsen, 1951, The ceratioid fishes: Ontogeny, taxonomy, distribution, and biology, *Dana Report* 39; T. Pietsch, 1976, Dimorphism, parasitism and sex: Reproductive strategies among deep sea ceratioid anglerfishes, *Copeia* 781–93.

2. B. Saemundsson, 1922, Zoologiske meddelelser fra Island: XIV. ll Fiske, nye for Island, og supplerende om andre, tidligere kendte, *Vidensk. Medd. fra Dansk Naturh. Foren. Bd.* 74:159–201.

3. C. Regen, 1925, Dwarfed males parasitic on the females in oceanic anglerfishes (Pediculati Ceratioidea), *Proc. R. Soc. Lond.,* ser. B, 97:386–400.

4. A. Vincent, I. Ahnesjö, A. Berglund, and G. Rosenqvist, 1992, Pipefishes and seahorses: Are they all sex role reversed? *Trends Ecol. Evol.* 7:237–41.

5. T. Clutton-Brock and A. C. J. Vincent, 1991, Sexual selection and the potential reproductive rates of males and females, *Nature* 351:58–60.

6. The sex ratio is meaningful when males and females are separate individuals. For hermaphroditic species, the analogous concept is sex allocation, which refers to how much of an individual's energy is placed into its female part and how much into its male part. For a theory explaining a fifty-fifty sex ratio species, see R. Fisher, 1927, *The Genetical Theory of Natural Selection,* Oxford University Press; for a fifty-fifty sex allocation, see E. Charnov, 1982, *The Theory of Sex Allocation,* Princeton University Press; and J. Roughgarden, 1991, The evolution of sex, *Amer. Natur.* 138:934–53.

7. A. C. J. Vincent and L. M. Sadler, 1995, Faithful pairbonds in wild seahorses, *Hippocampus whitei, Anim. Behav.* 50:1557–69.

8. S. T. Emlen, P. H. Wrege, and M. S. Webster, 1998, Cuckoldry as a cost of polyandry in the sex-role reversed wattled jacana, *Jacana jacana, Proc. R. Soc. Lond.,* ser. B, 265:2359–64. See also C. Yoon, 1998, In this battle of the sexes, the females win, *New York Times,* Dec. 22, 1998, p. D3; and photos at http://www.nwf.org/intlwild/2000/jacanaja.html.

9. See: D. J. Delehanty, R. C. Fleisher, M. A. Colwell, and L. W. Oring, 1996, Sex-role reversal and the absence of extra-pair fertilization in Wilson's phalarope, *Anim. Behav.* 55: 995–1002; L. W. Oring, J. M. Reed, and S. J. Maxson, 1994, Copulation patterns and mate guarding in the sex-role reversed, polyandrous spotted sandpiper *Actitis macularia, Anim. Behav.* 47:1065–72.

5: TWO-GENDER FAMILIES

1. L. Greenhouse, 2000, Case on visitation rights hinges on defining family, *New York Times,* Jan. 4, p. A11.

2. Louis Farrakhan, quoted in F. Clines, 2000, Families arrive in Washington for march called by Farrakhan, *New York Times,* Oct. 16.

3. M. Janofsky, 2002, Custody case in California paves way for "fathers," *New York Times,* June 8.

4. Ibid.

5. M. Frye, 1983, *The Politics of Reality: Essays in Feminist Theory,* Crossing Press, rpt. C. Gold, ed., *Key Concepts in Critical Theory: Gender,* Humanities Press, pp. 91–102.

6. Discussion based on P. Sherman, 1989, Mate guarding as paternity insurance in Idaho ground squirrels, *Nature* 338:418–20.

7. See T. R. Birkhead, 2000, Defining and demonstrating post-copulatory female choice—again, *Evolution* 54:1057–60; T. R. Birkhead, 2000, She knows what she wants, *New Scientist* 2244:28–31.

8. B. Smuts and R. Smuts, 1993, Male aggression and sexual coercion of females in nonhuman primates and other mammals: Evidence and theoretical implications, *Adv. Study Behav.* 22:1–63.

9. Ibid.

10. J. Mitani, 1985, Mating behavior of male orangutans in the Kutai Reserve, *Anim. Behav.* 33:392–402.

11. See: T. Clutton-Brock and G. Parker, 1995, Sexual coercion in animal societies, *Anim. Behav.* 49:1345–65; among birds, sexual coercion by males occurs in ducks and in the colorful long-billed South African white-fronted bee-eater *(Merops bullockoides).* F. McKinney, S. R. Derrickson, and P. Mineau, 1983, Forced copulation in waterfowl, *Behavior* 86:250–94; S. T. Emlen and P. H. Wrege, 1986, Forced copulations and intra-specific parasitism: Two costs of living in the white fronted bee-eater, *Ethology* 71:2–29.

12. T. Clutton-Brock, 1989, Review lecture: Mammalian mating systems, *Proc. R. Soc. Lond.,* ser. B, 236:339–72.

13. Discussion based on S. M. Smith, 1991, *The Black-Capped Chickadee: Behavioral Ecology and Natural History,* Comstock Publishing; K. Ratcliffe and L. Ratcliffe, 1996, Female initiated divorce in a monogamous songbird: Abandoning mates for males of higher quality, *Proc. R. Soc. Lond.,* ser. B, 263:351–54.

14. Ibid.

15. M. Lindén, 1991, Divorce in great tits—chance or choice? An experimental approach, *Amer. Natur.* 138:1039–48.

16. F. Cezilly and R. G. Nager, 1995, Comparative evidence for a positive association between divorce and extra-pair paternity in birds, *Proc. R. Soc. Lond.,* ser. B, 262:7–12.

17. Ibid.

18. T. Clutton-Brock, 1989, Review lecture: Mammalian mating systems, *Proc. R. Soc. Lond.,* ser. B, 236:339–72.

19. See: P. Johnsgard, 1997, *The Avian Brood Parasites: Deception at the Nest,* Oxford University Press.

20. S. Vehrencamp, 1978, The adaptive significance of communal nesting in groove-billed Anis, *Behav. Ecol. Sociobiol.* 4:1–33; B. S. Bowen, R. R. Koford, and S. L. Vehrencamp, Breeding roles and pairing patterns within communal groups of groove-billed Anis, *Anim. Behav.* 34:347–66.

21. Discussion based on J. Terborgh and A. Goldizen, 1985, On the mating system of the cooperatively breeding saddle-backed tamarin *Saguinus fuscicollis, Behav. Ecol. Sociobiol.* 16:293–99.

22. B. Keane, P. Waser, N. Creel, L. Elliott, and D. Minchella, 1994, Subordinate reproduction in dwarf mongooses, *Anim. Behav.* 47:65–75.

23. L. Whittingham, P. Dunn, and R. Magrath, 1997, Relatedness, polyandry and extra-group paternity in the cooperatively-breeding white-browed scrubwren *(Sericornis frontalis), Behav. Ecol. Sociobiol.* 40:261–70; H. Gibbs, C. Bullough, and A. Goldizen, 1994, Parentage analysis of multi-male social groups of Tasmanian native hens *(Tribonyx mortierii)*: Genetic evidence for monogamy and polyandry, *Behav. Ecol. Sociobiol.* 35:363–71; J. Faaborg, P. Parker, L. DeLay, T. de Vries, J. Bednarz, M. Paz, J. Naranjo, and T. Waite, 1995, Confirmation of cooperative polyandry in the Galapagos hawk *(Buteo galapagoensis), Behav. Ecol. Sociobiol.* 36:83–90; T. Burke, N. Davies, M. Bruford, and B. Hatchwell, 1989, Parental care and mating behavior of polyandrous dunnocks *Prunella modularis* related to paternity by DNA fingerprinting, *Nature* 338:249–51; I. Jamieson, J. Quinn, P. Rose, and B. White, 1994, Shared paternity is a result of an egalitarian mating system in a communally breeding bird, the pukeko, *Proc. R. Soc. Lond.* 257:271–77; W. Piper and G. Slater, 1993, Polyandry and incest avoidance in the cooperative stripe-backed wren of Venezuela, *Behavior* 124:227–47.

24. C. Packer and A. Pusey, 1997, Divided we fall: Cooperation among lions, *Scientific American* (May), 52–59.

25. S. Lewis and A. Pusey, 1997, Factors influencing the occurrence of communal care in plural breeding mammals, pp. 335–62 in N. G. Solomon and J. A. French, eds., *Cooperative Breeding in Mammals,* Cambridge University Press.

26. C. J. Manning, D. A. Dewsbury, E. K. Wakeland, and W. K. Potts, 1995, Communal nesting and communal nursing in house mice *(Mus musculus domesticus), Anim. Behav.* 50:741–51.

27. G. Wilkonson, 1990, Food sharing in vampire bats, *Scientific American* (February), 76–82.

28. Sometimes cooperation is distinguished from altruism, as altruism is giving up something to help someone else, whereas cooperation entails simply not competing with someone else. I'm using cooperation in the wider sense to include both helping and not hurting. See R. Trivers, 1984, *Social Evolution,* Benjamin-Cummings.

29. T. H. Clutton-Brock and G. A. Parker, 1995, Punishment in animal societies, *Nature* 373:209–15.

30. This suggestion extends the notion of public information now demonstrated to be involved in the choice of breeding sites. See B. Doligez, E. Danchin, and J. Clobert, 2002, Public information and breeding habitat selection in a wild bird population, *Science* 297:1168–69.

31. See: J. Jarvia, M. O'Riain, N. Bennett, and P. Sherman, 1994, Mammalian eusociality: A family affair, *Trends Ecol. Evol.* 9:47–51; E. Lacey and P. W. Sherman, 1997, Cooperative breeding in naked mole-rats: Implications for vertebrate and in-

vertebrate sociality, pp. 267–300 in Solomon and French, eds., *Cooperative Breeding in Mammals*.

32. Lacey and Sherman, 1997, Cooperative breeding in naked mole-rats.

33. Discussion based on C. Brown and M. Bomberger Brown, 1996, *Coloniality in the Cliff Swallow: The Effect of Group Size on Social Behavior,* University of Chicago Press.

34. Ibid.

35. S. Vehrencamp, 1983, Optimal degree of skew in cooperative societies, *Amer. Zool.* 23:327–35; H. Reeve, S. T. Emlen, and L. Keller, 1998, Reproductive sharing in animal societies: Reproductive incentives or incomplete control by dominant breeders? *Behav. Ecol.* 9:267–78; S. Emlen, 1995, An evolutionary theory of the family, *Proc. Nat. Acad. Sci. (USA)* 92:8092–99.

36. W. D. Hamilton, 1964, The genetical theory of social behavior, *J. Theor. Biol.* 7:1–52.

37. S. Vehrencamp, 1983, A model for the evolution of despotic versus egalitarian societies, *Anim. Behav.* 31:667–82.

38. L. Whittingham, P. Dunn, and R. Magrath, 1997, Relatedness, polyandry and extra-group paternity in the cooperatively-breeding white-browed scrubwren *(Sericornis frontalis)*, *Behav. Ecol. Sociobiol.* 40:261–70.

39. T. H. Clutton-Brock, 1998, Reproductive skew, concessions and limited control, *Trends Ecol. Evol.* 13:288–92; T. H. Clutton-Brock, P. N. M. Brotherton, A. F. Russell, M. J. O'Riain, D. Gaynor, R. Kansky, A. Griffin, M. Manser, L. Sharpe, G. M. McIlrath, T. Small, A. Moss, and S. Monfort, 2001, Cooperation, control, and concession in meerkat groups, *Science* 291:478–81.

40. L. Keller and H. K. Reeve, 1994, Partitioning of reproduction in animal societies, *Trends Ecol. Evol.* 9:98–102.

6: MULTIPLE-GENDER FAMILIES

1. R. D. Howard, 1978, The evolution of mating strategies in bullfrogs, *Rana catesbeiana, Evolution* 32:859–71; R. D. Howard, 1981, Sexual dimorphism in bullfrogs, *Ecology* 62:303–10; R. D. Howard, 1984, Alternative mating behaviors in young male bullfrogs, *Amer. Zool.* 24:397–406.

2. S. L. Lance and K. D. Wells, 1993, Are spring peeper satellite males physiologically inferior to calling males? *Copeia* 1162–66.

3. A. Bass, 1992, Dimorphic male brains and alternative reproductive tactics in a vocalizing fish, *TINS* 15:139–45; J. L. Goodson and A. H. Bass, 2000, Forebrain peptides modulate sexually polymorphic vocal circuitry, *Nature* 403:769.

4. M. R. Gross, 1985, Disruptive selection for alternative life histories in salmon, *Nature* 313:47–48.

5. F. F. Darling, 1937, *A Herd of Red Deer,* Oxford University Press.

6. G. Lincoln, R. Youngson, and R. Short, 1970, The social and sexual behavior of the red deer stag, *J. Reproduction and Fertility,* suppl. 11:71–103.

7. For Lake Opinicon studies, see: M. R. Gross, 1982, Sneakers, satellites and parentals: Polymorphic mating strategies in North American sunfishes, *Z. Tierpsychol.* 60:1–26; M. R. Gross, 1991, Evolution of alternative reproductive strategies: Frequency-dependent sexual selection in male bluegill sunfish, *Phil. Trans. R. Soc. Lond.,* ser. B, 332:59–66. For Lake Cazenovia studies, see: W. J. Dominey, 1980, Fe-

male mimicry in bluegill sunfish—a genetic polymorphism? *Nature* 284:546–48; W. J. Dominey, 1981, Maintenance of female mimicry as a reproductive strategy in bluegill sunfish *(Lepomis macrochirus), Environ. Biol. Fishes* 6:59–64.

8. Gross, 1991, Evolution of alternative reproductive strategies.

9. Dominey, 1981, Maintenance of female mimicry as a reproductive strategy in bluegill sunfish.

10. M. Taborsky, B. Hudde, and P. Wirtz, 1987, Reproductive behavior and ecology of *Symphodus (Crenilabrus) ocellatus,* a European wrasse with four types of male behavior, *Behavior* 102:82–118.

11. S. Alonzo, M. Taborsky, and P. Wirtz, 2000, Male alternative reproductive behaviours in a Mediterranean wrasse, *Symphodus ocellatus*: Evidence from otoliths for multiple life-history pathways, *Evol. Ecol. Research* 2:997–1007.

12. See: G. Barlow, 2000, *The Cichlid Fishes: Nature's Grand Experiment in Evolution,* Perseus.

13. R. Oliveira and V. Almada, 1998, Mating tactics and male-male courtship in the lek-breeding cichlid *Oreochromis mossambicus, J. Fish Biology* 52:1115–29.

14. M. Moore, D. Hews, and R. Knapp, 1998, Hormonal control and evolution of alternative male phenotypes: Generalizations of models for sexual differentiation, *Amer. Zool.* 38:133–51.

15. The low progesterone group had 0.1–1.0 ng/ml; the high progesterone group, 10–100 ng/ml.

16. D. DeNardo and B. Sinervo, 1994, Effects of corticosterone on activity and home-range size of free-ranging male lizards, *Hormones and Behavior* 28:53–65 and 28:273–87.

17. J. Kopachena and J. Falls, 1993, Aggressive performance as a behavioral correlate of plumage polymorphism in the white-throated sparrow *(Zonotrichia albicollis), Behavior* 124:249–66. See also: J. Lowther, 1961, Polymorphism in the white-throated sparrow *Zonotrichia albicollis* (Gmelin), *Can. J. Zool.* 39:281–92; H. Thorneycroft, 1975, A cytogenetic study of the white-throated sparrow, *Zonotrichia albicollis, Evolution* 29:611–21; A. Houtman and J. Falls, 1994, Negative assortative mating in the white-throated sparrow *Zonotrichia albicollis:* The role of mate choice and intra-sexual competition, *Anim. Behav.* 48:377–83; T. DeVoogd, A. Houtman, and J. Falls, 1995, White-throated sparrow morphs that differ in song production rate also differ in the anatomy of some song-related brain areas, *Neurobiology* 28:202–13.

18. B. Sinervo and C. M. Lively, 1996, The rock-paper-scissors game and the evolution of alternative male strategies, *Nature* 380:240–43; B. Sinervo, E. Sevensson, and T. Comendant, 2000, Density cycles and an offspring quantity and quality game driven by natural selection, *Nature* 406:985–88. See also photographs and movie clips on the web at http://www.biology.ucsc.edu/barrylab/lizardland/male_lizards .overview.html.

The paper cites a carrying capacity *K* for orange females as 0.70 and for yellow females as 1.18 without giving units. I'm guessing the appropriate unit of area for the rocky outcropping on which they live is 1 square meter.

19. Sinervo website, http://www.biology.ucsc.edu/barrylab/lizardland/male_ lizards.overview.html.

20. Ibid.

21. The orange female has been termed an "*r*-strategist" and the yellow female a "*K*-strategist." See: R. H. MacArthur, 1962, Some generalized theorems of natural selection, *Proc. Nat. Acad. Sci. (USA)* 231:123–38; J. Roughgarden, 1971, Density-dependent natural selection, *Ecology* 52:453–68.

22. Sinervo, Sevensson, and Comendant, 2000, Density cycles and an offspring quantity and quality game driven by natural selection.

23. R. Alatalo, L. Gustafsson, and A. Lundberg, 1994, Male coloration and species recognition in sympatric flycatchers, *Proc. Roy. Soc. Lond.*, ser. B, 256: 113–18.

24. T. Slagsvold and G. Sætre, 1991, Evolution of plumage color in male pied fly-catchers *(Ficedula hypoleuca)*: Evidence for female mimicry, *Evolution* 45:910–17. See also: E. Huhta and R. Alatalo, 1993, Plumage colour and male-male interactions in the pied flycatcher, *Anim. Behav.* 45: 511–18.

25. G. Sætre and T. Slagsvold, 1996, The significance of female mimicry in male contests, *Amer. Natur.* 147:981–95.

26. H. Hakkarainen, E. Korpimäki, E. Huhta, and P. Palokangas, 1993, De-layed maturation in plumage color: Evidence for the female-mimicry hypothesis in the kestrel, *Behav. Ecol. Sociobiol.* 33:247–51. (Quotations below from this source.)

27. See: R. Mason and D. Crews, 1985, Female mimicry in garter snakes, *Nature* 316:59–60; R. Shine, D. O'Connor, and R. Mason, 2000, Female mimicry in garter snakes: Behavioural tactics of "she-males" and the males that court them, *Can. J. Zool.* 78:1391–96; R. Shine, P. Harlow, M. Lemaster, I. Moore, and R. Mason, 2000, The transvestite serpent: Why do garter snakes court (some) other males? *Anim. Behav.* 59:349–59; for pictures and general information see http://www.naturenorth.com/spring/creature/garter/Fgarter.html.

28. Quotations from Shine, Harlow, Lemaster, Moore, and Mason, 2000, The transvestite serpent; Shine, O'Connor, and Mason, 2000, Female mimicry in garter snakes.

29. For example, see D. Hilton, 1987, A terminology for females with color pat-terns that mimic males, *Ent. News* 98:221–23.

30. M. Taborsky, 1994, Sneakers, satellites and helpers: Parasitic and cooperative behavior in fish reproduction, *Adv. Study Behav.* 23:1–100; M. Taborsky, 1998, Sperm competition in fish: 'Bourgeois' males and parasitic spawning, *Trends Ecol. Evol.* 13:222–27.

31. V. Brawn, 1961, Reproductive behavior of the cod *(Gadus callarias L)*, *Behavior* 18:177–98; C. Wedekind, 1996, Lek-like spawning behaviour and different female mate preferences in roach *(Rutilus rutilus)*, *Behavior* 102:82–118.

32. G. D. Constanz, 1985, Alloparental care in the tesselated darter: *Etheostoma olmstedi* (Pisces: Percidae), *Environ. Biol. Fishes* 14:175–83.

33. M. Taborsky and D. Limberger, 1981, Helpers in fish, *Behav. Ecol. Sociobiol.* 8:143–45; D. Limberger, 1983, Pairs and harems in a cichlid fish, *Lamprologus brichardi, Z. Tierpsychol.* 62:115–44; M. Taborsky, 1984, Broodcare helpers in the cichlid fish *Lamprologus brichardi*: Their costs and benefits, *Anim. Behav.* 32:1236–52.

34. K. McKaye and N. McKaye, 1977, Communal care and kidnapping of young by parental cichlids, *Evolution* 31:674–81; J. Coyne and J. Sohn, 1978, Interspecific

brood care in fishes: Reciprocal altruism or mistaken identity? *Amer. Natur.* 112:447–50.

35. R. Dawkins and J. Krebs, 1978, Animal signals: Information or manipulation? pp. 282–309 in J. R. Krebs and N. B. Davies, eds., *Behavioral Ecology—an Evolutionary Approach,* Blackwell; Taborsky, 1994, Sneakers, satellites, and helpers.

36. See: P. Ehrlich and J. Roughgarden, 1987, *The Science of Ecology,* Macmillan, pp. 310–18; W. Wickler, 1968, *Mimicry in Plants and Animals,* McGraw-Hill.

37. R. Bleiweiss, 1992, Widespread polychromatism in female sunangel hummingbirds (*Heliangelus:* Trochilidae), *Biol. J. Linnean Soc.* 45:291–314.

38. R. Bleiweiss, 2001, Asymmetrical expression of transsexual phenotypes in hummingbirds, *Proc. R. Soc. Lond.,* ser. B, 268:639–46.

39. R. Bleiweiss, 1999, Joint effects of feeding and breeding behavior on trophic dimorphism in hummingbirds, *Proc. R. Soc. Lond.,* ser. B, 266:2491–97.

40. T. Amundsen, 2000, Why are female birds ornamented? *Trends Ecol. Evol.* 15:149–55.

41. E. Morton, 1989, Female hooded warbler plumage does not become more male-like with age, *Wilson Bulletin* 101:460–62.

42. D. Niven, 1993, Male-male nesting behavior in hooded warblers, *Wilson Bulletin* 105:190–93.

43. See also: B. Stutchbury and J. Howlett, 1995, Does male-like coloration of female hooded warblers increase nest predation? *Condor* 97:559–64; B. Stutchbury, 1998, Extra-pair mating effort of male hooded warblers, *Wilsonia cirtina, Anim. Behav.* 55:553–61.

7: FEMALE CHOICE

1. E. Forsgren, 1997, Female sand gobies prefer good fathers over dominant males, *Proc. R. Soc. Lond.,* ser. B, 264:1283–86.

2. Discussion based on R. Warner, F. Lejeune, and E. van den Berghe, 1994, Dynamics of female choice for parental care in a fish species where care is facultative, *Behav. Ecol.* 6:73–81. (Quotation below from this source.)

3. Discussion based on R. Knapp and R. Sargent, 1989, Egg-mimicry as a mating strategy in the fantail darter, *Etheostoma flabellare:* Females prefer males with eggs, *Behav. Ecol. Sociobiol.* 321–26.

4. Discussion based on N. B. Davies, I. R. Hartley, B. J. Hatchwell, and N. E. Langmore, 1996, Female control of copulations to maximize male help: A comparison of polygynandrous alpine accentors, *Prunella collaris,* and dunnocks, *P. modularis, Anim. Behav.* 51:27–47.

5. T. Birkhead and A. Moller, 1992, *Sperm Competition in Birds,* Academic Press.

6. M. Rodriguez-Girones and M. Enquist, 2001, The evolution of female sexuality, *Anim. Behav.* 61:695–704.

7. Galán, 2000, Females that imitate males: Dorsal coloration varies with reproductive stage in female *Podarcis bocagei* (Lacertidae), *Copeia* 819–25.

8. G. Watkins, 1997, Inter-sexual signalling and the functions of female coloration in the tropidurid lizard *Microlophus occipitalis, Anim. Behav.* 53:843–52.

9. W. Cooper Jr. and N. Greenberg, 1992, Reptilian coloration and behavior, pp. 298–422 in C. Gans and D. Crews, eds., *Biology of the Reptilia: Hormones, Brain, and Behavior,* University of Chicago Press.

10. See: C. Johnson, 1975, Polymorphism and natural selection in ischnuran damselflies, *Evol. Theory* 1:81–90; Robin Corcoran, 1995, "Intraspecific sexual mimicry in insects," available at http://www.colostate.edu/Depts/Entomology/courses/en507/student_papers_1995/corcoran.html.

11. Discussion based on H. Robertson, 1985, Female dimorphism and mating behaviour in a damselfly, *Ischnura ramburi*: Females mimicking males, *Anim. Behav.* 33:805–9.

12. P. S. Corbet, 1980, Biology of the Odonata, *Ann. Rev. Entomol.* 25:189–217.

13. B. Hinnekint, Population dynamics of *Ischnura e. elegans* (Vander Linden) (Insects: Odonata) with special reference to morphological colour changes, female polymorphism, multiannual cycles and their influence on behavior, *Hydrobiologia* 146:3–31.

14. M. R. L. Forbes, 1991, Female morphs of the damselfly *Enallagma boreale* Selys (Odonata: Coenagrionidae): A benefit for androchromatypes, *Can. J. Zool.* 69:1969–70; M. R. L. Forbes, 1994, Tests of hypothesis for female-limited polymorphism in the damselfly, *Enallagma boreale* Selys, *Anim. Behav.* 47:724–26.

15. See D. J. Thompson, 1989, Lifetime reproductive success in andromorph females of the damselfly *Coenagrion puella* (L.) (Zygoptera: Coenagrionidae), *Odonatologica* 18:209–13; A. Cordero, 1992, Density-dependent mating success and colour polymorphism in females of the damselfly *Ischnura graellsii* (Odonata: Coenagrionidae), *J. Anim. Ecol.* 61:769–80; O. M. Fincke, 1994, On the difficulty of detecting density-dependent selection on polymorphic females of the damselfly *Ischnura graellsii*: Failure to reject the null hypothesis, *Evol. Ecol.* 8:328–29; O. M. Fincke, 1994, Female colour polymorphism in damselflies: Failure to reject the null hypothesis, *Anim. Behav.* 47:1249–66. On the genetics of color differences among female damselflies, see: C. Johnson, 1966, Genetics of female dimorphism in *Ischnura demorsa*, *Heredity* 21:453–59. On a masculine female phenotype in butterflies, see: R. I. Vane-Wright, 1979, Towards a theory of the evolution of butterfly colour patterns under directional and disruptive selection, *Biol. J. Linnean Soc., Lond.* 11:141–52; C. Clarke, F. Clarke, S. Collins, A. Gill, and J. Turner, 1985, Male-like females, mimicry and transvestism in butterflies (Lepidoptera: Papilionidae), *Systematic Entomology* 10:257–83; see: "female-limited Batesian mimicry" in T. Belt, 1874, *The Naturalist in Nicaragua*, Murray; R. Krebs and D. West, 1988, Female mate preference and the evolution of female-limited Batesian mimicry, *Evolution* 42:1101–4; S. E. Cook, J. G. Vernon, M. Bateson, and T. Guilford, 1994, Mate choice in the polymorphic African swallowtail butterfly, *Papllio dardanus*: Male-like females may avoid sexual harassment, *Anim. Behav.* 47:389–97; R. Vane-Wright, 1984, The role of pseudosexual selection in the evolution of butterfly colour patterns, in R. Vane-Wright and P. Ackerly, eds., *The Biology of Butterflies*, Academic Press.

16. D. Scott, 1986, Sexual mimicry regulates the attractiveness of mated *Drosophila melanogaster* females, *Proc. Nat. Acad. Sci. (USA)* 83:8429–33.

17. L. Gilbert, 1976, Postmating female odor in *Heliconius* butterflies: A male-contributed antiaphrodesiac, *Science* 193:419–20; C. Wiklund and J. Forsberg, 1985, Courtship and male discrimination between virgin and mated females in the orange tip butterfly *Anthocharis cardamines*, *Anim. Behav.* 34:328–32.

18. T. Amundsen, 2000, Why are female birds ornamented? *Trends Ecol. Evol.* 15:149–55.

19. See: A. Craig, 1996, The annual cycle of wing moult and breeding in the wattled starling *Creatophora cinera, Ibis* 138:448–54; W. Dean, 1978, Plumage, reproductive condition, and moult in non-breeding wattled starlings, *Ostrich* 49:97–101.

20. For the white-tailed deer, see: J. C. Donaldson and J. Doutt, 1965, Antlers in female white-tailed deer: A four-year study, *J. Wildlife Management* 29:699–705; W. Wishart, 1985, Frequency of antlered white-tailed does in Camp Wainright, Alberta, *J. Mammalogy* 35:486–88; G. Wislocki, 1954, Antlers in female deer, with a report on three cases in *Odocoileus, J. Mammalogy* 35:486–95; G. Wislocki, 1956, Further notes on antlers in female deer of the genus *Odocoileus, J. Mammalogy* 37:231–35.

For the black-detailed deer, see: I. McT. Cowan, 1946, Antlered doe mule deer, *Canadian Field-Naturalist* 60:11–12; B. Wong and K. Parker, 1988, Estrus in black-tailed deer, *J. Mammalogy* 69:168–71.

21. E. Reimers, 1993, Antlerless females among reindeer and caribou, *Can. J. Zool.* 71:1319–25.

22. A. Hogen-Warburg, 1966, Social behavior of the ruff, *Philomachus pugnax* (L.), *Ardea* 54:109–229; J. van Rhijn, 1991, *The Ruff,* Poyser.

23. J. Högland and R. Alatalo, 1995, *Leks,* Princeton University Press.

24. D. Lank and C. Smith, 1987, Conditional lekking in ruff *(Philomachus pugnax), Behav. Ecol. Sociobiol.* 20:137–45.

25. D. Hugie and D. Lank, 1997, The resident's dilemma: A female choice model for the evolution of alternative mating strategies in lekking male ruffs *(Philomachus pugnax), Behav. Ecol.* 8:218–25.

26. Discussion based on C. Groot and L. Margolis, 1991, *Pacific Salmon Life Histories,* University of British Columbia Press; I. A. Fleming, 1998, Pattern and variability in the breeding system of Atlantic salmon *(Salmo salar),* with comparisons to other salmonids, *Canadian J. Fisheries and Aquatic Sciences* 55 (suppl. 1): 59–76. See also: M. A. Elgar, 1990, Evolutionary compromise between a few large and many small eggs: Comparative evidence in teleost fish, *Oikos* 59: 283–87; I. A. Fleming and M. R. Gross, 1990, Latitudinal clines: A trade-off between egg number and size in Pacific salmon, *Ecology* 71: 1–11; G. S. Su, L. E. Liljedahl, and G. A. E. Gall, 1997, Genetic and environmental variation in female reproductive traits in rainbow trout *(Oncorhynchus mykiss), Aquaculture* 154:115–24; S. Einum and I. A. Fleming, 2000, Highly fecund mothers sacrifice offspring survival to maximise fitness, *Nature* 405:565–67.

27. D. Lack, 1947, The significance of clutch size: I. Intraspecific variation, *Ibis* 89:302–52; S. C. Kendeigh, T. C. Kramer, and F. Hamerstrom, 1956, Variations in egg characteristics of the House Wren, *Auk* 73:42–65; F. C. Rohwer, 1988, Inter- and intraspecific relationships between egg size and clutch size in waterfowl, *Auk* 105:161–76. See also: W. C. Kerfoot, 1974, Egg size cycle of a cladoceran, *Ecology* 55:1259–70; S. Kölding and T M. Fenchel, 1981, Patterns of reproduction in different populations of five species of the amphipod genus *Gammarus, Oikos* 37:167–72; S. A. H. Geritz, 1995, Evolutionary stable seed polymorphism and small-scale spatial variation in seedling density, *Amer. Natur.* 146:685–707; T. A. Mousseau and C W. Fox, 1998, The adaptive significance of maternal effects, *Trends Ecol. Evol.* 13: 403–7.

28. Discussion based on R. Wagner, 1992, The pursuit of extra-pair copulations

by monogamous female razorbills: How do females benefit? *Behav. Ecol. Sociobiol.* 29:455–64. (Quotation below from this source.)

29. J. Lifjeld and R. Robertson, 1992, Female control of extra-pair fertilization in tree swallows, *Behav. Ecol. Sociobiol.* 31:89–96. (All quotations on tree-swallow study from this source.)

30. S. Hrdy, 1977, *The Langurs of Abu: Male and Female Strategies of Reproduction,* Harvard University Press.

31. I have no ready explanation for why a "faithful" mate remains so after her pair-mate has died. Further work would be needed to confirm this report, and to ascertain how long she continued to use the pair-mate's sperm and whether courtship with other males involved any costs.

8: SAME-SEX SEXUALITY

1. G. Barlow, 2000, *The Cichlid Fishes,* Perseus, p. 145.

2. B. Bagemihl, 1999, *Biological Exuberance: Animal Homosexuality and Natural Diversity,* St. Martin's Press. Bagemihl, largely unknown to biologists, received his Ph.D. in linguistics from the University of British Columbia.

3. D. Crews and K. Fitzgerald, 1980, "Sexual" behavior in parthenogenetic lizards *(Cnemidophorus), Proc. Nat. Acad. Sci. (USA)* 77:499–502; also see illustrations of how courtship takes place in D. Crews, 1987, Courtship in unisexual lizards: A model for brain evolution, *Scientific American* 257 (6):116–21.

4. L. Young and D. Crews, 1995, Comparative neuroendocrinology of steroid receptor gene expression and regulation: Relationship to physiology and behavior, *Trends in Endocrinology and Metabolism* 6:317–23.

5. D. Crews, M. Grassman, and J. Lindzey, 1986, Behavioral facilitation of reproduction in sexual and unisexual whiptail lizards, *Proc. Nat. Acad. Sci. (USA)* 83:9547–50.

6. C. Cole and C. Townsend, 1983, Sexual behaviour in unisexual lizards, *Anim. Behav.* 31:724–28.

7. D. Crews and L. Young, 1991, Pseudocopulation in nature in a unisexual whiptail lizard, *Anim. Behav.* 42:512–14.

8. B. Leuck, 1982, Comparative burrow use and activity patterns of parthenogenetic and bisexual whiptail lizards (*Cnemidophorus:* Teiidae), *Copeia* 416–24; B. Leuck, 1985, Comparative social behavior of bisexual and unisexual whiptail lizards (Cnemidophorus), *J. Herpetology* 19:492–506.

9. The other asexual lizards mentioned earlier, the geckoes of Hawaii, also engage in same-sex courtship. Although not studied in detail, same-sex copulations have been photographed. Y. Werner, 1980, Apparent homosexual behavior in an all-female population of a lizard, *Lepidodactylus lugubris,* and its probable interpretation, *Z. Tierpsychol.* 54:144–50; M. J. McCoid and R. A. Hensley, 1991, Pseudocopulation in *Lepidodactylus lugubris, Herpetological Review* 22:8–9.

10. See G. K. Noble and H. T. Bradley, 1933, The mating behavior of lizards: Its bearing on the theory of sexual selection, *Annals of the New York Academy of Sciences* 35:25–100; R. L. Trivers, 1976, Sexual selection and resource-accruing abilities in the *Anolis garmani, Evolution* 30:253–69; and Bagemihl, 1999, *Biological Exuberance,* pp. 657–58, with references.

11. J. Terry, 2000, "Unnatural acts" in nature, *GLQ* 6:151–203, esp. 192 n. 59.

12. I. Jamieson and J. Craig, 1987, Male-male and female-female courtship and copulation behaviour in a communally breeding bird, *Anim. Behav.* 35:1251–53.

13. J. Craig, 1977, The behavior of the pukeko, *New Zealand J. Zoology* 4:413–33.

14. Jamieson and Craig, 1987, Male-male and female-female courtship and copulation behaviour in a communally breeding bird.

15. See: M. P. Harris, 1970, Territory limiting the size of the breeding population of the oystercatcher *(Haematopus ostralegus)*—a removal experiment, *J. Anim. Ecol.* 39:707–13; B. J. Ens, K. B. Briggs, U. N. Safriel, and C. J. Smut, 1996, pp. 186–218 in J. D. Goss-Custard, ed., *The Oystercatcher: From Individuals to Populations,* Oxford University Press.

16. D. Heg and R. van Treuren, 1998, Female-female cooperation in polygynous oystercatchers, *Nature* 391:687–91. The studies were carried out at the Natuurmonumenten reserve on Schiermonnikoog.

17. Bagemihl, 1999, *Biological Exuberance,* pp. 479–655.

18. R. Huber and M. Martys, 1993, Male-male pairs in greyleg geese *(Anser anser), J. Ornithologie* 134:155–64.

19. L. W. Braithwaite, 1981, Ecological studies of the black swan: III. Behavior and social organization, *Australian Wildlife Research* 8:135–46.

20. R. H. Wagner, 1996, Male-male mountings by a sexually monomorphic bird: Mistaken identity or fighting tactic? *J. Avian Biology* 27:209–14.

21. M. Fujioka and S. Yamagishi, 1981, Extramarital and pair copulations in the cattle egret, *Auk* 98:134–44; C. Ramo, 1993, Extra-pair copulations of gray herons nesting at high densities, *Ardea* 81:115–20; D. F. Werschkul, 1982, Nesting ecology of the little blue heron: Promiscuous behavior, *Condor* 84:381–84.

22. The respective scientific names for these birds are *Larus delawarensis, Larus canus, Larus occidentalis, Rissa tridactyla, Larus novaehollandiae, Larus argentatus, Larus ridibundus, Larus atricilla, Pagophila eburnea, Sterna caspia*, and *Sterna dougallii.*

23. M. R. Conover and G. L. Hunt Jr., 1984, Female-female pairings and sex ratios in gulls: A historical perspective, *Wilson Bulletin* 68:232–38; M. R. Conover and G. L. Hunt Jr., 1984, Experimental evidence that female-female pairs in gulls result from a shortage of males, *Condor* 86:472–76; M. R. Conover, D. E. Miller, and G. L. Hunt Jr., 1979, Female-female pairs and other unusual reproductive associations in ring-billed and California gulls, *Auk* 96:6–9; M. R. Conover, 1984, Occurrence of supernormal clutches in the Laridae, *Wilson Bulletin* 96:249–67; G. L. Hunt Jr. and M. W. Hunt, 1977, Female-female pairing in western gulls *(Larus occidentalis)* in Southern California, *Science* 196:1466–67; I. Nisbet and W. Drury, 1984, Supernormal clutches in herring gulls in New England, *Condor* 86:87–89; M. A. Fitch and G. W. Shugart, 1983, Comparative biology and behavior of monogamous pairs and one male–two female trios of herring gulls, *Behav. Ecol. Sociobiol.* 14:1–7; G. W. Shugart, M. A. Fitch, and G. A. Fox, 1987, Female floaters and nonbreeding secondary females in herring gulls, *Condor* 89:902–6; J. van Rhijn, 1985, Black-headed gull or black-headed girl? Advantage of concealing sex by gulls and other colonial birds, *Neth. J. Zool.* 35:87–102; J. van Rhijn and T. Groothuis, 1985, Biparental care and the basis for alternative bond-types among gulls, with special reference to black-headed gulls, *Ardea* 73:159–74; J. Hatch, 1993, Parental behavior of roseate terns:

Comparisons of male-female and multi-female groups, *Colonial Waterbird Society Bulletin* 17:43; I. Nisbet, 1989, The roseate tern, pp. 478–98 in *Audubon Wildlife Report 1989/1990,* Academic Press; I. Nisbet, J. Spendelow, J. Hatfield, J. Zingo, and G. Gough, 1998, Variations in growth of roseate tern chicks: II. Early growth as an index of parental quality, *Condor* 100:305–15.

24. M. Lombardo, R. Bosman, C. Faro, S. Houtteman, and T. Kluisza, 1994, Homosexual copulations by male tree swallows, *Wilson Bulletin* 106:555–57.

25. J. Blakey, 1996, Nest-sharing by female blue tits, *British Birds* 89:279–80.

26. O. Buchanan, 1966, Homosexual behavior in wild orange-fronted parakeets, *Condor* 68:399–400; P. C. Arrowood, 1988, Duetting, pair bonding, and agonistic display in parakeet pairs, *Behavior* 106:129–57.

27. A. I. Dagg, 1984, Homosexual behaviour and female-male mounting in mammals—a first survey, *Mammal Rev.* 14:155.

28. Bagemihl, 1999, *Biological Exuberance*, pp. 269–476.

29. V. Geist, 1971, *Mountain Sheep: A Study in Behavior and Evolution,* University of Chicago Press; J. T. Hogg, 1984, Mating in bighorn sheep: Multiple creative male strategies, *Science* 225:526–29; J. T. Hogg, 1987, Intrasexual competition in and mate choice in rocky mountain bighorn sheep, *Ethology* 75:119–44.

30. J. Berger, 1985, Instances of female-like behaviour in a male ungulate, *Anim. Behav.* 33:333–35.

31. A. Perkins and J. Fitzgerald, 1992, Luteinizing hormone, testosterone, and behavioral response of male-oriented rams to estrous ewes and rams, *J. Anim. Sci.* 70:1787–94.

32. C. Hulet, G. Alexander, and E. Hafez, 1975, The behavior of sheep, pp. 246–94 in E. Hafez, ed., *The Behaviour of Domestic Animals,* 3d ed., Bailliere-Tindall; J. Zenchak and G. Anderson, 1980, Sexual performance levels of rams *(Ovis aries)* as affected by social experience, *J. Anim. Sci.* 50:167; E. Price, L. Katz, S. Wallach, and J. Zenchak, 1988, The relationship of male-male mounting to sexual preferences of young rams, *Appl. Anim. Behav. Sci.* 21:347; L. Katz, E. Price, S. Wallach, and J. Zenchak, 1988, Sexual performance of rams reared with or without females after weaning, *J. Anim. Sci.* 11:1166.

33. G. Silver and E. Price, 1986, Effects of individual vs. group rearing on the sexual behavior of prepubertal bull beefs: Mount orientation and sexual responsiveness, *Appl. Anim. Behav. Sci.* 15:287.

34. A. Perkins and J. Fitsgerald, 1990, Is your ram a dud or a stud? Knowing the difference pays off, *Sheep!* (July), 4–5.

35. A. Perkins and J. Fitsgerald, 1993, Sexual behavior of rams: Biological perspectives to flock management, *Sheep Research J.* 9:51–58.

36. Bagemihl, 1999, *Biological Exuberance*, pp. 269–476.

37. The respective scientific names are *Odocoileus virginianus, Odocoileus hemionus, Cervus elaphus, Rangifer tarandus, Alces alces, Giraffa camelopardalis, Antilocapra americana, Kobus kob, Kobus ellipsiprymnus, Antilope cervicapra, Gazella thomsoni, Gazella granti, Ovibos moschatus, Oreamnos americanus, Bison bison, Equus zebra, Equus quagga, Placochoerus aethiopicus, Tayassu tajacu, Vicugna vicugna, Loxodonta africana,* and *Elephas maximus.*

38. The respective scientific names are *Panthera leo, Acinonyx jubatus, Vulpes vulpes, Canis lupis, Ursus arctos, Ursus americanus,* and *Crocuta crocuta.*

39. The respective scientific names are *Macropus giganteus, Macropus rufogriseus, Macropus parryi, Aepyprymnus rufescens, Dendrolagus dorianus, Dendrolagus matschiei, Phascolarctos cinereus, Sminthopsis crassicaudata,* and *Dasyurus hallucatus.*

40. The respective scientific names are *Tamiasciurus hudsonicus, Sciurus carolinensis, Tamias minimus, Marmota olympus, Marmota caligata, Microcavia autralis, Galea nusteloides, Cavia aperea, Hemiechinus auritus, Pteropus poliocephalus, Pteropus livingstonii,* and *Desmodus rotundus.*

41. The respective scientific names are *Inia geoffrensis, Tursiops truncatus, Stenella longirostris, Orcinus orca, Eschrichtius robustus, Balaena mysticetus, Balaena glacialis, Halichoerus grypus, Mirounga angustirostris, Phoca vitulina, Neophoca cinerea, Phocarctos hookeri, Callorhinus ursinus, Odobenus rosmarus,* and *Trichechus manatus.*

42. R. Connor, M. Heithaus, and L. Barre, 2001, Complex social structure, alliance stability and mating success in a bottlenose dolphin "super-alliance," *Proc. R. Soc. Lond.,* ser. B, 268:263–67; R. Connor, R. Wells, J. Mann, and A. Read, 2000, The bottlenose dolphin: Social relationships in a fission-fusion society, pp. 91–126 in J. Mann, R. Connor, P. Tyack, and H. Whitehead, eds., *Cetacean Societies: Field Studies of Whales and Dolphins,* University of Chicago Press; R. Connor, A. Richards, R. Smolker, and J. Mann, 1996, Patterns of female attractiveness in Indian Ocean bottlenose dolphins, *Behavior* 133:37–69; R. Connor and R. Smolker, 1995, Seasonal changes in the stability of male-male bonds in Indian Ocean bottlenose dolphins (*Tursiops* sp.), *Aquatic Mammals* 21:213–16; R. Connor, R. Smolker, and A. Richards, 1992, Dolphin alliances and coalitions, pp. 415–43 in A. Harcourt and F. de Waal, eds., *Coalitions and Alliances in Humans and Other Animals,* Oxford University Press.

43. P. Vasey, B. Chapais, and C. Gauthier, 1998, Mounting interactions between female Japanese macaques: Testing the influence of dominance and aggression, *Ethology* 104:387–98.

44. P. Vasey, 1998, Female choice and inter-sexual competition for female sexual partners in Japanese macaques, *Behavior* 135:579–97.

45. P. Vasey, 1996, Interventions and alliance formation between female Japanese macaques, *Macaca fuscata,* during homosexual consortships, *Anim. Behav.* 52:539–51.

46. P. Vasey and C. Gauthier, 2000, Skewed sex ratios and female homosexual activity in Japanese macaques: An experimental analysis, *Primates* 41:17–25.

47. E.g., L. Fairbanks, M. McGuire, and W. Kerber, 1977, Sex and agression during rhesus monkey group formation, *Aggr. Behav.* 3:241–49.

48. P. Vasey, 1998, Intimate sexual relations in prehistory: Lessons from the Japanese macaques, *World Archaeology* 29:407–25.

49. D. Futuyma and S. Risch, 1984, Sexual orientation, sociobiology, and evolution, *J. Homosexuality* 9:157–68.

50. Vasey, 1998, Intimate sexual relations in prehistory. See also: P. Abramson and S. Pinkerton, 1995, *With Pleasure: Thoughts on the Nature of Human Sexuality,* Oxford University Press.

51. See: S. Shafir and J. Roughgarden, 1998, Testing predictions of foraging theory for a sit-and-wait forager, *Anolis gingivinus, Behav. Ecol.* 9:74–84.

52. I thank Prof. Judy Stamps, of the University of California, Davis, for first suggesting this observation.

53. A trait is adaptive if the strength of natural selection, measured as the selection coefficient for the trait, exceeds the reciprocal of the population size by a factor of 10 or more. If the population size is 50, the reciprocal of the population size is 1/50, or 0.02. If the average number of offspring left by a homosexual female over her life is 2.5, and the average number of offspring left by a female who doesn't participate in STRs is 2.0, then the selection coefficient is (2.5/2.0–2.0/2.0), or 0.25. Since 0.25 is greater than 0.02 by a factor of 10 or more, this hypothetical degree of advantage would qualify homosexuality as adaptive.

54. F. de Waal, 1995, Bonobo sex and society, *Scientific American* (March), 82–88.

55. A. Parish, 1996, Female relationships in bonobos *(Pan paniscus)*: Evidence for bonding, cooperation, and female dominance in a male-philopatric species, *Human Nature* 7:61–96.

56. A. Richard, 1992, Aggressive competition between males, female-controlled polygyny, and sexual monomorphism in a Malagasy primate, *Propithecus verreauxi, J. Human Evolution* 22:395–406; A. Richard, 1974, Patterns of mating in *Propithecus verreauxi verreauxi,* pp. 49–74 in R. Martin, G. Doyle, and A. Walker, eds., *Prosimian Biology,* Duckworth.

57. G. Talmage-Riggs and S. Anschel, 1973, Homosexual behavior and dominance in a group of captive squirrel monkeys *Saimiri sciureus, Folia Primatologica* 19:61–72; S. Mendoza and W. Mason, 1991, Breeding readiness in squirrel monkeys: Female-primed females are triggered by males, *Physiology & Behavior* 49:471–79; C. Mitchell, 1994, Migration alliances and coalitions among adult male South American squirrel monkeys *Saimiri sciureus, Behavior* 130:169–90.

58. J. Manson, S. Perry, and A. Parish, 1997, Nonconceptive sexual behavior in bonobos and capuchins, *Int. J. Primatology* 18:767–86; S. Perry, 1998, Male-male social relationships in wild white-faced capuchins, *Cebus capucinus, Behavior* 135:139–72.

59. J. Akers and C. Conaway, 1979, Female homosexual behavior in *Macaca mulatta, Archives of Sexual Behavior* 8:63–80; J. Erwin and T. Maple, 1976, Ambisexual behavior with male-male anal penetration in male rhesus monkeys, *Archives of Sexual Behavior* 5:9–14; V. Reinhardt, A. Reinhardt, F. Bercovitch, and R. Goy, 1986, Does intermale mounting function as a dominance demonstration in rhesus monkeys? *Folia Primatologica* 47:55–60.

60. S. Chevalier-Skolnikoff, 1974, Male-female, female-female, and male-male sexual behavior in the stumptail monkey, with special attention to female orgasm, *Archives of Sexual Behavior* 3:95–116; S. Chevalier-Skolnikoff, 1976, Homosexual behavior in a laboratory group of stumptail monkeys *(Macaca arctoides)*: Forms, contexts, and possible social functions, *Archives of Sexual Behavior* 5:511–27; D. Goldfoot, H. Westerborg-van Loon, W. Groeneveld, and A. Slob, 1980, Behavioral and physiological evidence of sexual climax in the female stump-tailed macaque *(Macaca arctoides), Science* 208:1477–79.

61. R. Noë, 1992, Alliance formation among male baboons: Shopping for profitable partners, pp. 284–321 in A. Harcourt and F. de Waal, eds., *Coalitions and Alliances in Humans and Other Animals,* Oxford University Press.

62. G. Agoramoorthy and S. Mohnot, 1988, Infanticide and juvenilicide in Hanuman langurs *(Presbytis entellus)* around Jodhpur, India, *Human Evolution* 3:279–96; S. Mohnot, 1980, Intergroup infant kidnapping in Hanuman langur, *Folia Primatologica* 34:259–77; S. Hrdy, 1978, Allomaternal care and abuse of infants among Hanuman langurs, *Recent Advances in Primatology* 1:169–72.

63. S. Hrdy, 1977, *The Langurs of Abu: Male and Female Strategies of Reproduction,* Harvard University Press.

64. A. Edwards and J. Todd, 1991, Homosexual behavior in wild white-handed gibbons *(Hylobates lar), Primates* 32:231–36; R. Palombit, 1996, Pair bonds in monogamous apes: A comparison of the siamang *Hylobates syndactylus* and the white-handed gibbon *Hylobates lar, Behavior* 133:321–56; U. Reichard, 1995, Extra-pair copulation in a monogamous gibbon *(Hylobates lar), Ethology* 100:99–112.

65. R. Fischer and R. Nadler, 1978, Affiliative, playful and homosexual interactions of adult female lowland gorillas, *Primates* 19:657–64; D. Fossey, 1984, Infanticide in mountain gorillas *(Gorilla gorilla beringei)* with comparative notes on chimpanzees, pp. 217–35 in G. Hausfater and S. Hrdy, eds., *Infanticide: Comparative and Evolutionary Perspectives,* Aldine; M. Robbins, 1996, Male-male interactions in heterosexual and all-male wild mountain gorilla groups, *Ethology* 102:942–65.

66. P. Vasey, 1995, Homosexual behavior in primates: A review of evidence and theory, *Int. J. Primatology* 16:173–204.

67. F. de Waal, 1995, Bonobo sex and society, *Scientific American* (March), 82–88.

68. M. Zuk, 2002, *Sexual Selections: What We Can and Can't Learn about Sex from Animals,* University of California Press, p. 143.

9: THE THEORY OF EVOLUTION

1. C. Darwin, 1962 [1860], *The Voyage of the Beagle,* Anchor Books. Quotations here and below from pp. 393, 394, 398.

2. See: L. Margulis, 1996, Archaeal-eubacterial mergers in the origin of Eukarya: Phylogenetic classification of life, *Proc. Nat. Acad. Sci. (USA)* 93:1071–76; T. Cavalier-Smith, 1998, A revised six-kingdom system of life, *Biol. Rev.* 73:203–66; L. Margulis and K. V. Schwartz, 1988, *Five Kingdoms: An Illustrated Guide to the Phyla of Life on Earth,* 2d ed., W.H. Freeman; see also: Charting the evolutionary history of life, special issue of *Science* 300:1691–1709.

3. C. Darwin, 1967 [1859], *The Origin of Species by Means of Natural Selection or the Preservation of Favored Races in the Struggle for Life,* Atheneum, p. 63.

4. See: L. Margulis, 1993, *Symbiosis in Cell Evolution,* 2d ed., W.H. Freeman.

5. For Hrdy, see: S. Hrdy, 1974, Male-male competition and infanticide among the langurs *(Presbytis entellus)* of Abu, Rajasthan, *Folia Primatologica* 22:19–58; S. Hrdy, 1979, Infanticide among animals: A review, classification, and examination of the implications for the reproductive strategies of females, *Ethology and Sociobiology* 1:13–40; S. Hrdy, 1997, Raising Darwin's consciousness: Female sexuality and the prehominid origins of patriarchy, *Human Nature* 8:1–49.

For Gowaty, see: P. Gowaty, 2003, Power asymmetries between the sexes, mate preferences, and components of fitness, pp. 61–86 in C. Tarvis, 2003, *Evolution, Gender and Rape,* MIT Press; P. Gowaty, 2004, Sex roles, contests for the control of reproduction, and sexual selection, in P. Kappeler and C. van Schaik, eds., *Sexual Selection in Primates: New and Comparative Perspectives,* Cambridge University Press (in press); P. Gowaty, 2003, Sexual natures: How feminism changed evolutionary biology, *Signs* 28:901–22.

6. C. Darwin, 1871, *The Descent of Man, and Selection in Relation to Sex,* Princeton University Press (facsimile edition). Quotations here and below from pp. 218–22, 228–35, 449, 498–99.

7. In all fairness, a recent paper that models the outcome of male-female negotiation for whether a male should "desert" or remain at the nest to provide parental care is a decided step forward. See: M. Wade and S. Shuster, 2002, The evolution of parental care in the context of sexual selection: A critical reassessment of parental investment theory, *Amer. Natur.* 160:285–92.

8. M. Small, 1993, *Female Choices: Sexual Behavior of Female Primates,* Cornell University Press, p. 106

9. D. Buss, 1994, *The Evolution of Desire,* Basic Books, p. 20.

10. N. Malamuth, 1996, The confluence model of sexual aggression: Feminist and evolutionary perspectives, pp. 269–95 in D. Buss and N. Malamuth, eds., *Sex, Power, Conflict: Evolutionary and Feminist Perspectives,* Oxford University Press, p. 275.

11. J. Coyne, 2003, Of vice and men: A case study in evolutionary psychology, pp. 171–89 in C. Travis, ed., *Evolution, Gender, and Rape,* MIT Press. Quotations here and below from this source.

12. R. Thornhill and C. Palmer, 2000, *A Natural History of Rape: Biological Bases of Sexual Coercion,* MIT Press.

13. S. Shuster quoted in V. Gewin, 2003, Joan Roughgarden profile: A plea for diversity, *Nature* 422:368–69.

14. J. Crook, 1965, The adaptive significance of avian social organization, *Symp. Zool. Soc. Lond.* 14:181–218; J. Crook and J. Gartlan, 1966, Evolution of primate societies, *Nature* 210:1200–3; J. Crook, 1970, Social organization and environment: Aspects of contemporary social ethology, *Anim. Behav.* 18:197–209; R. Wrangham, 1987, Evolution of social structure, pp. 282–96 in B. Smuts, D. Cheney, R. Seyfarth, R. Wrangham, and T. Struhsaker, eds., *Primate Societies,* University of Chicago Press; T. H. Clutton-Brock, 1989, Review lecture: Mammalian mating systems, *Proc. R. Soc. Lond.,* ser. B, 236:339–72; N. Davies, 1991, Mating systems, pp. 263–300 in J. R. Krebs and N. B. Davies, eds., *Behavioral Ecology: An Evolutionary Approach,* Blackwell Scientific Publications.

10: AN EMBRYONIC NARRATIVE

1. R. Dawkins, 1990, *The Selfish Gene,* Oxford University Press.

2. Narrative based on: S. Gilbert, 2000, *Developmental Biology,* 6th ed., Sinauer. See also: M. Ginsberg, M. Snow, and A. McLaren, 1990, Primordial germ cells in the mouse embryo during gastrulation, *Development* 110:521–28; L. Pinsky, R. Erickson, and N. Schimke, 1999, *Genetic Disorders of Human Sexual Development,* Oxford University Press, p. 4; and A. Byskov, 1986, Differentiation of mammalian embryonic gonad, *Physiol. Rev.* 66:71–117.

3. The time between ovulation is 29.5 days on the average; Gilbert, 2000, *Developmental Biology,* p. 610.

4. E. Davidson, 1986, *Gene Activity in Early Development,* 3d ed., Academic Press. Maternal genes are expressed for the first two cell divisions after fertilization; see ibid., p. 356.

5. M. Profet, 1993, Menstruation as a defense against pathogens transported by sperm, *Q. Rev. Bio.* 68:335–85.

6. See: B. Storey, 1995, Interactions between gametes leading to fertilization: The sperm's eye view, *Reprod. Fertil. Dev.* 7:927–42.

7. S. Gilbert, 2000, *Developmental Biology,* 6th ed., Sinauer, p. 196.

8. See http://www.visembryo.com/baby/index.html; and http://anatomy.med .unsw.edu.au/cbl/embryo/Embryo.htm. An animation of human development in QuickTime movie format is available at http://www.uphs.upenn.edu/meded/ public/berp/overview.mov.

11: SEX DETERMINATION

1. See: A. McLaren, 1995, Germ cells and germ cell sex, *Phil. Trans. R. Soc. Lond.,* ser. B, 350:229–33; D. Whitworth, 1998, XX germ cells: The difference between an ovary and a testis, *TEM* 9:2–6.

2. P. Burgoyne, A. Thornhill, S. Bourdrean, C. Darling, C. Bishop, and E. Evans, 1995, The genetic basis of XX-XY differences present before gonadal sex differentiation in the mouse, *Proc. R. Soc. Lond.,* ser. B, 350:253–60.

3. M. Renfree and R. Short, 1988, Sex determination in marsupials: Evidence for a marsupial-eutherian dichotomy, *Phil. Trans. R. Soc. Lond.,* ser. B, 322:41–53.

4. M. Watanabe, A. Zinn, D. Page, and T. Nishimoto, 1993, Functional equivalence of human X- and Y-encoded isoforms of ribosomal protein S4 consistent with a role in Turner syndrome, *Nature Genetics* 4:268–71.

5. See, for example, C. Haqq and P. Donahoe, 1998, Regulation of sexual dimorphism in mammals, *Physiol. Rev.* 78:1–33; E. Vilain and R. McCabe, 1998, Mammalian sex determination: From gonads to brain, *Mol. Genet. Metab.* 65:74–84; J. Marshall Graves, 1998, Evolution of the mammalian Y chromosome and sex-determining genes, *J. Exp. Zool.* 281:472–81; A. Swain and R. Lovell-Badge, 1999, Mammalian sex determination: A molecular drama, *Genes and Development* 13: 755–67.

6. P. Koopman, J. Gubbay, N. Vivian, P. Goodfellow, and R. Lovell-Badge, 1991, Male development of chromosomally female mice transgenic for *Sry, Nature* 351:117–21.

7. E. Eicher and L. Washburn, 1986, Genetic control of primary sex determination in mice, *Ann. Rev. Genet.* 20:327–60; J. Gubbay, J. Collignon, P. Koopman, et al., 1990, A gene mapping to the sex-determining region of the mouse Y chromosome is a member of a novel family of embryonically expressed genes, *Nature* 346:245–50. See also: McLaren, 1995, Germ cells and germ cell sex.

8. D. Whitworth, G. Shaw, and M. B. Renfree, 1996, Gonadal sex reversal of the developing marsupial ovary *in vivo* and *in vitro, Development* 122:4057–63; see also: M. Renfree, J. Harry, and G. Shaw, 1995, The marsupial male: A role model for sexual development, *Proc. Trans. R. Soc. Lond.,* ser. B, 350:243–51.

9. S. Morais da Silva, A. Hacker, V. Harley, P. Goodfellow, A. Swain, and R. Lovell-Badge, 1996, SOX9 expression during gonadal development implies a con-

served role for gene in testis differentiation in mammals and birds, *Nature Genetics* 14:62–68.

10. K. McElreavey and M. Fellous, 1999, Sex determination and the Y chromosome, *Amer. J. Med. Genet. (Semin. Med. Genet.)* 89:176–85.

11. See: J. Marshall Graves, 1995, The origin and function of the mammalian Y chromosome and Y-borne genes—an evolving understanding, *Bioessays* 17:311–21; J. Marshall Graves, 1995, The evolution of mammalian sex chromosomes and the origin of sex determining genes, *Phil. Trans. R. Soc. Lond.,* ser. B, 350:305–12.

12. S. Edmands and R. Burton, 1999, Cytochrome *c* oxidase activity in interpopulational hybrids of a marine copepod: A test for nuclear-nuclear or nuclear-cytoplasmic coadaptation, *Evolution* 53:1972–78.

13. Both males and females have the same chromosomal makeup with an unpaired X, called XO. See: W. Just, W. Rau, W. Vogel, M. Akhverdian, K. Fredga, J. Marshall Graves, and E. Lyapunova, 1995, Absence of *Sry* in species of the vole *Ellobius, Nature Genetics* 11:117–18; A. Baumstark, M. Akhverdyan, A. Schulze, I. Reisert, W. Vogel, and W. Just, 2001, Exclusion of SOX9 as the testis determining factor in *Ellobius lutescens*: Evidence for another testis determining gene besides SRY and SOX9, *Mol. Genet. Metab.* 72 (1):61–66.

14. Just et al., 1995, Absence of *Sry* in species of the vole *Ellobius.*

15. N. Bianchi, M. Bianchi, G. Bailliet, and A. Chapelle, 1993, Characterization and sequencing of the sex determining region Y gene *(Sry)* in *Akadon* (Cricetidae) species with sex reversed females, *Chromosoma* 102:389–95.

16. R. Jiménez, M. Burgos, A. Sánchez, A. Sinclair, F. Alarcón, J. Marin, E. Ortega, and R. D. de la Guardia, 1993, Fertile females of the mole *Talpa occidentalis* are phynotypic intersexes with ovotestes, *Development* 118:1303–11.

17. A. Sánchez, M. Mullejos, M. Burgos, et al., 1996, Females of four mole species of genus *Talpa* (Insectivora, Mammalia) are true hermaphrodites with ovotestes, *Mol. Reprod. Dev.* 44:289–94.

18. See: H. Willard, 2000, The sex chromosomes and X chromosome inactivation, in C. Scriver, A. Beaudet, W. Sly, D. Valle, B. Childs, and B. Vogelstein, eds., *The Metabolic and Molecular Bases of Inherited Disease,* 8th ed., McGraw Hill, cited in T. Wizemann and M. L. Pardue, eds., 2001, *Exploring the Biological Contributions to Human Health: Does Sex Matter?* National Academy Press, pp. 29–32.

19. See: L. Carrel, A. Cottle, K. Goglin, and H. Willard, 1999, A first-generation X-inactivation profile of the human genome, *Proc. Nat. Acad. Sci. (USA)* 96:14440–44.

20. See: A. Arnold, 1996, Genetically triggered sexual differentiation of brain and behavior, *Hormones and Behavior* 30:495–505. See also: C. Smith and J. Joss, 1994, Sertoli cell differentiation and gonadogenesis in *Alligator mississippiensis, J. Exp. Zool.* 270:57–70; M. Thorne and B. Sheldon, 1993, Triploid intersex and chimeric chickens: Useful models for studies of avian sex determination, pp. 201–8 in K. Reed and J. Marshall Graves, *Sex Chromosomes and Sex Determining Genes,* Harwood Academic; I. Lagomarsino and D. Conover, 1993, Variation in environmental and genotypic sex-determining mechanisms across a latitudinal gradient in the fish, *Menidia menidia, Evolution* 47:487–94.

21. On reptiles generally, see: D. Deeming and M. Ferguson, 1988, Environmental regulation of sex determination in reptiles, *Phil. Trans. R. Soc. Lond.,* ser. B,

322:19–39; V. Lance, 1994, Introduction: Environmental sex determination in reptiles: Patterns and processes, *J. Exp. Zool.* 270:1–2; J. Spotila, L. Spotila, and N. Kaufer, 1994, Molecular mechanisms of TSD in reptiles: A search for the magic bullet, *J. Exp. Zool.* 270:117–27; T. Wibbels, J. Bull, and D. Crews, 1994, Temperature-dependent sex determination: A mechanistic approach, *J. Exp. Zool.* 270:71–78.

More specifically, on turtles, see: M. Ewert, D. Jackson, and C. Nelson, 1994, Patterns of temperature-dependent sex determination in turtles, *J. Exp. Zool.* 270:3–15. On crocodiles, see: J. W. Lang and H. Andrews, 1994, Temperature-dependent sex determination in crocodilians, *J. Exp. Zool.* 270:28–44; C. Smith and J. Joss, 1993, Gonadal sex differentiation in *Alligator mississippiensis*, a species with temperature-dependent sex determination, *Cell Tissue Res.* 273:149–62. On lizards, see: B. Viets, M. Ewert, L. Talent, and C. Nelson, 1994, Sex-determining mechanisms in squamate reptiles, *J. Exp. Zool.* 270:45–56.

22. See: C. Johnston, M. Barnett, and P. Sharpe, 1995, The molecular biology of temperature-dependent sex determination, *Phil. Trans. R. Soc. Lond.*, ser. B, 350: 297–304; Wibbles, Bull, and Crews, 1994, Temperature-dependent sex determination.

23. See: B. Schlinger, 1998, Sexual differentiation of avian brain and behavior: Hormone-dependent and independent mechanisms, *Ann. Rev. Physiol.* 60:407–29.

12: SEX DIFFERENCES

1. According to recent studies of the human genome, a human being's DNA contains about 3 billion (3×10^9) base pairs. This DNA encompasses about 30,000 genes (3×10^4), i.e., segments of DNA that encode a protein. Therefore, about 100,000 (10^5) DNA positions yield a gene ($3 \times 10^9/3 \times 10^4$). The exact sequence that encodes the protein is much shorter than this, but this average allows for all the intervening DNA, believed to be inactive. The genomes of individuals differ at non-sex chromosomes by 4 to 6 million (4–6×10^6) base positions, which corresponds to 60 genes ($6 \times 10^6/1 \times 10^5$). See: T. Wizemann and M. L. Pardue, eds., 2001, *Exploring the Biological Contributions to Human Health: Does Sex Matter?* National Academy Press, p. 25.

2. The X chromomome has about 160 million (160×10^6) DNA base pairs. (See ibid., p. 29.) Using the figure of one gene on the average per 10^5 base pairs, we obtain 1,600 genes, which may be rounded to 1,500 genes. Assuming that people differ by 0.2 percent of their genes, any two people differ by about 3 genes on their X chromosomes.

3. See: J. Belmont, 1996, Genetic control of X inactivation and processes leading to X-inactivation skewing, *Amer. J. Hum. Genet.* 58:1101–8; K. Wareham, M. Lyon, P. Glenister, and E. Williams, 1987, Age related reactivation of an X-linked gene, *Nature* 327:725–27; H. Willard, 2000, The sex chromosomes and X chromosome inactivation, in C. Scriver, A. Beaudet, W. Sly, D. Valle, B. Childs, and B. Vogelstein, eds., *The Metabolic and Molecular Bases of Inherited Disease*, 8th ed., McGraw-Hill; cited in ibid., pp. 30–31.

4. See Wizemann and Pardue, eds., 2001, *Exploring the Biological Contributions to Human Health*, p. 118.

5. XIST, see: L. Carrel and H. Willard, 1999, Heterogeneous gene expression from the inactive X chromosome: An X-linked gene that escapes X inactivation in some human cell lines but is inactivated in others, *Proc. Nat. Acad. Sci. (USA)* 96:7364–69.

6. L. Carrel, A. Cottle, K. Goglin, and H. Willard, 1999, A first-generation X-inactivation profile of the human X chromosome, *Proc. Nat. Acad. Sci. (USA)* 96:14440–44.

7. The Y chromosome is 60×10^6 base pairs long. One gene per 10^5 base pairs leads to an estimated 600 genes, which we can round to 500. If people differ on the average by 0.2 percent of their genes, $0.002 \times 500 = 1$ gene is the expected difference between two XY people because of their Y chromosomes alone.

8. M. Delbridge and J. Marshall Graves, 1999, Mammalian Y chromosome evolution and the male-specific functions of Y chromosome-borne genes, *Rev. Reprod.* 4:101–9. The sequence of the human Y chromosome is now available, see: H. Skaletsky et al., 2003, The male-specific region of the human Y chromosome is a mosaic of discrete sequence classes, *Nature* 423:825–37.

9. L. Whitfield, R. Lovell-Badge, and P. Goodfellow, 1993, Rapid sequence evolution of the mammalian sex-determining gene SRY, *Nature* 364:713–15; R. Tucker and B. Lundrigan, 1993, Rapid evolution of the sex determining locus in Old World mice and rats, *Nature* 364:715–17; E. Payten and C. Cotinot, 1994, Sequence evolution of SRY gene within *Bovidae* family, *Mammalian Genome* 5:723–25; P. Tucker and B. Lundrigan, 1995, The nature of gene evolution on the mammalian Y chromosome: Lessons from Sry, *Phil. Trans. R. Soc. Lond.,* ser. B, 350:221–27.

10. C. Nagamine, Y. Nishioka, K. Moriwaki, P. Boursot, F. Bonhomme, and Y. F. Lau, 1992, The musculus-type Y chromosome of the laboratory mouse is of Asian origin, *Mammalian Genome* 3:84–91; P. Tucker, B. Lee, B. Lundrigan, and E. Eicher, 1992, Geographic origin of the Y chromosome in "old" inbred strains of mice, *Mammalian Genome* 3:254–61; C. Nagamine, 1994, The testis-determining gene, SRY, exists in multiple copies in Old World rodents, *Genet. Res. Camb.* 64:151–59; K. Miller, B. Lundrigan, and P. Tucker, 1995, Length variation of CAG repeats in Sry across populations of *Mus domesticus, Mammalian Genome* 6:206–8.

11. P. Pamilo and R. Waugh O'Neill, 1997, Evolution of the Sry genes, *Mol. Biol. Evol.* 14:49–55.

12. P. Coward, K. Nagai, D. Chen, H. Thomas, C. Nagamine, and Y. F. Lau, 1994, Polymorphism of a CAG trinucleotide repeat within Sry correlates with B6Y[Dom] sex reversal, *Nature Genetics* 6:245–50; K. Miller, B. Lundrigan, and P. Tucker, 1995, Length variation of CAG repeats in Sry across populations of *Mus domesticus, Mammalian Genome* 6:206–8.

13. C. Clepet, A. Schafer, A. Sinclair, M. Palmer, R. Lovell-Badge, and P. Goodfellow, 1993, The human SRY transcript, *Hum. Mol. Genet.* 2:2007–12; J. Harry, P. Koopman, F. Brannan, J. Graves, and M. Renfree, 1995, Widespread expression of the testis-determining SRY in a marsupial, *Nature Genetics* 11:347–49; G. Lahr, S. Maxson, A. Mayer, W. Just, C. Pilgrim, and I. Reisert, 1995, Transcription of the Y chromosomal gene, Sry, in adult mouse brain, *Mol. Brain Res.* 33:179–82; E. Payen, E. Pailhous, R. About Merhi, L. Gianquinto, M. Kirszenbaum, A. Locatelli, and C. Cotinot, 1996, Characterization of ovine SRY transcript and developmental expression of genes involved in sexual differentiation, *Int. J. Devel. Biol.* 40:567–75.

14. Indeed this calculation may even understate the potential for genetic difference between XX- and XY-bodied people. David Page of the Whitehead Institute in Cambridge, Mass., has said, "Men and women differ by 1 to 2 percent of their genomes, which is about the same as the difference between a male man and a male chimpanzee

or between a woman and a female chimpanzee. . . . The reality is that the genetic difference between males and females absolutely dwarfs all other differences in the human genome." See also: N. Wade, 2003, Y chromosome depends on itself to survive, *New York Times,* June 19.

15. For the selfish gene alternative, see: L. Hurst, 1994, Embryonic growth and the evolution of the mammalian Y chromosome: I. The Y as an attractor for selfish genes, *Heredity* 73:223–32; L. Hurst, 1994, Embryonic growth and the evolution of the mammalian Y chromosome: II. Suppression of selfish Y-linked growth factors may explain escape from X-inactivation and rapid evolution of Sry, *Heredity* 73:233–43. Later work falsified this hypothesis. See: Pamilo and Waugh O'Neill, 1997, Evolution of the Sry genes.

16. F. Lillie, 1939, *Sex and Internal Secretions,* National Research Council; cited in A. Fausto-Sterling, 2000, *Sexing the Body: Gender Politics and the Construction of Sexuality,* Basic Books, p. 178.

17. Fausto-Sterling, 2000, *Sexing the Body,* p. 182.

18. See, for example: C. Migeon and A. Wisniewski, 1998, Sexual differentiation: From genes to gender, *Hormone Research* 50:245–51.

19. F. S. vom Saal, 1989, Sexual differentiation in litter-bearing mammals: Influence of sex of adjacent fetuses in utero, *J. Anim. Sci.* 67:1824–40.

20. D. McFadden, 1993, A masculinizing effect on the auditory systems of human females having male co-twins, *Proc. Nat. Acad. Sci. (USA)* 90:11900–4.

21. See Wizemann and Pardue, eds., 2001, *Exploring the Biological Contributions to Human Health,* p. 39.

22. Ibid., p. 49.

23. M. Herman-Giddens, E. Slora, R. Wasserman, C. Bourdony, M. Bhapkar, G. Koch, and C. Hasemeier, 1997, Secondary sexual characteristics and menses in young girls seen in office practice: A study from the Pediatric Research in Office Settings Network, *Pediatrics* 99:505–12.

24. J. Bilezikian, A. Morishima, J. Bell, and M. Grumbach, 1998, Increased bone mass as a result of estrogen therapy in a man with aromatase deficiency, *New England J. Medicine* 339:599–603.

25. Fausto-Sterling, 2000, *Sexing the Body,* p. 149.

26. A. Sullivan, 2000, The He hormone, *New York Times Magazine,* April 2, p. 46ff. Quotations here and below from this source.

27. Sullivan pointedly refers to the man, Drew Seidman, by his previous female name, Susan, and uses female pronouns, exemplifying the studied obstinacy of conservative gay commentators in refusing to acknowledge the reality of trans people. Ditto the conservative lesbian columnist Norah Vincent, whose 1999 article on Drew in the *Village Voice* is headlined "Girl grows hair on chest" and concludes by describing Drew as "slightly too pretty." So smug. To her credit, though, Vincent does accord masculine pronouns to Drew.

28. Shoshanna Scholar, 2000, An interview with Pat Califia, *The Channel* TGSF (Sept.), taken from *nerve.com.*

29. J. Green, 2004, *Becoming a Visible Man,* Vanderbilt University Press.

30. SRY is expressed in the adult mouse hypothalamus and midbrain, where the transcript is linear (not circular as in the testes) and is presumably translated. See Lahr

et al., 1995, Transcription of the Y chromosome gene, Sry, in adult mouse brain. SRY is also expressed in both the fetal and adult brain in the tammar wallaby, a marsupial. See Harry et al., 1995, Widespread expression of the testis-determining gene SRY in a marsupial (see note 13 above).

31. R. Zann, 1990, Song and call learning in wild zebra finches in south-east Australia, *Anim. Behav.* 40:811–28.

32. F. Nottebohm and A. Arnold, 1976, Sexual dimorphism in vocal control areas of the songbird brain, *Science* 194:211–13; see also: E. Adkins-Regan and J. Watson, 1990, Sexual dimorphism in the avian brain is not limited to the song system of songbirds: A morphometric analysis of the brain of the quail *(Corurnix japoonica), Brain Res.* 514:320–26.

33. See: G. Panzica, N. Aste, C. Viglietti-Panzica, and M. Ottinger, 1995, Structural sex differences in the brain: Influence of gonadal steroids and behavioral correlates, *J. Endocrinol. Invest.* 18:232–52, esp. fig. 4.

34. B. Schlinger, 1998, Sexual differentiation of avian brain and behavior: Current views on gonadal hormone-dependent and independent mechanisms, *Ann. Rev. Physiol.* 60:407–29; A. Arnold, 1996, Genetically triggered sexual differentiation of brain and behavior, *Hormones and Behavior* 30:495–505.

35. See studies of wrens, most of whom are dueting, except the Carolina wren, in which only the males sing: P. Nealen and D. Perkel, 2000, Sexual dimorphism in the song system of the Carolina Wren *Thryothorus ludovicianus, J. Comp. Neurology* 418:346–60.

36. S. MacDougall-Shackleton and G. Ball, 1999, Comparative studies of sex differences in the song-control system of songbirds, *TINS* 22:432–36.

37. Testosterone can induce female canaries to sing, but not female zebra finches. See: F. Nottebohm and A. Arnold, 1976, Sexual dimorphism in vocal control areas of the songbird brain, *Science* 194:211–13.

38. F. Nottebohm, 1981, A brain for all seasons: Cyclical anatomical changes in song control nuclei of the canary brain, *Science* 214:1368; J. Balthazard, O. Tlemcani, and G. Ball, 1996, Do sex differences in the brain explain sex differences in hormonal induction of reproductive behavior? What twenty-five years of research on the Japanese Quail tells us, *Hormones and Behavior* 30:627–61.

39. B. Schlinger and A. Arnold, 1995, Estrogen synthesis and secretion by the songbird brain, pp. 297–323 in P. Micevych and R. Hammer Jr., eds., *Neurobiological Effects of Sex Steroid Hormones,* Cambridge University Press.

40. B. Schlinger and G. Callard, 1989, Aromatase in quail brain: Correlations with aggressiveness, *Endocrinology* 124:437–43.

41. E. Adkins-Regan, M. Ottinger, and J. Park, 1995, Maternal transfer of estradiol to egg yolks alters sexual differentiation of avian offspring, *J. Exp. Zool.* 271:466–70; H. Schwabl, 1993, Yolk is a source of maternal testosterone for developing birds, *Proc. Nat. Acad. Sci. (USA)* 90:11446–50.

42. E. Adkins-Regan and M. Ascenzi, 1987, Social and sexual behavior of male and female zebra finches treated with oestradiol during the nestling period, *Anim. Behav.* 35:1100–12; V. Mansukhani, E. Adkins-Regan, and S. Yang, 1996, Sexual partner preference in female zebra finches: The role of early hormones and social environment, *Hormones and Behavior* 30:506–13.

43. T. DeVoogd, A. Houtman, and J. Falls, 1995, White-throated sparrow morphs that differ in song production rate also differ in the anatomy of some song-related brain areas, *Neurobiology* 28:202–13.

44. S. Torbet and I. Hanna, 1997, Ontogeny of sex differences in the mammalian hypothalamus and preoptic area, *Cellular and Molecular Neurobiology* 17:565–601; S. M. Breedlove, 1994, Sexual differentiation of the human nervous system, *Ann. Rev. Psychol.* 45:389–418; S. Torbet, D. Sahniser, and M. Baum, 1986, Differentiation in male ferrets of a sexually dimorphic nucleus of the preoptic/anterior hypothalamic area requires prenatal estrogen, *Neuroendocrinology* 44:299–308; R. Gorski, 1988, Hormone-induced sex differences in hypothalamic structure, *Bull. TMIN* 16:67–90; D. Commins and P. Yahr, 1984, Acetylcholinesterase activity in the sexually dimorphic area of the gerbil brain: Sex differences and influence of adult gonadal steroids, *J. Comp. Neurol.* 224:123–31; M. Hines, F. Davis, A. Coquelin, R. Goy, and R. Gorski, 1985, Sexually dimorphic regions in the medial preoptic area and the bed nucleus of the stria terminalis of the guinea pig brain: A description and an investigation of their relationship to gonadal steroids in adulthood, *J. Neurosci.* 5:40–47.

45. K. Dohler, J. Hancke, S. Srivastava, C. Hofmann, J. Shryne, and R. Gorski, 1984, Participation of estrogens in female sexual differentiation of the brain: Neuroanatomical, neuroendocrine and behavioral evidence, *Prog. Brain Res.* 61:99–117; E. Davis, P. Popper, and R. Gorski, 1996, The role of apoptosis in sexual differentiation of the rat sexually dimorphic nucleus of the preoptic area, *Brain Res.* 734:10–18.

46. F. DeJonge, A. Louwerse, M. Ooms, P. Evers, E. Endert, and N. Van de Poll, Lesions of the SDN-POA inhibit sexual behavior of male Wistar rats, *Brain Res. Bull.* 23:483–92.

47. A. Del Abril, S. Segovia, and A. Guillamon, 1987, The bed nucleus of the stria terminalis in the rat: Regional sex differences controlled by gonadal steroids early after birth, *Brain Res.* 429:295–300; A. Guillamon, S. Segovia, and A. Del Abril, 1988, Early effects of gonadal steroids on the neuron number in the medial posterior region and the lateral division of the bed nucleus of the stria terminalis in the rat, *Brain Res. Dev.* 44:281–90.

48. N. Forger, L. Hodges, S. Roberts, and S. Breedlove, 1992, Regulation of motoneuron death in the spinal nucleus of the bulbocavernosus, *J. Neurobiol.* 23:1192–1203; N. Forger, S. Roberts, V. Wong, and S. Breedlove, 1993, Ciliary neurotrophic factor maintains motoneurons and their target muscles in developing rats, *J. Neurosci.* 13.

49. D. Anderson, R. Rhees, and D. Fleming, 1985, Effects of prenatal stress on differentiation of the sexually dimorphic nucleus of the preoptic area SDN-POA of the rat brain, *Brain Res.* 332:113–18; W. Grisham, M. Kerchner, and I. Ward, 1991, Prenatal stress alters sexually dimorphic nuclei in the spinal cord of male rats, *Brain Res.* 551:126–31.

50. C. Moore, H. Dou, and J. Juraska, 1992, Maternal stimulation affects the number of motoneurons in a sexually dimorphic nucleus of the lumbar spinal cord, *Brain Res.* 572:52–56.

51. Recent data come from magnetic resonance imaging (MRI). See: N. Lange, J. Giedd, F. Castellanos, A. Vaituzis, and J. Rapoport, 1997, Variability of human brain

structure size: Ages four–twenty years, *Psychiatry Research: Neuroimaging Section* 74:1–12, esp. fig. 2, p. 8.

52. A. Dekaban and D. Sadawsky, 1978, Changes in brain weights during the span of human life: Relation of brain weights to body heights and body weights, *Ann. Neurol.* 44:345–56.

53. The difference is only 25 percent of the standard deviation of brain weight within either sex. In comparison, the difference between mean height for men and women is 200 percent of the standard deviation of height within either sex.

54. N. Forger and S. Breedlove, 1986, Sexual dimorphism in human and canine spinal cord: Role of early androgen, *Proc. Nat. Acad. Sci. (USA)* 83: 7527–31.

55. See: D. Swaab, 1995, Development of the human hypothalamus, *Neurochemical Research* 20:509–19, esp. fig. 5, p. 513; D. Swaab and E. Fliers, 1985, A sexually dimorphic nucleus in the human brain, *Science* 228:1112–15; D. Swaab and M. Hofman, 1988, Sexual differentiation of the human hypothalamus: Ontogeny of the sexually dimorphic nucleus of the preoptic area, *Dev. Brain Res.* 44:314–18; M. Hofman and D. Swaab, 1989, The sexually dimorphic nucleus of the preoptic area in the human brain: A comparative morphometric study, *J. Anat.* 164:55–72.

56. L. Allen and R. Gorski, 1990, Sex difference in the bed nucleus of the stria terminalis of the human brain, *J. Comp. Neurol.* 302:697–706; J.-N. Zhau, M. Hofman, L. Gooren, and D. Swaab, 1997, A sex difference in the human brain and its relation to transsexuality, *Nature* 378:68–70; F. Kruijver, J.-N. Zhou, C. Pool, M. Hofman, L. Gooren, and D. Swaab, 2000, Male-to-female transsexuals have female neuron numbers in a limbic nucleus, *J. Clin. Endocrin. Metab.* 85:2034–41.

57. R. Heath, ed., 1964, *The Role of Pleasure in Behavior*, Harper and Row; C. Sem-Jacobsen, 1968, *Depth-Electrographic Stimulation of the Human Brain and Behavior*, Charles C. Thomas.

58. B. Miller, J. Cummings, H. McIntyre, G. Ebers, and M. Grode, 1986, Hypersexuality or altered sexual preference following brain injury, *J. Neurology, Neurosurgery, and Psychiatry* 49:867–73; D. Gorman and J. Cummings, 1992, Hypersexuality following septal injury, *Arch. Neurol.* 49:308–10.

59. D. Swaab, J.-N. Zhou, T. Ehlhart, and M. Hofman, 1994, Development of vasoactive intestinal polypeptide neurons in the human suprachiasmatic nucleus in relation to birth and sex, *Dev. Brain Res.* 79:249–59; D. Swaab and M. Hofman, 1990, An enlarged suprachiasmatic nucleus in homosexual men, *Brain Res.* 537:141–48.

60. J. McGlone, 1980, Sex differences in human brain asymmetry: A critical survey, *Behav. Brain Sci.* 3:215–63.

61. J. Levy and W. Heller, 1992, Gender differences in human neuropsychological function, pp. 245–73 in A. Gerall, H. Moltz, and I. Ward, eds., 1992, *Handbook of Behavioral Neurobiology: Sexual Differentiation*, Plenum.

62. J. Wada, R. Clarke, and A. Hamm, 1975, Cerebral hemisphere asymmetry in humans: Cortical speech zones in one hundred adult and one hundred infant brains, *Arch. Neurol.* 32:239; M. de Lacoste, D. Horvath, and D. Woodward, 1991, Possible sex differences in the developing human fetal brain, *J. Clin. Exp. Neuropsychol.* 13:831–46.

63. L. Allen, M. Richey, Y. Chai, and R. Gorski, 1991, Sex differences in the corpus callosum of the living human being, *J. Neurosci.* 11:933–42.

64. G. de Courten-Myers, 1999, The human cerebral cortex: Gender differences in structure and function, *J. Neuropath. and Exp. Neurol.* 58:217–26.

65. E. Davis, P. Popper, and R. Gorski, 1996, The role of apoptosis in sexual differentiation of the rat sexually dimorphic nucleus of the preoptic area, *Brain Res.* 734:10–18; T. Rabinowicz, G. de Courten-Myers, J. Petetot, et al., 1996, Human cortex development: Estimates of neuronal numbers indicate major loss late during gestation, *J. Neuropath. and Exp. Neurol.* 55:320–28.

66. M. Laubach, J. Wessberg, and M. Nicolelis, 2000, Cortical ensemble activity increasingly predicts behaviour outcomes during learning of a motor task, *Nature* 405:567–71; J. Sanes and J. Donoghue, 2000, Plasticity and primary motor cortex, *Ann. Rev. Neuroscience* 23:393–415.

67. J. Hyde and M. Linn, 1988, Gender differences in verbal ability: A meta-analysis, *Psychological Bull.* 104:53–69; E. Hampson and D. Kimura, 1992, Sex differences and hormonal influences on cognitive function in humans, pp. 357–98 in J. Becker, S. Breedlove, and D. Crews, eds., *Behavioral Endocrinology*, MIT Press.

68. F. Patterson, C. Holts, and L. Saphire, 1991, Cyclic changes in hormonal, physical, behavioral, and linguistic measures in a female lowland gorilla, *Amer. J. Primatology* 24:181–94.

69. E. Hampson, 1990, Variations in sex-related cognitive abilities across the menstrual cycle, *Brain and Cognition* 14:26–43; E. Hampson, 1990, Estrogen-related variations in human spatial and articulatory-motor skills, *Psychoneuroendocrinology* 15:97–111; M. Moody, 1997, Changes in scores on the Mental Rotations Test during the menstrual cycle, *Perception and Motor Skills* 84:955–61; K. Phillips and I. Silverman, 1997, Differences in the relationship of menstrual cycle phase to spatial performance on two- and three-dimensional tasks, *Hormones and Behavior* 32:167–75; M. McCourt, V. Mark, K. Radonovich, S. Willison, and P. Freeman, 1997, The effects of gender, menstrual phase, and practice on the perceived location of the midsagittal plane, *Neuropsychologia* 35:717–24.

70. B. Sherwin, 1997, Estrogen effects on cognition in menopausal women, *Neurology* 48:S21–S26; K. Yaffe, D. Grady, A. Pressman, and S. Cummings, 1998, Serum estrogen levels, cognitive performance, and risk of cognitive decline in older community women, *J. Amer. Geriatric Soc.* 46:816–21; S. Shaywitz, B. Shaywitz, K. Pugh, R. Fulbright, P. Skudlarski, W. Mencl, R. Constable, F. Naftolin, S. Palter, K. Marchione, L. Katz, D. Shankweiler, J. Fletcher, C. Lacadie, M. Keltz, and J. Gore, 1999, Effect of estrogen on brain activation patterns in postmenopausal women during working memory tasks, *J. Amer. Med. Assoc.* 281:1197–1202; B. McEwen and S. Alves, 1999, Estrogen actions in the central nervous system, *Endocrine Reviews* 20:279–307.

71. S. Shaywitz, 1996, Dyslexia, *Scientific American* (Nov.), 98–103; S. Shaywitz, B. Shaywitz, K. Puch, R. Fulbright, R. Constable, W. Mencl, D. Shankweiler, A. Liberman, P. Skudlarski, J. Fletcher, L. Katz, K. Marchione, C. Lacadie, C. Gatenby, and J. Gore, 1998, Functional disruption in the organization of the brain for reading in dyslexia, *Proc. Nat. Acad. Sci. (USA)* 95:2636–41.

72. M. Linn and A. Petersen, 1985, Emergence and characterization of sex differences in spatial ability: A meta-analysis, *Child Devel.* 56:1479–98; C. Benbow, 1988, Sex differences in mathematical reasoning ability in intellectually talented preadolescents: Their nature, effects, and possible causes, *Behav. Brain Sci.*

11:169–232; H. Stumpf and J. Stanley, 1998, Standardized tests: Still gender based? *Current Directions in Psychological Sci.* 7:192–96; A. Gallagher, R. De Lisi, P. Holst, A. McGillicuddy-DeLisi, M. Morely, and C. Cahalan, 2000, Gender differences in advanced mathematical problem solving, *J. Exp. Child Psychol.* 75:165–90; D. Halpern, 2000, *Sex Differences in Cognitive Abilities*, 3d ed., Lawrence Erlbaum.

73. See: B. Shaywitz, S. Shaywitz, K. Pugh, R. Constable, P. Skudlarski, R. Fulbright, R. Bronen, J. Fletcher, D. Shankweiler, L. Katz, and J. Gore, 1995, Sex differences in the functional organization of the brain for language, *Nature* 373:607–9, esp. the color photograph on p. 608.

74. G. Gron, A. Wunderlich, M. Spitzer, R. Tomczak, and M. Riepe, 2000, Brain activation during human navigation: Gender-different neural networks as substrate of performance, *Nature Neuroscience* 3:404–8; N. Sandstrom, J. Kaurman, and S. Huettel, 1998, Males and females use different distal cues in a virtual environment navigation task, *Brain Res.: Cognitive Brain Res.* 6:351–60.

75. G. Miller, 2000, *The Mating Mind: How Sexual Choice Shaped the Evolution of Human Nature*, Random House.

76. Ibid., p. 2. Subsequent quotations are from pp. 3, 4, 7, 29, 103, 104, 92, 98.

77. Ibid., pp. 217–18.

78. National Center for Health Statistics, 2000, *Health, United States, 2000, with Health and Aging Chartbook* (PHS-2000-1232), cited in Wizemann and Pardue, eds., 2001, *Exploring the Biological Contributions to Human Health*, p. 60.

79. R. Kaplan and P. Erickson, 2000, Gender differences in quality-adjusted survival using a Health-Utilities Index, *Amer. J. Preventive Med.* 18:77–82; cited in ibid., p. 61.

80. Wizeman and Pardue, 2001, *Exploring the Biological Contributions to Human Health*, p. 60.

81. For the math about how natural selection influences life-history evolution, see: J. Roughgarden, 1998, *Primer of Ecological Theory*, Prentice Hall, pp. 203–16; and J. Roughgarden, 1996 [1979], *Theory of Population Genetics and Evolutionary Ecology: An Introduction*, Prentice Hall, pp. 347–70.

82. R. Michael, J. Gagnon, E. Laumann, and G. Kolata, 1995, *Sex in America: A Definitive Survey*, Warner Books, fig. 6, p. 90.

83. Ibid., table 8, p. 116.

13: GENDER IDENTITY

1. T. Elbert, C. Pantev, C. Wienbruch, B. Rockstroh, and E. Taub, 1995, Increased cortical representation of the fingers of the left hand in string players, *Science* 270:305–6.

2. J.-N. Zhau, M. Hofman, L. Gooren, and D. Swaab, 1997, A sex difference in the human brain and its relation to transsexuality, *Nature* 378:68–70; F. Kruijver, J.-N. Zhou, C. Pool, M. Hofman, L. Gooren, and D. Swaab, 2000, Male-to-female transsexuals have female neuron numbers in a limbic nucleus, *J. Clin. Endocrin. Metab.* 85:2034–41.

3. Kruijver et al., 2000, Male-to-female transsexuals have female neuron numbers in a limbic nucleus, pp. 2037, 2039.

4. A. Del Abril, S. Segovia, and A. Guillamon, 1987, The bed nucleus of the stria terminalis in the rat: Regional sex differences controlled by gonadal steroids early

after birth, *Brain Res.* 429:295–300; A. Guillamon, S. Segovia, and A. Del Abril, 1988, Early effects of gonadal steroids on the neuron number in the medial posterior region and the lateral division of the bed nucleus of the stria terminalis in the rat, *Brain Res. Dev. Brain Res.* 44:281–90.

5. Kruijver et al., 2000, Male-to-female transsexuals have female neuron numbers in a limbic nucleus, p. 2034.

6. L. Gooren, 1990, The endocrinology of transsexualism: A review and commentary, *Psychoneuroendocrinology* 15:3–14, esp. p. 4.

7. Kruijver et al., 2000, Male-to-female transsexuals have female neuron numbers in a limbic nucleus, p. 2041.

8. D. Abramowich, I. Davidson, A. Longstaff, and C. Pearson, 1987, Sexual differentiation of the human mid trimester brain, *Eur. J. Obst. Gyn. Reprod. Biol.* 25:7–14.

9. D. Sandberg, H. Meyer-Bahlburg, G. Aranoff, J. Sconzo, and T. Hensle, 1989, Boys with hypospadias: A survey of behavioral difficulties, *J. Pediat Psychol.* 14:491–514; D. Sandberg, H. Meyer-Bahlburg, T. Yager, T. Hensle, S. Levitt, S. Kogan, and E. Reda, 1995, Gender development in boys born with hypospadias, *Psychoneuroendocrinology* 20:693–709.

10. See: K. Zucker, 1999, Intersexuality and gender identity differentiation, *Ann. Rev. Sex Res.* 10:1–69, esp. references on p. 20.

11. Ibid., pp. 28–29.

12. Ibid., pp. 14–15; L. Pinsky, R. Erickson, and R. Schimke, 1999, *Genetic Disorders of Human Sexual Development*, Oxford University Press, pp. 233–38.

13. R. Goy, F. Bercovitch, and M. McBrair, 1988, Behavioral masculinization is independent of genital masculinization in prenatally androgenized female rhesus macaques, *Hormones and Behavior* 22:552–71.

14. M. Diamond and H. Sigmundson, 1997, Sex reassignment at birth, long-term review and clinical implications, *Arch. Pediatr. Adolesc. Med.* 151:298–304.

15. J. Money, cited in J. Colapinto, 1997, The true story of John/Joan, *Rolling Stone,* Dec. 11, pp. 55ff.

16. M. Diamond and H. Sigmundson, 1997, Sex reassignment at birth, long-term review and clinical implications, *Arch. Pediatr. Adolesc. Med.* 151:298–304.

17. J. Colapinto, 2001, *As Nature Made Him: The Boy Who Was Raised As a Girl,* Harper.

18. S. Bradley, G. Oliver, A. Chernick, and K. Zucker, 1998, Experiment of nurture: Ablatio penis at two months, sex reassignment at seven months and a psychosexual followup in young adulthood, *Pediatrics* 102:e9 (abstract).

14: SEXUAL ORIENTATION

1. D. Swaab and M. Hofman, 1995, Sexual differentiation of the human hypothalamus in relation to gender and sexual orientation, *Trends in Neuroscience* 18:264–70; D. Swaab, L. Gooren, and M. Hofman, 1995, Brain research, gender, and sexual orientation, *J. Homosexuality* 28:283–301; D. Swaab and M. Hofman, 1990, An enlarged suprachiasmatic nucleus in homosexual men, *Brain Res.* 537:141–48.

2. This difference is manifest from ten to thirty years of age. See: D. Swaab, 1995, Development of the human hypothalamus, *Neurochemical Research* 5:509–19, esp. fig. 4.

3. Results based on a comparison of brains from eighteen homosexual males be-

tween twenty-two and seventy-four years of age who died of AIDS with the brains of ten heterosexual males between twenty-five and forty-three years of age who also died of AIDS. See: Swaab, Gooren, and Hofman, 1995, Brain research, gender, and sexual orientation, esp. fig. 3.

4. See: S. LeVay, 1991, A difference in hypothalamic structure between heterosexual and homosexual men, *Science* 253:1034–37, esp. fig. 2.

5. D. McFadden and E. Pasanen, 1998, Comparison of the auditory systems of heterosexuals and homosexuals: Click-evoked otoacoustic emissions, *Proc. Nat. Acad. Sci. (USA)* 95:2709–13.

6. D. McFadden, E. Pasanen, and N. Callaway, 1998, Changes in otoacoustic emissions in a transsexual male during treatment with estrogen, *J. Acoust. Soc. Am.* 104:1555–58.

7. R. Pillard and J. Weinrich, 1986, Evidence of familial nature of male homosexuality, *Arch. Gen. Psychiatry* 43:808–12.

8. J. Bailey and D. Benishay, 1993, Familial aggregation of female sexual orientation, *Am. J. Psychiatry* 150:272–77; J. Bailey and A. Bell, 1993, Familiality of female and male homosexuality, *Behav. Genetics* 23:313–22; M. Pattatucci and D. Hamer, 1995, Development and familiality of sexual orientation in females, *Behav. Genetics* 25:407–20.

9. J. Bailey and R. Pillard, 1991, A genetic study of male sexual orientation, *Arch. Gen. Psychiatry* 48:1089–96.

10. F. Whitam, M. Diamond, and J. Martin, 1993, Homosexual orientation in twins: A report on sixty-one pairs and three triplet sets, *Arch. Sexual Behavior* 22:187–206.

11. J. Bailey, R. Pillard, M. Neale, and Y. Agyei, 1993, Heritable factors influence sexual orientation in women, *Arch. Gen. Psychiatry* 50:217–23.

12. M. King and E. McDonald, 1992, Homosexuals who are twins: A study of forty-six probands, *Brit. J. Psychiatry* 160:407–9.

13. J. Bailey and N. Martin, 1995, A twin registry study of sexual orientation, paper presented at the annual meeting of the International Academy of Sex Research, Provinceton, Mass., 1995.

14. E. Eckert, T. Boouchard, J. Bohlen, and L. Heston, 1986, Homosexuality in monozygotic twins reared apart, *Brit. J. Psychiatry* 148:421–25.

15. Bailey and Pillard, 1991, A genetic study of male sexual orientation.

16. D. Hamer, S. Hu, V. Magnuson, N. Hu, and A. Pattatucci, 1993, A linkage between DNA markers on the X chromosome and male sexual orientation, *Science* 261:321–27. For biographical information on Hamer, see: D. Hamer and P. Copeland, 1994, *The Science of Desire*, Simon and Schuster; C. Burr, *A Separate Creation*, Hyperion. For biographical information on Pattatucci, see: K. Brandt, 1993, Doctor Angela Pattatucci: Not your typical government scientist, *Deneuve* (Dec.), 44–46; S. Levay and E. Nonas, 1995, *City of Friends: A Portrait of the Gay and Lesbian Community in America*, MIT Press, pp. 194–95.

17. See: W. Byne and B. Parsons, 1993, Human sexual orientation: The biologic theories reappraised, *Arch. Gen. Psychiatry* 50:228–39; W. Byne, 1994, The biological evidence challenged, *Scientific American* (May), 50–55.

18. R. Kirkpatrick, 2000, The evolution of human homosexual behavior, *Current Anthropology* 41:385–413.

19. Burr, 1996, *A Separate Creation,* pp. 179–80.

20. Ibid., pp. 235–36.

21. Ibid., pp. 173–74.

22. See: C. Golden, 1966, What's in a name? Sexual self-identification among women, pp. 229–47 in R. Savin-Williams and K. Cohen, eds., *The Lives of Lesbians, Gays, and Bisexuals,* Harcourt Brace.

23. See: Hamer et al., 1993, A linkage between DNA markers on the X chromosome and male sexual orientation, fig. 2, p. 322.

24. S. Hu, A. Pattatucci, C. Patterson, L. Li, D. Fulker, S. Cherny, L. Kruglyak, and D. Hamer, 1995, Linkage between sexual orientation and chromosome Xq28 in males but not in females, *Nature Genetics* 11:248–56.

25. J. Bailey, R. Pillard, K. Dawood, M. Miller, L. Farrer, S. Trivedi, and R. Murphy, 1999, A family history study of male sexual orientation using three independent samples, *Behav. Genetics* 29:79–86.

26. G. Rice, C. Anderson, N. Risch, and G. Ebers, 1999, Male homosexuality: Absence of linkage to microsatellite markers at Xq28, *Science* 284:665–67. Data and quotations from this source.

27. Burr, 1996, *A Separate Creation,* p. 236.

28. Quotations here and below cited in ibid., pp. 200–5.

29. L. Peplau, L. Spalding, T. Conley, and R. Veniegas, 1999, The development of sexual orientation in women, *Ann. Rev. Sex Res.* 10:70–99; see also M. Dickemann, 1993, Reproductive strategies and gender construction: An evolutionary view of homosexualities, *J. Homosexuality* 24:55–71.

30. C. Daskalos, 1998, Changes in the sexual orientation of six heterosexual male-to-female transsexuals, *Arch. Sex. Behav.* 27:605–14.

31. Cassandra Smith of Boston University, cited in Burr, 1996, *A Separate Creation,* p. 243.

32. Philip Reilly of the Shriver Center for Mental Retardation, cited in ibid., p. 244.

33. Ibid., p. 245.

34. Ibid.

35. S. Essock-Vitale and M. McGuire, 1988, What seventy million years hath wrought: Sexual histories and reproductive success of a random sample of American women, pp. 221–35 in M. Betzig, B. Mulder, and P. Turke, eds., *Human Reproductive Behavior: A Darwinian Perspective,* Cambridge University Press.

36. *A Yankelovich MONITOR Perspective on Gays and Lesbians,* 1994, Yankelovich.

37. R. Baker and M. Bellis, 1995, *Human Sperm Competition: Copulation, Masturbation, and Infidelity,* Chapman and Hall.

38. *A Yankelovich MONITOR Perspective on Gays and Lesbians,* 1994.

39. S. Isomura and M. Mizogami, 1992, The low rate of HIV infection in Japanese homosexual and bisexual men: An analysis of HIV seroprevalence and behavioral risk factors, *AIDS* 6:501.

40. L. Dean, I. Meyer, K. Robinson, R. Sell, R. Sember, V. Silenzio, D. Bowen, J. Bradford, E. Rothblum, M. Scout, J. White, P. Dunn, A. Lawrence, D. Wolfe, and J. Xavier, 2000, Lesbian, gay, bisexual, and transgender health: Findings and concerns, *J. Gay Lesbian Med. Assoc.* 4:101–51.

41. E. O. Wilson, 1975, *Sociobiology: The New Synthesis,* Harvard University Press, see esp. p. 555 for a discussion of homosexuality; and E. Wilson, 1979, Gay as normal: Homosexuality and human nature: A sociobiological view, *Advocate* (May 3), pp. 15, 18.

42. S. LeVay, 1996, *Queer Science: The Use and Abuse of Research into Homosexuality,* MIT Press, p. 193.

43. See also: J. Weinrich, 1995, Biological research on sexual orientation: A critique of the critics, *J. Homosexuality* 28:197–213.

44. Hamer and Copeland, 1994, *The Science of Desire,* esp. chap. 11.

45. Burr, 1996, *A Separate Creation,* pp. 257–59.

46. F. Muscarella, 2000, The evolution of homoerotic behavior in humans, *J. Homosexuality* 40:51–77.

47. R. Kirkpatrick, 2000, The evolution of human homosexual behavior, *Current Anthropology* 41:385–413. (Subsequent quotations from this source.)

48. N. Barber, 1998, Ecological and psychosocial correlates of male homosexuality: A cross-cultural investigation, *J. Cross-Cultural Psychology* 29:387–401.

49. See: Dickemann, 1993, Reproductive strategies and gender construction.

15: PSYCHOLOGICAL PERSPECTIVES

1. J. Myerowitz, 2002, *How Sex Changed: A History of Transsexuality in the United States,* Harvard University Press.

2. P. Califia, 1997, *Sex Changes: The Politics of Transgenderism,* Cleis Press.

3. M. Brown and C. Rounsley, 1996, *True Selves: Understanding Transsexualism for Families, Friends, Coworkers, and Helping Professionals,* Jossey-Bass.

4. Ibid., pp. 30, 31, 42.

5. Ibid., pp. 46, 36.

6. Ibid., pp. 47, 37.

7. Ibid., p. 43.

8. Ibid., p. 44.

9. Ibid., pp. 51, 52.

10. Ibid., p. 73.

11. Ibid., p. 53.

12. S. Corbett, 2001, When Debbie met Christina, who then became Chris, *New York Times Magazine,* Oct. 14, pp. 84–87.

13. Brown and Rounsley, 1996, *True Selves,* p. 61.

14. Ibid., p. 71.

15. Ibid., p. 88.

16. S. Ramet, 1996, Introduction, *Gender Reversals and Gender Cultures,* Routledge.

17. T. Sellers, 1992, *The Correct Sadist: The Memoirs of Angel Stern,* Temple Press, p. 98, cited in ibid., p. 13.

18. S. Stryker, 1998, The transgender issue: An introduction, *GLQ* 4:145–58, esp. 150.

19. A. Lawrence, http://www.annelawrence.com/autogynephiliaoriginal.html.

20. The 1996 study is available at http://www.ren.org/Lawrence1.html; Lawrence's site (http://www.annelawrence.com/nwc.html) includes a summary of the data from the 1996 study. The 1998 study appears at http://www.annelawrence.

com/autogynephiliaoriginal.html. Quotations in subsequent paragraphs are from these sites.

21. M. Schroder, 1995, New women: Sexological outcomes of gender reassignment surgery, Ph.D. diss., Institute for Advanced Study of Human Sexuality, San Francisco.

22. A. Lawrence, http://www.annelawrence.com/aghistories.html; http://www.annelawrence.com/agnarratives.html.

23. R. Michael, J. Gagnon, E. Laumann, and G. Kolata, 1995, *Sex in America: A Definitive Survey,* Warner Books, table 14, p. 157.

24. Brown and Rounsley, 1996, *True Selves,* p. 71.

25. M. Martino, 1977, *Emergence: A Transsexual Autobiography,* Crown Publishers, cited in Califia, 1997, *Sex Changes,* p. 38.

26. Califia, 1997, *Sex Changes,* p. 39.

27. N. Vincent, 1999, A real man: Brooklyn girl grows hair on chest, *Village Voice,* Nov. 17–23.

28. J. Green, 2004, *Becoming a Visible Man,* Vanderbilt University Press.

29. R. Michael, J. Gagnon, E. Laumann, and G. Kolata, 1995, *Sex in America: A Definitive Survey,* Warner Books, table 8, p. 116.

30. Brown and Rounsley, 1996, *True Selves,* p. 79.

31. Ibid., p. 81.

32. Ibid., p. 82.

33. Ibid., p. 100.

34. A full guide to transition appears in G. Isreal and D. Trarver II, 1997, *Transgender Care: Recommended Guidelines, Practical Information and Personal Accounts,* Temple University Press.

35. Brown and Rounsley, 1996, *True Selves,* p. 126.

36. Ibid., p. 127.

37. Ibid., p. 154.

38. M. Rottnek, ed., 1999, *Sissies and Tomboys: Gender Nonconformity and Homosexual Childhood,* New York University Press.

39. M. Lassell, 1999, Boys don't do that, pp. 245–62 in ibid.

40. K. Chernin, 1999, My life as a boy, pp. 199–208 in ibid.

16: DISEASE VERSUS DIVERSITY

1. See, for example, the medical dictionaries available on MedLine at http://www.nlm.nih.gov/medlineplus/dictionaries.html.

2. G. Maranto, 1998, On the fringes of the bell curve: The evolving quest for normality, *New York Times,* May 26, p. B13; A. Dreger, 1998, When medicine goes too far in pursuit of normality, *New York Times,* July 28. Also see J. Terry and J. Urla, 1995, *Deviant Bodies,* Indiana University Press, which looks at history's changing definitions of normalcy by looking at its flip side, deviance.

3. At the mutation-selection equilibrium, a dominant mutant gene's frequency in the population, p, is v/s, and a recessive mutant gene's frequency is $\sqrt{v/s}$, where v is the aggregate recurrent mutation rate from the remaining alleles to the mutant allele (typically, say, 10^{-6}), and s is the selection coefficient that measures how bad the mutant allele is relative to the nonmutants. Specifically, if the Darwinian fitness of the nonmutant genotypes is standardized to 1, then the Darwinian fitness of a mutant genotype is expressed as $1 - s$. Thus s times 100 can be thought of as the percentage

by which the mutant's Darwinian fitness is less than the nonmutant's, and s varies from 1 for a lethal mutation down to 0 for a mutation that is not deleterious at all.

For a worst-case scenario, where the mutant gene is lethal, the gene frequency, p, varies from 10^{-6} if dominant to 10^{-3} if recessive. The frequency of births of individuals with the mutant genotype is about $2p$ if the mutant gene is dominant, because the mutant gene is primarily expressed in heterozygotes, and about p^2 if recessive, because the mutant gene is expressed only in homozygotes. These birth frequencies are then $2v/s$ for a dominant and v/s for a recessive. In both cases, as well as intermediate cases, the birth frequency b is summarized within a factor of two, by the approximate formula $b \approx v/s$. For a dominant lethal gene with $s = 1$ and $v = 10^{-6}$, this formula works out to be about one birth per million. Lethal genetic defects are therefore very rare, with numbers like one in a million births being typical.

If a trait is more common than one in a million, then something else is going on too—the gene is either not so deleterious or is deleterious only when combined with certain other genes or when expressed in a particular environment—or it is ameliorated in some other way. We may take a figure of 1 in 50,000 as an approximate criterion for traits less severe than lethal to qualify as genetic diseases. More specifically, if we rearrange the summary formula of $b \approx v/s$ as $s \approx v/b$, we see that a b of 1 in 50,000 corresponds to a selection coefficient of 0.05. A selection coefficient of 0.05 means for every 100 net offspring produced by individuals not carrying the mutant genotype, only 95 are produced by those who do carry the mutant genotype, a 5 percent deficit in Darwinian fitness. A 5 percent deficit in fitness represents natural selection strong enough to stand out against the evolutionary noise of genetic drift in a population of 200 or more. A 5 percent deficit in fitness may not seem like a lot, but if sustained in all environments at all times, it will lead to a birth frequency of only 1 in 50,000.

If the birth frequency is higher than 1 in 50,000, the fitness deficit is not as high as 5 percent. Thus traits whose birth frequency is as common as 1 in 1,000 to 1 in 100, the range of most LBGTI traits, are inconsistent with any but the weakest opposing selection, selection so weak as to be lost in the evolutionary noise of genetic drift for most human effective-population sizes.

The formulas for a mutation-selection balance can be found in any textbook on population genetics. See, for example, J. Crow and M. Kimura, 1970, *An Introduction to Population Genetics Theory,* Harper and Row, pp. 259–60; and J. Roughgarden, 1996, *Theory of Population Genetics and Evolutionary Ecology,* Prentice Hall, p. 47. The conversion from gene frequencies to birth frequencies uses the Hardy-Weinberg relations.

4. *Time,* January 11, 1999, pp. 57–58.

5. See also R. Nesse and G. Williams (a medical doctor and an evolutionary biologist), 1996, *Why We Get Sick: The New Science of Darwinian Medicine,* Vintage Books. See especially chapter 7, "Genes and Disease," which includes a table of benefits from genes considered diseases; work by J. Diamond, a medical doctor and ecologist, in B. Spyropoulos and J. Diamond, 1989, *Nature,* 331:666; *Discover,* November 1989, pp. 72–78; *Natural History,* June 1988, pp. 10–13, and February 1990, pp. 26–30. See also G. Kolata, 1993, *New York Times,* Nov 16, pp. C1 and C3, on cystic fibrosis; and N. Angier, 1994, *New York Times,* June 1, p. B9, on evolutionary aspects of genetic diseases.

6. The prevalence of gay males and of lesbians is between 10 percent and 1 percent, according to Alfred Kinsey's data and subsequent surveys. The numbers vary according to how being gay is defined. In any particular study, gay men appear to be as much as twice as prevalent as lesbian women. See: R. Michael, J. Gagnon, E. Laumann, and G. Kolata, 1994, *Sex in America: A Definitive Survey,* Warner Books, pp. 169–83; S. LeVay, 1996, *Queer Science,* MIT Press, pp. 47–49 and 60–64.

7. The Harry Benjamin International Gender Dysphoria Association's *Standards of Care for Gender Identity Disorders,* Fifth Draft, January 29, 1997, claims that in Holland 1 in 11,900 people are male-to-female transsexuals and 1 in 30,400 people are female-to-male transsexuals. The association was started through the efforts of a transgendered man, Reed Erickson. See: http://web.uvic.ca/erick123/.

8. See http://www.lynnconway.com/.

9. Http://ai.eecs.umich.edu/people/conway/TS/TS-II.html#anchor635615.

10. U.K. Home Office, July 2000, *Report of the Interdepartmental Working Group on Transsexual People,* available as a pdf file at http://www.homeoffice. gov.uk/ccpd/wgtrans.pdf, and as an html file at http://www.pfc.org.uk/workgrp/wgrp-all.htm.

11. See the figure of 5,000 used by Press for Change, http://www.pfc.org.uk/index.htm.

12. See: D. Kelly, 2001, Estimating of the prevalence of transsexualism in the United Kingdom, http://ai.eecs.umich.edu/people/conway/TS/UK-TSprevalence.html.

13. See: J. Fichtner, D. Filipas, A. Mottrie, G. Voges, and R. Hohenfellner, 1995, Analysis of meatal location in 500 men: Wide variation questions need for meatal advancement in all pediatric anterior hypospadias cases, *J. Urology* 154:833–34, cited in A. Fausto-Sterling, 2000, *Sexing the Body: Gender, Politics, and the Construction of Sexuality,* Basic Books, pp. 57–58.

14. A frequency of 0.5797/1,000 people is reported in M. Blackless, A. Charuvastra, A. Derryck, A. Fausto-Sterling, K. Lauzanne, and E. Lee, 2000, How sexually dimorphic are we? Review and synthesis, *Amer. J. Human Biol.* 12:151–66.

15. Fichtner et al., 1995, Analysis of meatal location in 500 men.

16. Blackless et al., 2000, How sexually dimorphic are we? See also: Fausto-Sterling, 2000, *Sexing the Body,* table 3.2, p. 53; L. Pinsky, R. Erickson, and R. N. Schimke, 1999, *Genetic Disorders of Human Sexual Development,* Oxford University Press, p. 236.

17. Pinsky et al., 1999, *Disorders of Human Sexual Development,* p. 234.

18. Blackless et al., 2000, How sexually dimorphic are we?

19. B. Gottlieb, H. Lehvaslaiho, K. Beitel, M. Trifiro, R. Lumbrosos, and L. Pinsky, 1998, The androgen receptor gene mutations database, *Nucleic Acids Res.* 26:234–38.

20. Pinsky et al., 1999, *Disorders of Human Sexual Development,* pp. 242–52.

21. K. Buckton, 1983, Incidence and some consequences of X-chromosome abnormalities in liveborn infants: Cytogenetics of the mammalian X chromosome, pp. 7–22 in *Part B: X Chromosome Abnormalities and Their Clinical Manifestations,* ed. Alan R. Liss, cited in Blackless et al., 2000, How sexually dimorphic are we?, p. 152.

22. L. Abramsky and J. Chapple, 1997, 47,XXY (Klinefelter syndrome) and 47,XYY: Estimated rates of and indication for postnatal diagnosis for prenatal coun-

seling, *Prenat. Diagn.* 17:363–68, cited in Blackless et al., 2000, How sexually dimorphic are we?, p. 152.

23. Blackless et al., 2000, How sexually dimorphic are we?; see also A. Danso and F. Nkrumah, 1992, The challenges of ambiguous genitalia, *Centr. Afr. J. Med.* 38:367–71; G. Krob, A. Braun, and U. Kuhnle, 1994, True hermaphroditism: Geographical distribution, clinical findings, chromosomes and gonadal histories, *Eur. J. Pediatr.* 152:2–10; U. Kuhnle, H. Schwarz, U. Lohrs, S. Stengel-Ruthkowski, H. Cleve, and A. Braun, 1993, Familial true hermaphroditism: Paternal and maternal transmission of true hermaphroditism (46,XX) and XX maleness in the absence of the Y-chromosome sequences, *Hum. Genet.* 92:571–76; S. Slaney, I. Chalmers, N. Affara, and L. Chitty, 1998, An autosomal or X-linked mutation results in true hermaphrodites and 46,XX males in the same family, *J. Med. Genet.* 35:17–22.

24. W. van Niekerk, 1976, True hermaphroditism: An analytic review with a report of three new cases, *Am. J. Obst. Gyn.* 126:890–907.

25. The examples described are drawn from LeVay, 1996, *Queer Science,* pp. 91–97.

26. E. Goode, 1998, On gay issue, psychoanalysis treats itself, *New York Times,* Dec. 12, p. A29.

27. R. Weller, 1998, Board nixes gay conversion therapy, Associated Press, Dec. 11, available on the web at http://www.skeptictank.org/hs/apanogay.htm. For information on the actual resolutions, see http://www.apa.org/pi/lgbpolicy/orient.html; and http://www.psych.org/pract_of_psych/copptherapyaddendum83100.cfm. For an overview, see http://psychology.ucdavis.edu/rainbow/html/facts_changing.html.

28. M. Lemonick, 1999, Designer babies, *Time,* Jan. 11, p. 66.

29. See: The Harry Benjamin International Gender Dysphoria Association, *Standards of Care for Gender Identity Disorders,* Fifth Draft, Jan. 29, 1997.

30. G. Rekers and O. Lovaas, 1974, Behavioral treatment of deviant sex-role behaviors in the male child, *J. Applied Behav. Analysis* 7:173–90; cited in K. Zucker and S. Bradley, 1995, *Gender Identity Disorder and Psychosexual Problems in Children and Adolescents,* Guilford Press, p. 271.

31. Zucker and Bradley, 1995, *Gender Identity Disorder,* pp. 281, 315.

32. Ibid., p. 281.

33. I. Haraldsen and A. Dahl, 2000, Symptom profiles of gender dysphoric patients of transsexual type compared to patients with personality disorders and healthy adults, *Acta Psychiatr. Scand.* 102:276–81.

34. Ibid.

35. P. Califia, 1997, *Sex Changes: The Politics of Trans-genderism,* Cleis Press.

36. See the web page for *GIDreform.org* at http://www.transgender.org/tg/gidr/, specifically K. Wilson, 1998, The disparate classification of gender and sexual orientation in American psychiatry, available on the web at http://www.transgender.org/tg/gidr/kwapa98.html.

37. N. Bartlett, P. Vasey, and W. Bukowski, 2000, Is gender identity disorder in children a mental disorder? *Sex Roles* 43:753–85.

38. Pinsky et al., 1999, *Disorders of Human Sexual Development,* p. 51.

39. P. Donahoe and J. Schnitzer, 1996, Evaluation of the infant who has ambiguous genitalia, and principles of operative management, *Seminars Pediatr. Surg.* 5:30–40.

40. D. Federman and P. Donahoe, 1995, Ambiguous genitalia—etiology, diagnosis, and therapy, *Adv. Endocrin. and Metab.* 6:91–116; quotations here and below from pp. 103, 108.

41. See Fausto-Sterling, 2000, *Sexing the Body*, diagram on p. 59.

42. For penis length, see E. Flatau, Z. Josefsberg, S. Reisner, O. Bialik, and Z. Laron, 1975, Penile size in the newborn infant, *J. Pediatr.* 87:663–64. For clitoral length, see S. Oberfield, A. Mondok, F. Shahrivr, J. Klein, and L. Leving, 1989, Clitoral size in full-term infants, *Amer. J. Perinat.* 6:453–54.

43. Fausto-Sterling, 2000, *Sexing the Body*, pp. 62–63.

44. For more specific information, consult http://www.the-penis.com/hypo spadias.html, which has links to more information as well as personal testimonials; and a hypospadias support group at http://www.hypospadias.co.uk/index2.htm. See also the home page of the Intersex Society of America, an intersex advocacy group, at http://isna.org, and their guide to hypospadias at http://isna.org/library/hypospa diasguide.html.

45. Federman and Donahoe, 1995, Ambiguous genitalia, p. 104.

46. See diagrams in Fausto-Sterling, 2000, *Sexing the Body,* pp. 61–63.

47. Pinsky et al., 1999, *Genetic Disorders of Human Sexual Development*, p. 235. For a booklet for parents of children with CAH, see http://www.rch.unimelb.edu.au /publications/cah_book/index.html.

48. See Fausto-Sterling, 2000, *Sexing the Body,* p. 65.

49. Pinsky et al., 1999, *Genetic Disorders of Human Sexual Development*, p. 252.

50. See the website for the AIS Support Group at http://www.medhelp.org/www /ais/; a publication is available at http://www.rch.unimelb.edu.au/publications /CAIS.html.

51. Donahoe and Schnitzer, 1996, Evaluation of the infant who has ambiguous genitalia, p. 30.

52. S. Jahoda, Theatres of madness, pp. 251–76 in Terry and Urla, 1995, *Deviant Bodies.*

53. R. Maines, 1999, *The Technology of Orgasm: "Hysteria," the Vibrator, and Women's Sexual Satisfaction,* Johns Hopkins University Press; reviewed by N. Angier, *New York Times,* Feb. 13, 1999.

54. Study finds dysfunction in sex lives is widespread, 1999, *New York Times,* Feb. 10, p. A14.

55. T. Abate, 1999, Vivus patents sex cream for women, *San Francisco Chronicle,* March 9, pp. E1, E4.

56. D. Stead, 1999, Circumcision's pain and benefits re-examined, *New York Times,* March 2, p. D6.

57. A. Raghunathan, 1999, A bold rush to sell drugs to the shy, *New York Times,* May 18, pp. C1, C6.

58. Robert Pear, 1999, Few seek to treat mental disorders, a U.S. study says, *New York Times,* Dec. 13.

59. C. Goldberg, 1999, Mental health of students gets new push at Harvard, *New York Times,* Dec. 14.

60. E. Eakin, 2000, Bigotry as mental illness, or just another norm, *New York Times,* Jan. 15, p. A21.

61. G. Kolata, 2002, More may not mean better in health care, studies find, *New York Times,* July 21, pp. A1, A20.

62. P. G. Allen, 1990, The woman I love is a planet, in I. Diamond and G. Ornstein, eds., *Reweaving the World: The Emergence of Ecofeminism,* Sierra Club Books, pp. 52–57.

17: GENETIC ENGINEERING VERSUS DIVERSITY

1. See R. Proctor, 1995, The destruction of "Lives Not Worth Living," pp. 170–96 in J. Terry and J. Urla, 1995, *Deviant Bodies,* Indiana University Press.

2. I. Wilmut, A. Schnieke, J. McWhir, A. Kind, and K. Campbell, 1997, Viable offspring from fetal and adult mammalian cells, *Nature* 385:810–14.

3. G. Kolata, 1998, Japanese scientists clone a cow, making eight copies, *New York Times,* Dec. 8, p. 1. For information about cloning from the web, start surfing at http://www.ornl.gov/hgmis/elsi/cloning.html and http://www.ri.bbsrc.ac.uk/library/research/cloning/. Books include the skeptical introduction by G. Kolata, 1998, *Clone: The Road to Dolly, and the Path Ahead,* William Morrow; and a positive account by G. Pence, 1998, *Who's Afraid of Human Cloning?* Rowman & Littlefield.

4. Dr. Y. Tsunoda of Kinki University in Nara, Japan, in Kolata, 1998, Japanese scientists clone a cow, p. 1.

5. See the U.S. government's infomercial web page on cloning: http://www.howstuffworks.com/cloning.htm.

6. Http://www.nature.com/nsu/020211/020211-13.html.

7. Http://news.bbc.co.uk/hi/english/ sci/tech/newsid_1820000/1820749.stm.

8. Quoted in Kolata, 1998, Japanese scientists clone a cow, p. 1. See: L. Silver, 1998, *Remaking Eden: Cloning and Beyond in a Brave New World,* Avon Books.

9. On the South Korean claim, see: S. WuDunn, 1998, South Korean scientists say they cloned a human cell, *New York Times,* Dec. 17; also: G. Kolata, 1998, Speed of cloning advances surprises US ethics panel, *New York Times,* Dec. 8. On U.S. claim, see: http://www.sciam.com/explorations/2001/112401ezzell.

10. Researcher promises human clone by next year, team plans delivery outside of the US, 2002, *San Francisco Chronicle,* May 16.

11. See: http://www.msnbc.com/news/681969.asp; G. Kolata, 2003, First mammal clone dies: Dolly made science history, *New York Times,* Feb. 15.

12. G. Kolata, 2001, In cloning, failure far exceeds success, *New York Times,* Dec. 11.

13. See also: Simerly et al., 2003, Molecular correlates of primate nuclear transfer failures, *Science* 300:297–98.

14. Kolata, 2001, In cloning, failure far exceeds success.

15. J. Whitfield, 2001, Cloned cows in the pink: Healthy cows buck the trend for sickly clones, Nov. 23, available at http://www.nature.com/nsu/011129/011129-1.html. (Quotations that follow are from this source.)

16. H. Gee, 1999, Dolly is not quite a clone, Aug. 31, available at http://www.nature.com/nsu/990902/990902-5.html.

17. See: T. Maniatis and R. Reed, 2002, An extensive network of coupling among gene expression machines, *Nature* 416: 499–506.

18. H. Pearson, 2002, Cause of sick clones contested: Tentative diagnosis of

clones' complaints, Jan. 11, available at http://www.nature.com/nsu/020107/020107 –10.html.

19. See http://www.globalchange.com/clonech.htm; http://www.humancloning. org/; http://www.clonaid.com/.

20. C. Hall, 2002, UCSF admits human clone research, *San Francisco Chronicle,* May 25.

21. Dr. H. Sweeney, of the University of Pennsylvania Medical Center, quoted in P. Recer, 1988, New genes may restore old muscles, study finds, *Santa Barbara News Press,* Dec. 15, p. A1.

22. Dr. W. Haseltine, chairman of Human Genome Sciences in Rockville, Md., quoted in N. Wade, 1998, Immortality of a sort beckons to biologists, *New York Times,* Nov. 17, p. D1.

23. L. Jaroff, 1999, Fixing the genes, *Time,* Jan. 11, pp. 68–70.

24. S. Stolberg, 2000, Gene therapy ordered halted at university, *New York Times,* Jan. 22, p. A1.

25. M. D. Lemonick, 1999, Designer babies, *Time,* Jan. 11, pp. 64–66.

26. Lance Morrow, 1999, Is this right? Who has the right to say? A mother of octuplets, one already gone, says God has blessed her, *Time,* Jan. 11, p. 41.

27. Dr. J. Botkin, a University of Utah pediatrics professor, quoted in F. Golden, 1999, Good eggs, bad eggs, *Time,* Jan. 11, pp. 56–59.

28. G. Kolata, 1999, $50,000 offered to tall, smart egg donor, *New York Times,* March 3, p. A10.

29. T. Abate, 1999, Artist proposes using jellyfish genes to create glow-in-the-dark dogs, *San Francisco Chronicle,* Oct. 18, p. B1.

30. S. Madoff, 2002, The wonders of genetics breed a new art, *New York Times,* May 26.

31. S. Begley, 2002, Cloned calves may offer cure for many ills, *Wall Street Journal,* Aug. 12.

32. N. Wade, 1999, Life is pared to basics: Complex issues arise, *New York Times,* Dec. 14.

33. James Watson, 1999, All for the good, *Time,* Jan. 11, p. 91.

34. N. Wade, 2002, Scientist reveals secret of genome: It's his, *New York Times,* April 27, p. A1. (Quotations and discussion that follow are from this source.)

35. A. Pollack, 2002, A jury orders Genentech to pay $300 million in royalties, *New York Times,* June 11.

36. A. Pollack and D. Johnson, 2002, Former chief of ImClone Systems is charged with insider trading, *New York Times,* June 13.

37. N. Wade, 2002, Human genome sequence has errors, scientists say, *New York Times,* June 11.

38. J. DeRisi, V. Iyer, and P. Brown, 1997, Exploring the Metabolic and Genetic Control of Gene Expression on a Genomic Scale, *Science* 278: 680–86. A wonderful animation of how the technique works is available on the web at http://www.bio. davidson.edu/courses/genomics/chip/chip.html. A slide show containing the pictures of whole banks of genes shutting down and starting up during transition from fermentation to respiration appears at http://industry.ebi.ac.uk/ālan/MicroArray/In troMicroArrayTalk/index.htm. For more information, a good place to begin search-

ing is the Stanford Microarray Database (SMD) at http://genome-www.stanford. edu/microarray: G. Sherlock et al., 2001, The Stanford Microarray Database, *Nucleic Acids Res.* 29:152–55.

39. D. Lockhart and E. Winzeler, 2000, Genomics, gene expression and DNA arrays, *Nature* 405:827–36.

40. See: http://www.monsantoafrica.com/reports/fieldpromise/page4.html.

41. D. Quist and I. Chapela, 2001, Transgenic DNA introgressed into traditional maize landraces in Oaxaca, Mexico, *Nature* 414:541–43.

42. See the back-and-forth exchange in *Nature* (advance online publication), April 4, 2002, available at www.nature.com, 1–3.

43. N. Wade, 2002, After scare, a gene therapy trial proceeds, *New York Times,* Jan. 8.

44. N. Wade, 1999, Ten drug makers join to find genetic roots of diseases, *New York Times,* April 15. Dr. F. Collins, director of the Human Genome Project at the National Institutes of Health, is quoted as saying, "If we are talking about building a genome map, it should be a real catalogue of human variation and not just northern Europeans." Furthermore, Wade (1999, The genome's combative entrepreneur, *New York Times,* May 18, pp. D1, D2) quotes Dr. Venter of Celera as intending to sequence genomes of "five individuals, to be selected from the major ethnic groups, including three men and two women."

45. See N. Wade, 1999, Tailoring drugs to fit the genes, *New York Times,* April 20. Wade reports that the ten drug companies that banded together to set up genome maps for the various races are planning to develop tailor-made drugs.

46. G. Kolata, 1999, Using gene testing to decide a patient's treatment, *New York Times,* Dec. 20, p. 1; Andrew Pollack, 1999, In the works: Drugs tailored to individual patients, *New York Times,* Dec. 20, p. C8.

47. See, for example: N. Wade, 2002, Genes help identify oldest human population, *New York Times,* Jan. 8.

48. W. Broad and J. Miller, 1999, Defector discloses germ weapons, *New York Times,* April 5.

49. N. Wade, 2002, Thrown aside, genome pioneer plots a rebound, *New York Times,* April 30.

50. See: T. Abate, 2002, Depressing times for biotech as key index hits three-year low, *San Francisco Chronicle,* May 13; A. Pollack, 2002, Biotechnology seeks end to slump, *New York Times,* May 19.

51. D. Levy, 2002, Scientists need to play a bigger role in national security, Drell says, *Stanford Report,* Feb. 14, available at http://www.stanford.edu/dept/news/report/news/february20/drell-a.html.

18: TWO-SPIRITS, MAHU, AND HIJRAS

1. W. Williams, 1992, *The Spirit and the Flesh: Sexual Diversity in American Indian Culture,* Beacon Press; W. Roscoe, 1998, *Changing Ones: Third and Fourth Genders in Native North America,* St. Martin's Press; W. Roscoe, 1988, *Living the Spirit: A Gay American Indian Anthology,* St. Martin's Press; S. Long, 1996, There is more than just women and men, pp. 183–96 in S. Ramet, ed., *Gender Reversals and Gender Cultures: Anthropological and Historical Perspectives,* Routledge; S. Lang,

1998, *Men As Women, Women As Men,* University of Texas Press; S. E. Jacobs, W. Thomas, and S. Lang, 1997, *Two-Spirit People: Native American Gender Identity, Sexuality, and Spirituality,* University of Illinois Press.

2. Williams, 1992, *The Spirit and the Flesh,* p. 19.

3. Roscoe, 1998, *Changing Ones,* p. 31. All quotations regarding Osh-Tisch are from this source.

4. Ibid., p. 35.

5. Ibid., p. 47. All quotations regarding Klah are from this source.

6. Ibid., p. 54.

7. Ibid., p. 31.

8. Ibid., p. 33.

9. Ibid., p. 75.

10. Ibid., p. 145.

11. S. Lang, 1999, Lesbians, men-women and two-spirits: Homosexuality and gender in Native American cultures, pp. 91–116 in E. Blackwood and S. Wieringa, eds., *Female Desires, Same-Sex Relations and Transgender Practices across Cultures,* Columbia University Press.

12. Williams, 1992, *The Spirit and the Flesh,* p. 24. The quotations that follow are from this source.

13. Roscoe, 1998, *Changing Ones,* p. 150.

14. Williams, 1992, *The Spirit and the Flesh,* p. 98.

15. Ibid., p. 137.

16. Roscoe, 1998, *Changing Ones,* p. 202

17. Ibid., p. 17.

18. See Williams, 1992, *The Spirit and the Flesh,* pp. 80, 219.

19. N. Besnier, 1994, Polynesian gender liminality through time and space, pp. 285–328 in G. Herdt, ed., *Third Sex, Third Gender: Beyond Sexual Dimorphism in Culture and History,* Zone Books. The quotation is from p. 289.

20. Ibid., p. 290.

21. D. Elliston, 1999, Negotiating transnational sexual economies: Female mahu and same-sex sexuality in "Tahiti and Her Islands," pp. 232–52 in Blackwood and Wieringa, eds., *Female Desires, Same-Sex Relations and Transgender Practices across Cultures.*

22. Besnier, 1994, Polynesian gender liminality, p. 291.

23. Ibid., pp. 296–98.

24. Ibid., p. 300.

25. Ibid., p. 298.

26. Elliston, 1999, Negotiating transnational sexual economies, p. 234.

27. Ibid., p. 237.

28. Ibid., p. 237.

29. Ibid., p. 241.

30. This account of the *hijras* is drawn primarily from the research of anthropologist Serena Nanda and linguist Kira Hall. See: S. Nanda, 1999, *Neither Man nor Woman: The Hijras of India,* 2d ed., Wadsworth; L. Cohen, 1995, The pleasures of castration: The postoperative status of hijras, jankhas, and academics, pp. 276–304 in P. R. Abramsan and S. D. Pinkerton, eds., *Sexual Nature, Sexual Culture,* University of Chicago Press; K. Hall, 1995, *Hijra/bijrin: Language and Gender Identity,*

Ph.D. diss., Department of Linguistics, University of California, Berkeley; K. Hall, 1997, "Go suck your husband's sugarcane!": Hijras and the use of sexual insult, pp. 430–60 in A. Livia and K. Hall, eds., *Queerly Phrased: Language, Gender and Sexuality,* Oxford University Press; and K. Hall, 2001, personal communication. See also photographs by Takeshi Ishikawa at http://home.interlink.or.jp/takeshii/index.htm.

31. B. Bearnak, 2001, A pox on politicians: A eunuch you can trust, *New York Times,* Jan. 19.

32. J. Karp, 1998, And she's a eunuch: Ms. Nehru goes far in Indian politics: Lower than the untouchables, 'hijras' begin to change some popular prejudices, *Wall Street Journal,* Sept. 24.

33. S. Nanda, 1999, *Neither Man nor Woman,* p. 17.

34. Ibid., p. 166.

35. Ibid., p. 6.

36. Ibid., p. 56.

37. Quotations from Meera here and below from ibid., pp. 81, 78–79, 82.

38. Ibid., p. 63.

39. Quotations from Sushila here and below from ibid., pp. 87, 95–96.

40. Quotations from Salima here and below from ibid., pp. 98–100.

41. Ibid., p. 28.

42. Ibid., pp. 66, 35.

43. The pejorative remarks just cited are quoted by Nanda in ibid.; see pp. 35–37.

44. All quotations here from ibid., see pp. 67, 15–16, 28.

45. Quotations from Meera from ibid., pp. 81, 156.

46. Quotation from Meera from ibid., p. 81; from Sushila, ibid., p. 94.

47. Ibid., p. 67.

48. Ibid., p. 11.

49. Ibid., p. 69.

50. Ibid., p. 160.

51. L. Cohen, 1995, The pleasures of castration: The postoperative status of hijras, jankhas, and academics, pp. 276–304 in P. R. Abramson and S. D. Pinkerton, eds., *Sexual Nature, Sexual Culture,* University of Chicago Press; Kira Hall, 2001, personal communication; S. Nanda, 1999, *Neither Man nor Woman,* p. 11.

52. G. Thadani, 1999, The politics of identities and languages: Lesbian desire in ancient and modern India, pp. 67–90 in Blackwood and Wieringa, eds., *Female Desires, Same-Sex Relations and Transgender Practices across Cultures.*

53. Jordy Jones, 2001, personal communication.

19: TRANSGENDER IN HISTORICAL EUROPE AND THE MIDDLE EAST

1. This section is drawn primarily from the work of historian Mathew Kuefler: M. Kuefler, 2001, *The Manly Eunuch: Masculinity, Gender Ambiguity, and Christian Ideology in Late Antiquity,* University of Chicago Press. For the quotations in this paragraph, see pp. 32–34.

2. Ibid., p. 34; see also J. Long, 1996, *Claudian's In Eutropium: Or, How, When, and Why to Slander a Eunuch,* University of North Carolina Press.

3. Kuefler, 2001, *The Manly Eunuch,* p. 62.

4. Ibid.

5. Ibid., p. 61.

6. Ibid., p. 252.

7. Ibid., p. 101.

8. Ibid., pp. 85, 64.

9. Catullus, considered the Roman equivalent of Shakespeare; see: Poem 63, translated by C. Sisson, 1967, *The Poetry of Catullus,* Orion; available at http://aztriad.com/catullus.html.

10. Kuefler, 2001, *The Manly Eunuch,* pp. 255–56.

11. Http://news.bbc.co.uk/hi/english/uk/england/newsid_1999000/1999734.stm. The tradition has been revived. See: http://www.gallae.com/ and http://aztriad.com/index.html.

12. Kuefler, 2001, *The Manly Eunuch,* p. 248.

13. Ibid., p. 250.

14. Ibid., pp. 250–51.

15. Ibid., p. 254.

16. Ibid., p. 249.

17. I thank the Bible Gateway of the Gospel Communications Network Online Christian Resources. Their website, http://bible.gospelcom.net/cgi-bin/bible, offers online access to several editions of the Bible.

18. Kuefler, 2001, *The Manly Eunuch*, pp. 137–39.

19. Ibid., p. 261.

20. Ibid., pp. 223, 231. See also: J. Welch, 1996, Cross-dressing and cross-purposes: Gender possibilities in the Acts of Thecla, pp. 66–78 in S. Ramet, ed., *Gender Reversals and Gender Cultures,* Routledge; K. Torjesen, 1996, Martyrs, ascetics, and gnostics, pp. 79–91 in S. Ramet, ed., *Gender Reversals and Gender Cultures,* Routledge.

21. Kuefler, 2001, *The Manly Eunuch,* pp. 261–62.

22. K. Ringrose, 1994, Living in the shadows: Eunuchs and gender in Byzantium, pp. 85–109 in G. Herdt, ed., *Third Sex: Beyond Sexual Dimorphism in Culture and History,* Zone Books, p. 99.

23. Kuefler, 2001, *The Manly Eunuch,* pp. 269, 280. Cassian spoke admiringly of a case of what we would call religious abuse, involving a father who entered a community of Egyptian monks with his eight-year-old son. To teach the father a lesson about renunciation of family, the abbot had the other monks persistently mistreat the boy in his father's presence, exposing the son to the "blows and slaps of many so that the father never saw him without his cheeks marked with the dirty tracks of tears." The father endured in silence the suffering that the maltreatment of his son caused him, and even willingly attempted to drown the boy when commanded to do so by the abbot, before being prevented from so doing, according to plan. When the abbot died, the father succeeded him. The point was clear: by ignoring the responsibilities of his biological paternity, the man made himself worthy to be father to the whole community of monks. Not a tradition to entrust with protecting family values.

24. Ibid., p. 274.

25. E. Rowson, 1991, The effeminates of early Medina, *J. Amer. Oriental Soc.* 111:671–93. All quotations in this section are from this source.

26. According to Rowson, both the Quran and the hadith strongly condemn ho-

mosexual activity, though. In the Quran, see 7:80f., 26:165f., 27:54f., 29:27f., 54:37.

27. See L. Feinberg, 1996, *Transgender Warriors: Making History from Joan of Arc to RuPaul,* Beacon Press, pp. 31–37. An extensive scholarly literature, including lengthy transcripts of the legal procedures, as well as popular dramatizations, now exist for Jehanne, and may be accessed over the web at http://www.smu.edu/ijas; http://www.stjoan-center.com; and http://perso.wanadoo.fr/jean-claude.colrat. See especially: B. Wheeler and C. T. Wood, eds., 1996, *Fresh Verdicts on Joan of Arc,* Garland; and S. Crane, 1996, Clothing and gender definition: Joan of Arc, *J. Medieval and Early Modern Studies* 28 (2):297–320.

28. Feinberg, 1996, *Transgender Warriors,* p. 34.

29. Ibid., p. 35.

30. Ibid., p. 36.

20: SEXUAL RELATIONS IN ANTIQUITY

1. This section draws on the research of K. J. Dover. K. Dover, 1989 [1978], *Greek Homosexuality,* Harvard University Press. See especially pp. 98, 101, 102, 106.

2. Ibid., p. 31.

3. Ibid., p. 36.

4. This section draws on the research of Tom Horner and Daniel Helminiak: T. Horner, 1978, *Jonathan Loved David,* Westminster Press; D. Helminiak, 2000, *What the Bible Really Says about Homosexuality* (millennium edition), Alamo Square Press, with a foreword by J. Spong. For further scholarship, see: J. Boswell, 1980, *Christianity, Social Tolerance and Homosexuality: Gay People in Western Europe from the Beginning of the Christian Era to the Fourteenth Century,* University of Chicago Press; W. Countryman, 1988, *Dirt, Greed and Sex: Sexual Ethics in the New Testament and Their Implications for Today,* Fortress Press; B. Brooten, 1996, *Love Between Women: Early Christian Responses to Female Homoeroticism,* University of Chicago Press. For ministry, see: R. Truluck, 2000, *Steps to Recovery from Bible Abuse,* Chi Rho Press, available online at http://www.chirhopress.com/products /gayandchristian.html; and links at http://www.truluck.com/.

5. Helminiak, 2000, *What the Bible Really Says about Homosexuality,* p. 83.

6. Ibid., p. 79.

7. Brooten, 1996, *Love Between Women.*

8. Helminiak, 2000, *What the Bible Really Says about Homosexuality,* pp. 85–86.

9. N. Wilson, 2000, *Our Tribe, Queer Folks, God, Jesus and the Bible* (millennium edition), Alamo Square Press, p. 100.

10. D. Halperin, 1989, Sex before sexuality: Pederasty, politics, and power in classical Athens, pp. 37–53 in M. Duberman, M. Vicinus, and G. Chauncey Jr., *Hidden from History: Reclaiming the Gay and Lesbian Past,* Meridian.

21: TOMBOI, VESTIDAS, AND GUEVEDOCHE

1. E. Blackwood, 1999, *Tonbois* in west Sumatra: Constructing masculinity and erotic desire, pp. 182–205 in E. Blackwood and S. Wieringa, eds., *Same-Sex Relations and Female Desires: Transgender Practices across Cultures,* Columbia University Press. Quotations here and below from p. 186.

2. S. Wieringa, 1999, Desiring bodies or defiant cultures: Butch-femme lesbians in Jakarta and Lima, pp. 206–32 in Blackwood and Wieringa, eds., *Same-Sex Relations and Female Desires*. All information on Java is based on this source. Quotations, in order, are from pp. 206, 209, 218, 217, 222.

3. For a definitive study of a butch/femme community in Buffalo, New York, see E. Kennedy and M. Davis, 1993, *Boots of Leather, Slippers of Gold,* Penguin Books.

4. Wieringa, 1999, Desiring bodies or defiant cultures, p. 226.

5. This section relies primarily on the work of Annick Prieur: A. Prieur, 1998, *Mema's House, Mexico City: On Transvestites, Queens, and Machos,* University of Chicago Press.

6. Ibid., pp. 49–50.

7. Quotations from Marta here and below from ibid., pp. 104–8, 178.

8. Ibid., p. 108.

9. Ibid., p. 109.

10. Ibid., p. 129.

11. Ibid., pp. 145–46, 154.

12. Ibid., p. 156.

13. Ibid., pp. 158–59.

14. Ibid., pp. 171, 174.

15. Ibid., p. 173.

16. This exposition relies on the work of Gilbert Herdt: G. Herdt, 1994, Mistaken sex: Culture, biology and the third sex in New Guinea, pp. 419–45 in G. Herdt, ed., *Third Sex, Third Gender: Beyond Sexual Dimorphism in Culture and History,* Zone Books.

22: TRANS POLITICS IN THE UNITED STATES

1. L. Feinberg, 1996, *Transgender Warriors: Making History from Joan of Arc to RuPaul,* Beacon Press, pp. 97–99; extensive gay political activism was taking place in San Francisco at this time and earlier, as documented in S. Stryker and J. Van Buskirk, 1996, *Gay by the Bay,* Chronicle Books.

2. See http://www.hrc.org.

3. *The Channel,* TGSF, 1999, 18 (2) (Feb):13; reprinted from *QUILL,* Gender-Pac Media Advisory, Jan. 14, 1999. All quotations regarding Lauryn Paige's murder are from this source.

4. Http://www.rememberingourdead.org.

5. D. France, 2000, An inconvenient woman, *New York Times Magazine,* May 28, pp. 24–29, and cover photograph. All quotations regarding Barry Winchell's murder are from this source. A dramatization of Winchell's murder and his romance with Calpernia Addams, entitled "Soldier's Girl," premiered on the premium cable movie channel Showtime on May 31, 2003.

6. C. Heredia, 2001, Gay, lesbian troops can serve openly—for now: Pentagon suspends discharges during conflict, *New York Times,* Sept. 19.

7. A. Livia, 1997, Disloyal to masculinity, pp. 349–68 in A. Livia and K. Hall, eds., *Queerly Phrased: Language, Gender, and Sexuality,* Oxford University Press.

8. H. Lee, 2003, Guilty plea in transgender killing, *San Francisco Chronicle,* Feb. 25.

9. K. St. John, 2003, Slain teen's last night recounted, *San Francisco Chronicle,* Feb. 20.

10. There's lots of credit to go around on this innovative legislation. According to Jamison Green, a trans activist and leader both nationally and in the San Francisco area, initial credit belongs to Mayor Willie Brown, a controversial yet effective figure with a long history of success in human rights. Mayor Brown appointed Melissa Welch as director of the city's Health Services Board, in part because of her support for addressing the medical needs of trans people. Dr. Welch crafted most of the proposal that eventually passed, with the involvement of many trans people.

11. See http://ai.eecs.umich.edu/people/conway/TSsuccesses/TSsuccesses.html; and http://ai.eecs.umich.edu/people/conway/TSsuccesses/TransMen.html.

12. K. Bornstein, 1994, *Gender Outlaw: On Men, Women, and the Rest of Us,* Vintage Books, quote taken from back cover.

13. See http://www.thesisters.org/sistory/spihistory.htm.

14. See discussion and photographs in J. Halberstam, 1998, *Female Masculinity,* Duke University Press; D. Volcano and J. Halberstam, 1999, *The Drag King Book,* Serpent's Tail.

APPENDIX: POLICY RECOMMENDATIONS

1. For discussion about the Hippocratic Oath today, see http://www.pbs.org/wgbh/nova/doctors/oath.html; for a possible Hippocratic Oath for scientists and engineers, see http://www.globalideasbank.org/BOV/BV-381.html.

2. C. Yoon, 1999, Reassessing ecological risks of genetically altered crops, *New York Times,* Nov. 3, p. 1.

3. *City and County of San Francisco Voter Information Pamphlet and Sample Ballot,* November 7, 2000, Supervisor-District 6, Ballot Type 03, p. 13.

Index

Indexer:	Andrew Joron
Compositor:	Binghamton Valley Composition
Text:	10/14 Sabon
Display:	Akzidenz Grotesk
Printer and binder:	Maple-Vail Manufacturing Group